Heavy Oils: Reservoir Characterization and Production Monitoring

Edited by

Satinder Chopra

Laurence R. Lines

Douglas R. Schmitt

Michael L. Batzle

SEG Geophysical Developments Series No. 13

Ian Jones, managing editor

Society of Exploration Geophysicists
The international society of applied geophysics

ISBN 978-0-931830-41-9 (Series)
ISBN 978-1-56080-222-8 (Volume)

Copyright 2010
Society of Exploration Geophysicists
P.O. Box 702740
Tulsa, OK U.S.A. 74170-2740

All rights reserved. No part of this book may be reproduced,
stored in a retrieval system, or transcribed in any form or by any means,
electronic or mechanical, including photocopying and recording, without
prior written permission of the publisher.

Published 2010

Printed in the United States of America

Library of Congress Cataloging-in-Publication Data

Heavy oils : reservoir characterization and production monitoring / edited by Satinder Chopra ... [et al.].
   p. cm. -- (SEG geophysical developments series ; no. 13)
Includes bibliographical references and index.
ISBN 978-1-56080-222-8 (volume : alk. paper) -- ISBN 978-0-931830-41-9 (series : alk. paper)
1. Oil sands. 2. Heavy oil. 3. Petroleum reserves. 4. Petroleum--Geology. 5. Oil fields--Production methods. I. Chopra, Satinder.
 TN870.54.H43 2010
 622'.3383--dc22
                    2010039030

To our families

Table of Contents

About the Editors xv
Preface xvii
Acknowledgments xx

Chapter 1: Heavy-oil Reservoirs: Their Characterization and Production 1
Satinder Chopra, Larry Lines, Douglas R. Schmitt, and Mike Batzle

    Heavy Oil as an Important Resource for the Future 1
    Oils and Rocks 5
        Chemical properties of conventional crude oil 6
        Classification of crude oil 8
    Mechanisms for the Formation of Heavy Oil 10
    General Phase Behavior of Hydrocarbons 11
        Single-component system 12
    Hydrocarbon Mixtures, Terminology, and Phase Diagrams 13
        Retrograde gas condensate 13
        Wet gas 14
        Dry gas 14
    Properties of Heavy Oil 14
        Shear properties 17
    Rocks with Heavy Oil 18
    Geology of Two Major Heavy-oil/Oil-sands Areas 21
        Venezuela 21
        Canada 22
    Heavy-oil Recovery 25
        Surface mining 25
        In situ recovery 27
    Geophysical Characterization of Heavy-oil Formations 32
        Elastic parameters from AVO analysis 33
        Multicomponent data for characterization of heavy-oil formations 37
    Enhanced Oil Recovery, Time-lapse Monitoring, and Crosswell Imaging 42
        Monitoring ISC fire front 43
        Steam flood monitoring 45
        Monitoring SAGD steam flooding 50
        Frequency attenuation 52
        Seismic tomography methods 56
    Challenges for Heavy-oil Production 60
        Land surface disturbance and reclamation 62
        Mine tailings disposal 62
        Water consumption 62
        Fuel consumption 62
        Upgrading of heavy oil 63

      Greenhouse gas emissions 64
      Future outlook 65
  References 65

Section 1: Rock Physics Aspects 71

Chapter 2: Seismic Properties of Heavy Oils — Measured Data* 73
De-hua Han, Jiajin Liu, and Michael Batzle

  Introduction 73
  Velocity Models for Light and Heavy Oils 74
  Factors Influencing Velocity 76
      Pressure effect 76
      Gas effect 76
      Temperature effect 77
      Liquid point 78
      Frequency effect 79
  Summary 79
  References 80

Chapter 3: Modeling Studies of Heavy Oil — In Between Solid and Fluid Properties* 81
Agnibha Das and Mike Batzle

  Introduction 81
  Uvalde Heavy-oil Rock and the Canadian Tar Sands 81
  Elastic Property Estimation Using HS Bounds 82
      HS modeling of the Uvalde heavy-oil rock 83
      Generalized Gassmann's equations for a solid infill of the pore space 84
      HS modeling of Canadian tar sand 86
  Conclusions 87
  Acknowledgments 87
  References 87

Chapter 4: Correlating the Chemical and Physical Properties of a Set of
Heavy Oils from around the World 89
Amy Hinkle, Eun-Jae Shin, Matthew W. Liberatore, Andrew M. Herring, and Mike Batzle

  Introduction 89
  Experimental 90
      Heavy-oil samples 90
      Rheology 90
      Ultrasonic measurements 91
      Pyrolysis-MBMS 91
  Results 91
      Rheology 91
      Ultrasonic measurements 93
      Pyrolysis-MBMS 95

*These chapters appeared in the September 2008 issue of THE LEADING EDGE and have been edited for inclusion in this volume.

Conclusions	96
Acknowledgments	96
References	96

Chapter 5: Measuring and Monitoring Heavy-oil Reservoir Properties* ... 99
Kevin Wolf, Tiziana Vanorio, and Gary Mavko

Introduction	99
Converted PSEI	99
PSEI inversion for reservoir characterization and monitoring	100
Data in PSEI space	101
Laboratory Measurements of Heavy-oil Reservoirs	102
Transducer design	103
Conclusions	105
References	105

Chapter 6: Seismic Rock Physics of Steam Injection in Bituminous-oil Reservoirs* ... 107
Evan Bianco, Sam Kaplan, and Douglas Schmitt

Introduction	107
Integrating Rock Physics and Reservoir Parameters for Improved Synthetic Modeling	109
Comparison with Real Seismic Data and 2D Time-lapse Imaging	111
Conclusions	112
Acknowledgments	112
References	112

Chapter 7: Prediction of Pore Fluid Viscosity Effects on P-wave Attenuation in Reservoir Sandstones ... 113
Angus Best, Clive McCann, and Jeremy Sothcott

Introduction	113
Laboratory Measurements	114
Comparison with BISQ Theory	114
Results	115
Implications for Seismic Monitoring of Heavy-oil Reservoirs	116
Conclusions	118
Acknowledgments	118
References	118

Chapter 8: Elastic Property Changes in a Bitumen Reservoir during Steam Injection* ... 121
Ayato Kato, Shigenobu Onozuka, and Toru Nakayama

Introduction	121
Pressure and Temperature Dependence	121
Application of the Gassmann Equation	123
Sequential Rock Physics Model	124
Velocity Dispersion	125

*These chapters appeared in the September 2008 issue of THE LEADING EDGE and have been edited for inclusion in this volume.

Sequential Elastic Property Changes	126
Conclusions	127
Acknowledgments	127
References	127

Section 2: Geologic/Geophysical Characterization — 129

Chapter 9: The Devonian Petroleum System of the Western Canada Sedimentary Basin — with Implications for Heavy-oil Reservoir Geology — 131
Hans G. Machel

Introduction	131
Basin Evolution and Structure	133
Sedimentation and Facies	136
Diagenetic Evolution of Carbonates	137
Source Rocks	139
Hydrology and Migration	139
Reservoirs	143
Dry gas reservoirs	144
Heavy-oil reservoirs	144
Light crude, sweet, and sour gas reservoirs	148
Conclusions	151
Acknowledgments	151
References	151

Chapter 10: Review of Geology of a Giant Carbonate Bitumen Reservoir, Grosmont Formation, Saleski, Alberta — 155
Kent R. Barrett and J. C. Hopkins

Introduction	155
Reservoir geology of the Grosmont C and D	155
Fracturing	157
Megaporosity zones	158
Origin of Breccia Zones	159
What was the source of the solubility contrast?	159
Exploitation of the Bitumen Resource	161
Why is the Grosmont an attractive SAGD candidate?	161
Conclusions	162
Acknowledgments	162
References	163

Chapter 11: Deterministic Mapping of Reservoir Heterogeneity in Athabasca Oil Sands Using Surface Seismic Data* — 165
Yong Xu and Satinder Chopra

| Rock Physics Analysis | 165 |
| Workflow for Mapping Reservoir Heterogeneity | 166 |

*These chapters appeared in the September 2008 issue of The Leading Edge and have been edited for inclusion in this volume.

Improved Three-term AVO Inversion.	166
Application to Real Data	167
Conclusions	171
Acknowledgments	171
References	171

Chapter 12: Imaging Oil-sands Reservoir Heterogeneities Using Wide-angle Prestack Seismic Inversion* **173**

Baishali Roy, Phil Anno, and Michael Gurch

Introduction	173
Angle Requirements and Difficulties of Wide-angle Inversion	174
Wide-angle processing and inversion for density	175
Anisotropic imaging	176
Wavelet stretch correction	176
Inelastic losses	177
Constrained prestack linear inversion	178
Results	179
Conclusions	181
Acknowledgments	181
References	181

Chapter 13: Characterization of Heavy-oil Reservoir Using V_P/V_S Ratio and Neural Networks Analysis. **183**

Carmen C. Dumitrescu and Larry Lines

Introduction	183
Method	185
Results	188
Conclusions	190
Acknowledgments	190
References	190

Chapter 14: Multicomponent Processing of Seismic Data at the Jackfish Heavy-oil Project, Alberta. **191**

Karen J. Pengelly, Larry R. Lines, and Don C. Lawton

Introduction	191
Seismic Survey Acquisition	191
Vertical Component Data Processing	192
Geometry	192
Radial filter	193
Gabor deconvolution	193
Near-surface static solution	193
Kirchhoff migration	194
Radial Component Data Processing	195
Radial component geometry	195
Radial filter	196

*These chapters appeared in the September 2008 issue of THE LEADING EDGE and have been edited for inclusion in this volume.

      Gabor deconvolution 196

      Near-surface statics solution 197

      Depth-variant stack 198

  Conclusions 199

  Acknowledgments 199

  References 199

Section 3: Reservoir Monitoring 201

Chapter 15: Geostatistical Reservoir Modeling Focusing on the Effect of Mudstone Clasts on Permeability for the Steam-assisted Gravity Drainage Process in the Athabasca Oil Sands 203

Koji Kashihara, Akihisa Takahashi, Takashi Tsuji, Takahiro Torigoe, Koji Hosokoshi, and Kenji Endo

  Introduction 203

  Environment of Deposition 203

  Permeability of Sand with Mudstone Clast Facies 204

  Seismic Attribute Selection for Facies Discrimination 206

  Reservoir Modeling Workflow 208

      Facies modeling 209

      Mudstone volume modeling 209

      Porosity and permeability estimations 211

  Discussion and Conclusions 212

  Acknowledgments 213

  References 213

Chapter 16: Monitoring an Oil-sands Reservoir in Northwest Alberta Using Time-lapse 3D Seismic and 3D P-SV Converted-wave Data* 215

Toru Nakayama, Akihisa Takahashi, Leigh Skinner, and Ayato Kato

  Time-lapse 3D Seismic Data 215

  Core Velocity Data and Rock Physics Model 215

  Seismic Calibration and Interpretation 215

  Spectral Decomposition 222

  P-SV Analysis and Interpretation 222

  Discussion 225

  Conclusions 225

  Acknowledgments 226

  References 226

Chapter 17: Oil-sands Reservoir Characterization for Optimization of Field Development 227

Akihisa Takahashi

  Introduction 227

  Geologic Background 227

  Geophysical Background 228

  Workflow 229

*These chapters appeared in the September 2008 issue of THE LEADING EDGE and have been edited for inclusion in this volume.

Evaluation of Facies Prediction	230
Horizontal Well Pair Planning	231
Conclusions	234
Acknowledgments	234
References	234

Section 4: Production . . . 235

Chapter 18: The Effects of Cold Production on Seismic Response* . . . 237
Fereidoon Vasheghani, Joan Embleton, and Larry Lines

Introduction	237
Model and Methodology	237
Seismic Modeling and Imaging	238
Conclusions	240
Acknowledgments	241
References	241

Chapter 19: Effects of Heavy-oil Cold Production on V_P/V_S Ratio . . . 243
Duojun (Albert) Zhang, Larry Lines, and Joan Embleton

Introduction	243
Fluid Substitution: Gassmann's Equation	243
Difference of Heavy-oil Physical Properties between Pre- and Postproduction	245
Effects of Heavy-oil Cold Production on V_P/V_S Ratio	245
Conclusions	248
Acknowledgments	248
References	248

Chapter 20: Collaborative Methods in Enhanced Cold Heavy-oil Production* . . . 251
Larry Lines, Hossein Agharbarati, P.F. Daley, Joan Embleton, Mathew Fay, Tony Settari, Fereidoon Vasheghani, Tingge Wang, Albert Zhang, Xun Qi, and Douglas Schmitt

Introduction	251
Wormholes, Foamy Oil, and Cold Production	251
Seismic Resolution of Cold Production Zones	251
Rock Physics of Cold Production of Heavy Oil	253
Reservoir Simulation of Heavy Oil Cold Production	255
Conclusions	256
Acknowledgments	256
References	256

Chapter 21: Crosswell Seismic Imaging — A Critical Tool for Thermal Projects . . . 259
Mark McCullum

| Introduction | 259 |
| Background | 259 |

*These chapters appeared in the September 2008 issue of THE LEADING EDGE and have been edited for inclusion in this volume.

Imaging Reservoir Features 260

Conclusions 263

References 263

Chapter 22: The Impact of Oil Viscosity Heterogeneity on Production from Heavy Oil and Bitumen Reservoirs: Geotailoring Recovery Processes to Compositionally Graded Reservoirs 265

Ian D. Gates, Jennifer J. Adams, and Steve R. Larter

Fluid Heterogeneity Is Caused by Biodegradation over Geologic Timescales 267

Impact of Fluid Property Variations on Recovery 268

Geotailoring Recovery Processes for Reservoirs with Viscosity Variations 269

J-well and Gravity Drainage Recovery Process: A Geotailored Process 270

References 273

Chapter 23: Using Time-lapse Seismic to Monitor the Toe-to-Heel-Air-Injection (THAI™) Heavy-oil Production Process 275

Rob Kendall

Introduction 275

Reservoir description at Whitesands (Leismer) 275

THAI and CAPRI 276

Conditioning of Seismic Data for Time-lapse Analysis 276

Velocity Anomalies 278

Time-lapse Results 279

PP Time-lapse Results 279

PS Time-lapse Results 282

Conclusions 283

Acknowledgments 283

References 284

Section 5: Geomechanical Aspects 285

Chapter 24: Geomechanics Effects in Thermal Processes for Heavy-oil Exploitation 287

Maurice B. Dusseault and Patrick M. Collins

Introduction 287

Thermally Induced Stress Changes 287

Shearing, Dilation, and Mechanical Damage of the Rock 288

Shear and dilation 288

What are the consequences for geophysics? 288

Effective stress 289

Shear at the shale caprock 289

Reservoir Deformations 289

Impact of Geomechanics on Seismic Monitoring 290

Suggestions 290

References 291

*These chapters appeared in the September 2008 issue of THE LEADING EDGE and have been edited for inclusion in this volume.

Chapter 25: Passive Seismic and Surface Monitoring of Geomechanical Deformation Associated with Steam Injection* **293**

Shawn C. Maxwell, Jing Du, and Julie Shemeta

    Introduction 293

    Case Study 294

    Microseismic Results 294

    Tiltmeter Results 297

    Discussion and Conclusions 298

    References 300

Chapter 26: Using Multitransient Electromagnetic Surveys to Characterize Oil Sands and Monitor Steam-assisted Gravity Drainage **301**

Folke Engelmark

    Introduction 301

    Steam Injection from the Pore Perspective 302

    Resistivity Monitoring of Thermal Recovery 304

    Conclusions 306

    References 307

Section 6: Environmental Aspects **309**

Chapter 27: Tar Sands: Key Geologic Risks and Opportunities* **311**

Jack R. Century

    Acknowledgments 313

Index **315**

*These chapters appeared in the September 2008 issue of THE LEADING EDGE and have been edited for inclusion in this volume.

About the Editors

Satinder Chopra received MSc and MPhil degrees in physics from Himachal Pradesh University, Shimla, India. He joined the Oil and Natural Gas Corporation Limited (ONGC) of India in 1984 and served there until 1997. In 1998, he joined CTC Pulsonic at Calgary, which later became Scott Pickford and Core Laboratories Reservoir Technologies. Currently, he is working as Chief Geophysicist (Reservoir), at Arcis Corporation, Calgary. In the last 26 years, Satinder has worked in regular seismic processing and interactive interpretation but has spent more time in special processing of seismic data involving seismic attributes including coherence, curvature and texture attributes, seismic inversion, AVO, VSP processing, and frequency enhancement of seismic data. His research interests focus on techniques that are aimed at characterization of reservoirs. He has published seven books and more than 190 papers and abstracts, and he likes to make presentations at any beckoning opportunity. He is the chief editor of the CSEG *Recorder*, a past member of THE LEADING EDGE Editorial Board, and a former chairman of the SEG Publications Committee.

Satinder Chopra received several awards at ONGC; recently, he has received the Best Oral Presentation Award for his paper titled "Delineating stratigraphic features via cross-plotting of seismic discontinuity attributes and their volume visualization," presented at the 2010 AAPG Annual Convention held in New Orleans; the Top 10 Paper Award for his poster titled "Extracting meaningful information from seismic attributes," presented at the 2009 AAPG Annual Convention held in Denver; the Best Poster Award for his paper titled "Seismic attributes for fault/fracture characterization," presented at the 2008 SEG Convention held in Las Vegas; the Best Paper Award for his paper titled "Curvature and iconic coherence attributes adding value to 3D seismic data interpretation," presented at the CSEG Technical Luncheon, Calgary, in January 2007; and the 2005 CSEG Meritorious Services Award. He and his colleagues have received the CSEG Best Poster Awards in successive years from 2002 to 2005.

He is a member of SEG; CSEG; CSPG; Canadian Heavy Oil Association; EAGE; AAPG; Association of Professional Engineers, Geologists, and Geophysicists of Alberta; and Texas Board of Professional Geoscientists.

Laurence "Larry" Lines received BSc (1971) and MSc (1973) geophysics degrees from the University of Alberta and a PhD (1976) in geophysics from the University of British Columbia. His industrial career included 17 years with Amoco in Calgary and Tulsa (1976–1993). Following a career in industry, Lines held the NSERC/Petro-Canada Chair in Applied Seismology at Memorial University of Newfoundland (1993–1997) and the Chair in Exploration Geophysics at the University of Calgary (1997–2002). From 2002 through 2007, he served as the Head, Department of Geology and Geophysics at the University of Calgary.

In professional service, Larry was the President of SEG in 2008–2009. Previous to that, he served SEG as GEOPHYSICS Editor (1997–99), Distinguished Lecturer, GEOPHYSICS Associate Editor, Translations Editor, Publications Chairman, and as a member of THE LEADING EDGE Editorial Board. He served the CSEG as Editor and Associate Editor. Larry and coauthors have won SEG's Best Paper in GEOPHYSICS Award twice (1988, 1995) and have twice won Honorable Mention for Best Paper (1986, 1998). Larry is an Honorary Member of SEG, CSEG, and the Geophysical Society of Tulsa. Additionally, he is a member of APEGGA, CGU, EAGE, and AAPG. Larry is married with two children, and he enjoys hobbies of choir, softball, and hiking with his Alaskan malamute.

Douglas R. Schmitt's research team, the Experimental Geophysics Group, carries out field- and laboratory-based geophysical measurements with a particular focus on understanding the rock physics of geological systems at different scales. Field studies focus on time-lapse and near-surface geophysics. He also is involved in numerous scientific drilling projects around the world having carried out onshore and offshore borehole seismic, logging, and hydraulic fracturing stress measurements in the Antarctic, the Arctic, Africa, Europe, and North America.

Current laboratory studies focus on seismic anisotropy, dielectric permittivity, and fundamental studies of wave propagation in porous media. He currently holds the Canada Research Chair in Rock Physics and is a professor of physics and geophysics at the University of Alberta since 1989. He obtained a BSc in physics from the University of Lethbridge and a PhD in geophysics from the Seismological Laboratory at the California Institute of Technology. Prior to graduate studies, he worked as an exploration geophysicist at Texaco Canada and upon graduation carried out postdoctoral research in the Department of Geophysics at Stanford University. He has been a visiting scientist at the Geophysikalisches Institute at the University of Karlsurhe as an A. vonHumboldt research fellow and was recently a visiting scientist at the Research School of Earth Sciences at the Australian National University, Canberra. He currently serves on advisory and research assessment panels for both the International Continental Drilling Program and the Integrated Ocean Drilling Program.

Mike Batzle holds the Baker Hughes Distinguished Chair of Petrophysics and Borehole Geophysics at the Colorado School of Mines, where he has been a member of the geophysics department for the past 17 years. Previously, he was a principal scientist at ARCO Oil and Gas Company in Plano, Texas. He has a BS in geology from the University of California, Riverside, and a PhD in geophysics from Massachusetts Institute of Technology. His main interests have been in rock properties research primarily for engineering and geophysical purposes. Mike's laboratory has a wide range of equipment to measure seismic and acoustic properties of rocks and fluids, including low frequency and low amplitude velocity, attenuation, and modulus measurements. He was awarded the Kauffman Gold Medal by SEG for his research with Zhijing Wang on fluid properties. In the past, he also has conducted borehole geophysics research and development, and he holds U. S. patents on tool designs.

Currently, his research focus is on seismic lithology and fluid identification, reservoir characterization, and time-lapse seismic monitoring. He also has conducted extensive research on rock strength and stability and their estimation from both laboratory measurements and well logs. At the School of Mines, he established and codirects the Rock Physics Laboratory (affectionately known as The Center for Rock Abuse).

Preface

Heavy oil is an important global resource with reserves comparable to those of conventional oil. As conventional resources get thinner, attention is being focused on heavy oil and bitumen, which hold the promise of becoming useful fuels. Already more than 1 million barrels of oil are being produced from the oil sands in Canada; heavy oil represents half of California's crude oil production in the United States and is a major production in Mexico. With demand for global energy soaring, heavy oil will undoubtedly be an important resource to be exploited in a big way in the near future.

The SEG Development and Production Committee held its Heavy Oil Forum in Edmonton, Alberta, in July 2007. This was a joint research forum cosponsored by the Canadian Society of Exploration Geophysicists (CSEG) and SEG and hosted by the University of Alberta. Preceding the forum, a field trip took the participants to the vast Athabasca Oil Sands region where they observed the outcrops, open pit mining, and steam injection operations, followed by a tour of the steam-assisted gravity drainage projects. Topics of the well-attended forum included the definition of heavy oil; where is heavy oil found; how it is produced; heavy-oil reservoir characterization; fluid and rock properties; electrical, tilt, and gravity techniques; borehole, surface seismic measurements; and microseismicity.

Although some of the primary engineering concerns such as water cycling or gas availability may appear to have little to do with geophysics, the forum consensus was that geophysics can play a critical role in reservoir characterization and production monitoring, but improvements in resolution, turnaround, and the investment involved are needed.

Heavy oil is different from conventional oil, and so its exploration and production will require special seismic strategies, tools, and rock physics models. This forum aimed to identify the issues and problems and their workarounds. Canadian heavy oil has an American Petroleum Institute (API) gravity range of 8°–12°. Found at depths of 760 m and more, much of it is too deep to mine and will require special production and monitoring strategies. For lighter oils (up to 20° API), such as are found in the North Sea, Indonesia, China, and Brazil, combinations of gravity and stimulated production can be effective.

Problems to be addressed include

- integration of borehole, seismic, nonseismic, and engineering techniques to solve enhanced oil recovery problems

- roles of service and oil companies, and different disciplines such as engineering, geology, and geophysics among others

- resolutions — core scale heterogeneity versus seismic resolution and how to map wormholes and other production artifacts

- case studies

Members of the field trip to the open-pit heavy-oil mines north of Fort McMurray.

- time lapse — what can we expect in
  - migrated time sections
  - impedance inversions
  - amplitude variation with offset (AVO) response
  - seismic traveltimes
  - attenuation
  - V_P/V_S and Poisson's ratios

Issues arise when we link models from various disciplines and interpret them in terms of rock and fluid properties. Currently we have good feedback between seismic interpretations and seismic modeling. However, our workflow needs to be broader to incorporate reservoir models and also to gain improved understanding of the rock physics behavior.

Estimates of heavy oil in place range from 9 to 13 trillion barrels. In the Western Hemisphere, heavy oil is poised to become the principal source of hydrocarbons, particularly in Canada and Venezuela. The broader geologic settings of the Alberta-Saskatchewan and Orinoco deposits are strikingly similar. Oil is sourced in deep basins adjacent to mountain belts. Lighter oils migrate up near the surface to be biodegraded into heavy, viscous fluids. Although the heavy oils of both areas have similar API gravities, Venezuelan oils have lower viscosity and are easier to produce. This is due partly to the Orinoco's higher in situ temperatures and to chemical variations.

Reservoir characterization is one of the fundamental problems facing engineering design. We need to map reservoir heterogeneity accurately and differentiate between lithologies. In fluvial environments, rock types can change rapidly from coarse sands to shales. Conventional stacked seismic data are useful for delineating structures and stratigraphy at a large reservoir scale. However, in many heavy-oil projects, we need spatial and temporal resolutions of 2–3 m. Shale breaks in particular can interfere with thermal and other production techniques. Currently, heavy oil carbonate reservoirs are becoming significant producers and may require entirely different characterization schemes.

Recovery processes vary drastically. For shallow deposits, mining is the common extraction method. For deeper reservoirs, the two main production schemes are cold heavy-oil production with sand (CHOPS), used mainly in eastern Alberta, and thermal stimulation, including cyclic steam and steam-assisted gravity drainage (SAGD), used in Alberta, California, and Indonesia. Time-lapse seismic data can be particularly effective in tracking steam migration. These production methods have individual advantages and drawbacks. Thermal recovery techniques may soon be limited by the availability of input resources (gas, water). Different oils and environments will require new or modified recovery schemes; for example, in situ combustion or vapor extraction in offshore production. We should be planning to apply our techniques to these new schemes.

For rock physics, we need to characterize different recovery methods and corresponding changes in reservoir and rock properties. Theoretical, laboratory, and field examples for heavy-oil interpretations must be developed. Where rock and fluid properties are concerned, numerous developments need to take place

- better rock physics models
- how to go beyond Gassmann
- shear modulus for frame and for fluid
- anelastic/viscoelastic/elastic parameters
- anomalous Q values in heavy oils themselves and rocks with heavy oil
- lithology discrimination
- shale measurements
- characterize heterogeneity

Seismic data play a significant role in reservoir characterization and production monitoring. As stated earlier, the prime challenges are mapping reservoir heterogeneity and differentiating lithologies. In the case of shallow reservoirs, higher frequencies (>200 Hz) at wide angles can be obtained, but statics and noise suppression are some of the problems. The workflow must include anisotropic imaging, wavelet stretching, and wide-angle inversion. AVO or with angle of incidence (AVA) analyses are important to derive density and Poisson's ratio information. Venezuela and Canada have significantly different signatures in this respect. Wider angle (60°) inversions give much better results, particularly for density. For monitoring, particularly on land, the repeatability of the data must be quantified and improved.

Multicomponent data are often collected, but the additional use of this shear information is limited. There appears to be no consensus on P-wave reflection (PP) data versus converted-wave (PS) data; for example, Chevron found that AVO gave good enough information for the Alba Field, whereas the use of PS data definitely improved interpretations for the Grane Field. Can multicomponent data be added without losing resolution? Because PS waves have higher attenuation, it is difficult to compare the resulting data. PP has higher resolution, but the two types of data give different information. So, although PS has lower apparent resolution, it is wise to not average but add information from both data sets. One problem is that even when PS information has been acquired, it is not delivered rapidly enough to be used.

For interpretation, there needs to be close interaction between geophysicists and engineers. We can provide probabilities from geophysical interpretations for engineering applications. Uncertainties can be estimated by incorporating geostatistics. We must perform joint inversions including reservoir and geologic models. For each different recovery method, we must document the corresponding changes in rock and reservoir properties. Can we go beyond Gassmann? Finally, any product must be furnished in a timely manner and in a form useful to the engineer.

Log data are already incorporated extensively in reservoir description and engineering planning. Initial assessments have always used the more standard tools such as gamma ray, density, resistivity, and acoustics. Now, other tools such as image logs or nuclear magnetic resonance (NMR) logs provide better descriptions of the lithologies and the oils themselves. Cross-borehole and time-lapse logging show great potential for monitoring recovery processes.

Other techniques such as tilt, electrical, and microseismicity have strong potential. Injecting steam causes an obvious deformation at the surface, easily resolved by tilt monitoring techniques. During steam injection, the exchange of fluids and fluid phases can result in significant changes in the electrical and electromagnetic signatures. Microseismicity is already extensively used. These data provide an image of fluid motion and stress changes. The microseismic events have indicated which wells in a cluster are taking steam, when breakthrough may be imminent, and if steam migrates out of zone.

In conclusion, geophysical methods can play a significant role in reservoir characterization and production monitoring of heavy-oil reservoirs. Some of the salient issues that have to be addressed are

- How to best model the oils and rocks that contain them.
- How to increase resolution in reservoir characterization.
- Are full three-dimensional monitoring surveys necessary?
- Are multicomponent data worth the acquisition and processing costs?
- How to best integrate seismic, nonseismic, and engineering data.

Currently, Canada is acting as a primary laboratory for developing and testing heavy-oil geophysical techniques. However, for application to other areas and conditions, tested techniques will need substantial modification and local calibration.

The authors who presented at the forum were encouraged to submit their work in the form of articles for the September 2008 special issue of THE LEADING EDGE on heavy oil. The response was so overwhelming that it was not possible to include them all, and so the idea of a book was conceived. As we organized the submissions into topical sections, a few more contributions were sought from experts in areas not fully covered. An introductory chapter to this compilation provides a comprehensive review on the subject for the convenience of the readers.

We hope that as many more advancements materialize in the years to come, this book will serve as a platform to build on.

Satinder Chopra
Calgary, Alberta, Canada
Larry Lines
Calgary, Alberta, Canada
Doug Schmitt
Edmonton, Alberta, Canada
Mike Batzle
Boulder, Colorado, United States

Acknowledgments

CSEG and SEG sponsored the SEG D&P Forum on Heavy Oil held at the University of Alberta in Edmonton in July 2007 and provided logistical and promotional support. Kristi Smith put in a vast amount of effort with the organization and registration of the field trip and forum as did several graduate student "slaves" from various universities. Dr. Murray Gingras and particularly graduate student Curtis Lettley taught the geology of the Athabasca deposits to forum delegates. ConocoPhillips and CGGVeritas provided financial support. All are gratefully acknowledged.

Special thanks go to *TLE* editor Dean Clark, who reviewed the 12 articles first published in the September 2008 issue of *TLE* also included in this book; Dolores Proubasta for copy editing chapter 1; and Jennifer Cobb, SEG manager of Geophysics and books, for her patience and professionalism in producing this book.

Nothing would have been possible without the authors' contributions, patience, and cooperation. The many useful discussions held with Paul MacKay were also instrumental in clarifying many doubts in the section on the geology of Athabasca Oil Sands We're grateful to Fotis Kalantzis for providing images for Figures 87 to 93, in Chapter 1.

We also thank our respective employers for the encouragement and support that justify the time and effort spent on such a writing venture.

Finally, the editors thank the following companies for permission to use the images/pictures to illustrate the text.

Arcis Corporation

Oil Sands Imaging

The Pembina Institute

Petrobank Energy

Suncor Canada Ltd.

Syncrude Canada Ltd.

The Editors

Chapter 1

Heavy-oil Reservoirs: Their Characterization and Production

Satinder Chopra,[1] Larry Lines,[2] Douglas R. Schmitt,[3] and Mike Batzle[4]

Heavy Oil as an Important Resource for the Future

With more than 87 million barrels of oil being consumed worldwide every day, oil has come to be the lifeblood of modern civilization. It is cheap, relatively easy to procure and use, and has become addictive in terms of its flexibility in enhancing our lives in multiple applications. First and foremost, we are dependent on oil for transportation because more than 90% of transportation energy comes from oil. In addition, oil provides a feedstock for pharmaceuticals, agriculture, plastics, clothing, mining, electricity, and several other products that we use in our everyday lives. Almost all goods are connected to oil in one way or another; we are all dependent on oil and gas more than any other resource, yet not many of us think about this dependence.

Oil exploration and production has fueled world economic growth over the last century, and it has reached a stage where the economy of several nations is dependent on the exports of oil to the international market. Global demand for oil is now outstripping supply growth and the importance of this crucial commodity is such that companies engaged in oil exploration and production or transportation have dwarfed those in every other commodities sector. Some important aspects to keep in mind are that oil and gas are absolutely critical to the operation of today's industrial society, essential for sustained economic growth in the industrialized world, and key to progress in nations working their way toward prosperity. This translates into a growing demand for oil and gas, much of it coming from developing nations with low levels of energy use per capita.

However, oil fields are not uniformly distributed in the world; some countries boast giant accumulations and others have none at all. Those that are not self-sufficient spend billions of dollars each year importing oil to satisfy their growing demand. The United States alone imports 65% of the oil it consumes. It is not possible for such countries to insulate themselves from the impact of global oil supply-and-demand imbalances. Also, oil is a finite resource bound to be exhausted one day. At year-end 2007, proven worldwide oil reserves were reported as 1238 billion barrels. In Figure 1, the proven reserves for some significant oil producers are shown, with Saudi Arabia topping the list. The production rate at the end of 2007 was 81.5 million barrels a day or 29.7 billion barrels per year. A simple calculation suggests that conventional crude oil supply could last for about 42 more years (at the 2007 production rate). Some energy experts argue that there is fallacy in using such numbers as reserves keep changing. Even if these figures are taken as conservative, or the reserves are taken as growing, the crude oil supply could last a little longer. But world oil demand is forecast to grow by 50% by 2025, and so the two could offset each other. Although these figures are based on assumptions that may not be strictly true (Campbell and Laherrère, 1998), this simple calculation forewarns us that conventional oil cannot last forever. Most would shudder to think of such a day, but it would be prudent to discuss and analyze a scenario wherein conventional crude oil supply does dwindle some day for us to evaluate what would be the next reliable resource.

Another significant reality to be considered is that many of the oil-producing nations have peaked their production and are on a decline. Figure 2 shows the production from the United States (which peaked in 1970), Mexico (production from the giant Cantarell Field is in decline), Venezuela, Norway, Russia, United Kingdom, and Egypt since 1985. The only exception seen on the graph is Russia, where production has increased since 2000 and continues to climb. North Sea crude oil production, not shown in

---

[1]Arcis Corporation, Calgary, Alberta, Canada
[2]University of Calgary, Department of Geoscience, Calgary, Alberta, Canada
[3]University of Alberta, Institute of Geophysical Research, Department of Physics, Edmonton, Alberta, Canada
[4]Colorado School of Mines, Golden, Colorado, U.S.A.

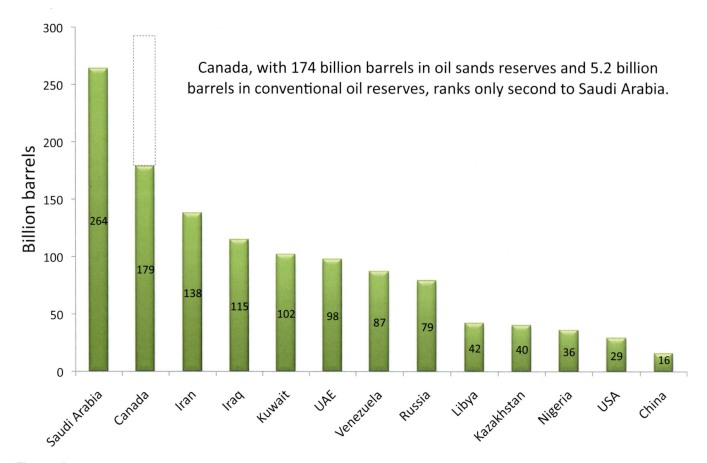

Figure 1. Proven combined reserves of countries shown here add up to 1225 billion barrels, which is close to the world's proven reserves of 1238 billion barrels at the end of 2007. Canada's National Energy Board estimates that domestic conventional oil reserves stand at 5 billion barrels. *The Oil and Gas Journal* pegs Canadian oil-sands reserves at 174 billion barrels, for a combined estimate of 179 billion barrels of conventional and oil-sands reserves. Adding to this figure unofficial estimates of 111 billion barrels of recoverable reserves (at a 35% recovery factor) from Alberta's Grosmont formation pushes Canadian oil reserves above Saudi Arabia's. From BP Statistical Review of World Energy, 2008.

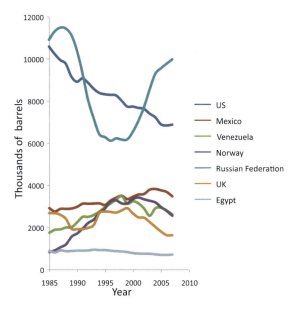

Figure 2. Daily oil production has been in decline for all countries listed except the Russian Federation. From BP Statistical Review of World Energy, 2008.

Figure 2, is also declining. Yet another important factor to note is that world oil discoveries have been steadily shrinking over the last few decades. In the last 25 years, no oil fields capable of producing more than 1 million barrels a day, such as Ghawar (discovered in 1948 in Saudi Arabia), Kirkuk (1938 in Iraq), Burgan Greater (1927 in Kuwait), or Cantarell (1976 in Mexico), have been discovered.

Since the early 1980s, world oil production has fallen short of the number of barrels consumed (Figure 3). As this shortfall increases so does the price per barrel. Higher prices in turn are taxing on the economy of developing countries. However, lately the demand of countries such as India and China has not diminished despite the spiking oil prices. At this writing, the 2008 global economic meltdown has dampened the demand for oil worldwide and prices have fallen. This could be a temporary phase followed by higher prices. Of course, we are talking about "conventional" oil.

Conventional oil consumption in the last three decades has increased in general and, as stated earlier, total world consumption stands at more than 87 million barrels per

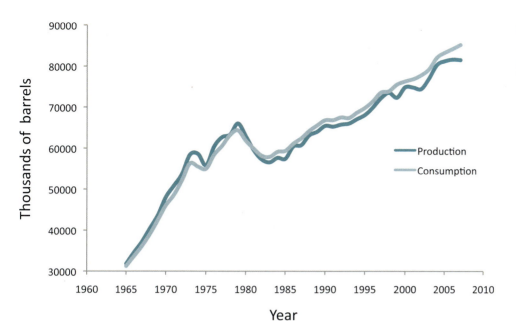

Figure 3. Daily oil production and consumption worldwide. From BP Statistical Review of World Energy, 2008.

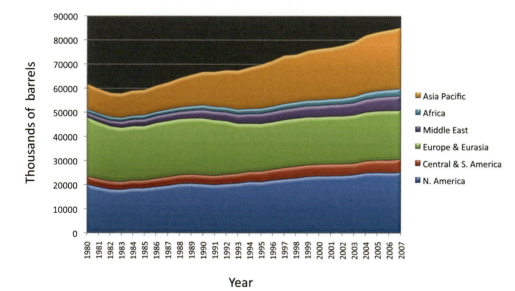

Figure 4. Daily oil consumption for various regions. From BP Statistical Review of World Energy, 2008.

day (b/d), despite the average price per barrel for 2008 being above $75 (Stonehouse, 2008). Oil demand has been projected to increase to 110 million b/d by 2015 and to 120 million b/d by 2025. As seen in Figure 4, after an initial lowering trend in the energy crisis of the early 1980s, there has been an overall increase of oil consumption in all regions of the world except Europe and Eurasia, with Asia Pacific showing significantly higher growth than others. If we analyze the data for some of the developed countries of the world, we again notice that oil consumption does not show significant growth over the last three decades (Figure 5a). However, two rapidly growing economies, China and India, have recorded a growth by a factor of 4 (Figure 5b) in the last three decades, and this growth has doubled in the last decade alone. This trend is expected to continue as these economies expand.

Production in China and India has not kept pace with consumption (Figure 6a and b). As with many other oil producing countries, Indonesia, which had been a net exporter of oil until recently, turned into a net importer of oil in 2005 (Figure 6c). This gap between production and consumption is widening, especially in developing countries. Figure 7 shows this gap as 10.5 million b/d for some developing nations, including China, India, Indonesia, Mexico, Pakistan, Philippines, Singapore, South Korea, and Taiwan. Two decades ago, the overall production in these countries as a whole matched their consumption.

Energy experts have suggested for some time that conventional oil production will peak this decade and decline thereafter never to rise again. Some say that world oil has already peaked, but others believe that Saudi oil supplies might decline earlier than expected (Simmons, 2005).

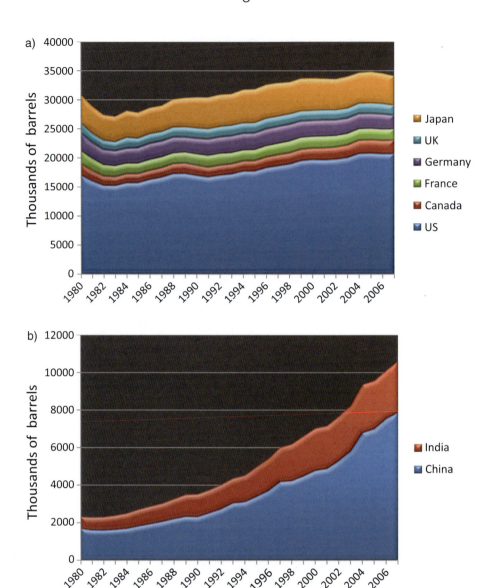

Figure 5. Daily oil consumption for (a) some developed countries, and (b) India and China. From BP Statistical Review of World Energy, 2008.

Either scenario acknowledges the fact that hydrocarbons are a finite resource. The sooner we accept it, the better we will handle the shortfall and avoid a crisis. Adding to our concerns is the geopolitical climate around the world that makes access to some oil-rich nations such as Russia, Venezuela, and others increasingly difficult. In such a global scenario, and with the price per barrel soaring past $140 as recently as 2008, it is natural to turn to alternative energy sources.

An assessment of the range of alternatives reveals that only unconventional hydrocarbon resources like oil sands, heavy oil, shale deposits (oil/gas), and coal liquids are sizeable enough to be considered. Globally, the recoverable reserves of heavy oil and natural bitumen are equal to the remaining reserves of conventional oil (Figure 8a). When the available data are analyzed, we find that although the Middle East dominates in terms of conventional oil (Figure 8b), South America, principally Venezuela, leads in heavy-oil reserves (Figure 8c), and bitumen reserves are abundant in North America, mainly in Canada (Figure 8d).

Conventional oil production in Canada has declined over the last few years, falling from 1.2 million b/d (including light and medium grades as well as heavy oil from Alberta and Saskatchewan) to 1 million b/d in 2006. This output could fall further to only 671,000 b/d by 2020. During the same five-year period, total production from mined oil sands and oil sands produced in situ by steam-assisted gravity drainage (SAGD) or other methods increased from 659,000 b/d in 2001 to 1.1 million b/d in 2006. This number is forecast to swell up to 4 million b/d in 2020 according to the Canadian Association of Petroleum Producers (CAPP). In 2008, total Canadian oil production including conventional and oil sands stood at 2.6 million b/d. This is expected to double by 2020 (Stell, 2008).

Russia has about 246.1 billion barrels of natural bitumen in place, of which only 33.7 billion barrels (approximately 14%) are recoverable. The remaining bitumen cannot

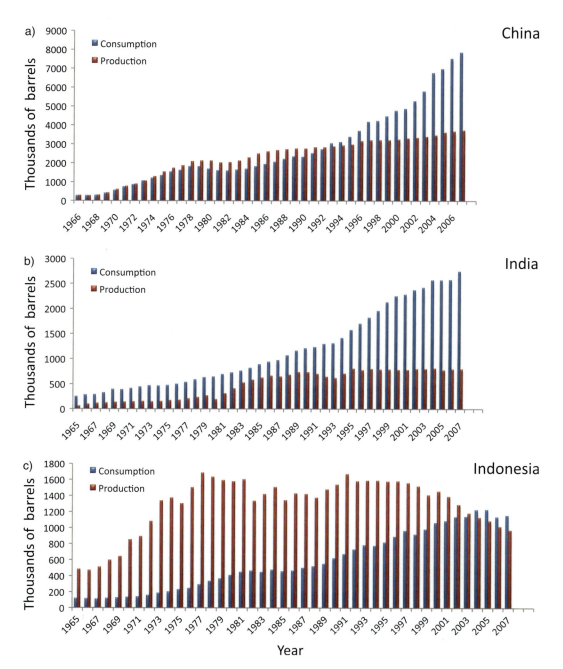

Figure 6. Daily oil consumption versus production for (a) China, (b) India, and (c) Indonesia. From BP Statistical Review of World Energy, 2008.

be realistically recovered as it exists either in remote areas or scattered in many small deposits. A large accumulation occurs in the Olenik Highland located in the Lena-Anabar Basin in eastern Siberia (Veazeay, 2006).

Forecasts vary because of the uncertainty associated with data sources and the level of optimism individuals may assign to their forecasts, which in turn is a function of their individual experience and interpretation of the data. Also, influencing such forecasts is the prevailing geopolitical climate. Yet the fact remains that for the foreseeable future, heavy oil could be contributing significantly to the world oil production.

Oils and Rocks

To be able to interpret the data collected during monitoring of heavy-oil production processes, we must have a thorough understanding of how the fluids and rocks behave under production conditions. Heavy oils are unique in the influence they have on the sands or consolidated rocks containing them.

- Heavy oils can act like a solid at low temperatures — they have a shear modulus.
- Heavy oils are strongly temperature dependent.

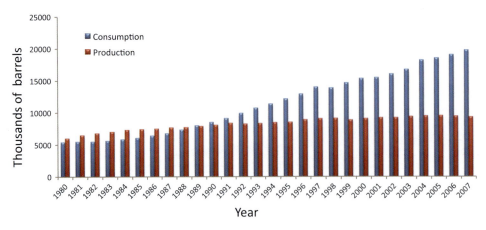

Figure 7. Combined daily oil consumption and production for some developing nations including China, India, Indonesia, Mexico, Thailand, (the following have no significant production) Pakistan, Philippines, Singapore, South Korea, and Taiwan. The gap between production and consumption, which was close to zero in 1989, widened to 5.8 million b/d in 1999 and to 10.5 million b/d by 2007. From BP Statistical Review of World Energy, 2008.

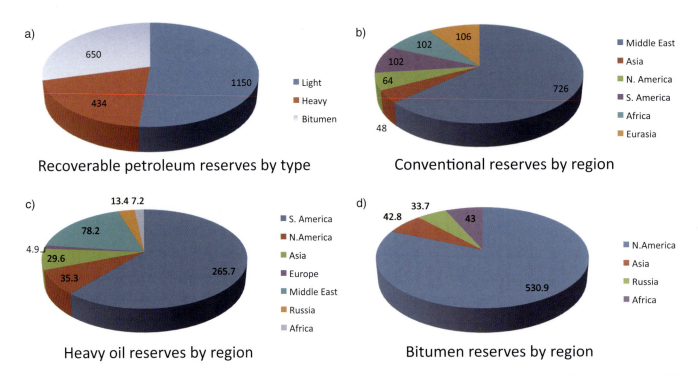

Figure 8. World's current recoverable petroleum reserves in billions of barrels. From Meyer and Attasi, 2003 and Schmitt, 2005.

- Heavy oil properties are frequency dependent.
- Gas coming out of solution (even smallest amounts) can produce a large geophysical signature.
- Heavy oils often act as a cementing agent in unconsolidated sands.
- During production, the reservoir rock matrix is often structurally changed.
- The physical properties of rocks containing heavy oils are temperature dependent.
- The physical properties of rocks containing heavy oils are frequency dependent.
- Simple Gassmann substitution will fail in heavy-oil reservoirs.

As a result of these complexities, we will elaborate on the characteristics of the oils themselves and the influence they have on reservoir rocks.

Chemical properties of conventional crude oil

Crude oil consists primarily of hydrocarbons or compounds comprising hydrogen and carbon only. Some elements such as sulfur, nitrogen, and oxygen are also present in small quantities and are generally combined with the carbon and hydrogen in complex molecules. As oils get heavier or more viscous, their composition becomes more complex and can contain chain and sheet structures with molecular weights in the thousands.

In fact, carbon and hydrogen can form hydrocarbons in several patterns depending on how the carbon and hydrogen atoms are attached to each other. The simplest pattern is the one that has a straight chain and represents the saturated hydrocarbons called normal paraffins. Examples are methane (CH_4, one carbon atom surrounded by four hydrogen atoms), ethane (C_2H_6, two carbon atoms surrounded by six hydrogen atoms), propane (C_3H_8), butane (C_4H_{10}), and so on.

Carbon atoms can also be attached to each other in a branched chain. Compounds with such a pattern are referred to as isoparaffins. If there are three carbon atoms they will form a straight chain compound, but four carbon atoms could also form a branched chain (Figure 9).

An additional pattern is when paraffins form a ring, referred to as cycloparaffins or naphtenes. An example of cyclohexane is shown in Figure 10.

Alternatively, multiple bonds between two carbon atoms are common. In the case of benzene, each carbon is bonded to a single hydrogen. The remaining bonds are shared among the other carbons. A common representation of benzene (Figure 11) is somewhat misleading because the carbon double bonds are not rigidly fixed between alternating carbons but are actually shared over all six carbons. Hence, the hexagon enclosing a circle is often used as a schematic representation. Numerous such rings can form together, growing into sheets. Such compounds are also called aromatics (because of their aromatic odor).

As oils become heavy and gain in molecular weight, the chains and sheets become larger, and identification of individual molecules becomes difficult. As a result, an alternative characterization is often used based on solubility.

SARA fractionation breaks the liquid into **S**aturates, **A**romatics, **R**esins, and **A**sphaltenes. We have discussed saturates and aromatics. Resins are usually defined as the propane-insoluble but pentane-soluble fraction and asphaltenes as soluble in carbon disulfide but insoluble in petroleum ether or *n*-pentane. Notice that resins and asphaltenes are not defined as specific molecular structures. In practice, these molecules have molecular weights in the thousands, are polar, and contain elements such as sulfur, nitrogen, oxygen, and heavy metals. They have no definite melting point (decompose between 300°C and 400°C); exist in a dispersed state in crude oils, but can aggregate to form precipitates; and often are composed of condensed aromatic rings in the form of a nonhomogeneous flat sheet. Examples of asphaltene structures are shown in Figure 12a and 12b.

Sulfur is present in all crude oil samples. Its content can vary from almost negligible content to 5%–6%. It may occur not only as free sulfur, but also in combined form as hydrogen sulfide or as organic sulfur compounds

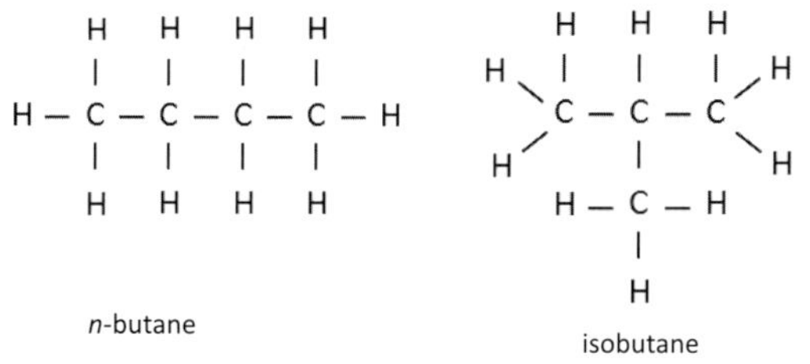

Figure 9. *n*-butane and isobutane.

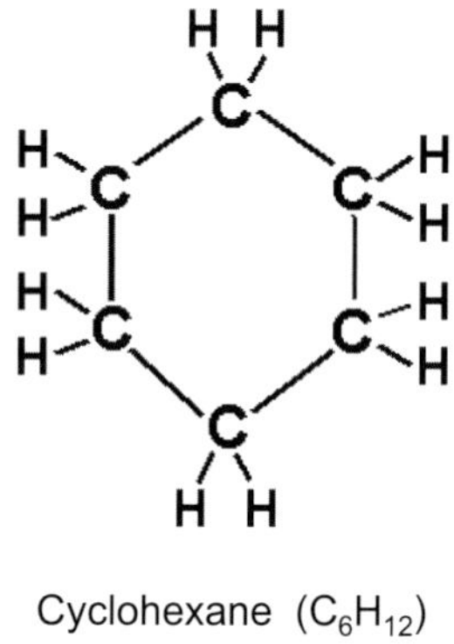

Figure 10. Cyclohexane.

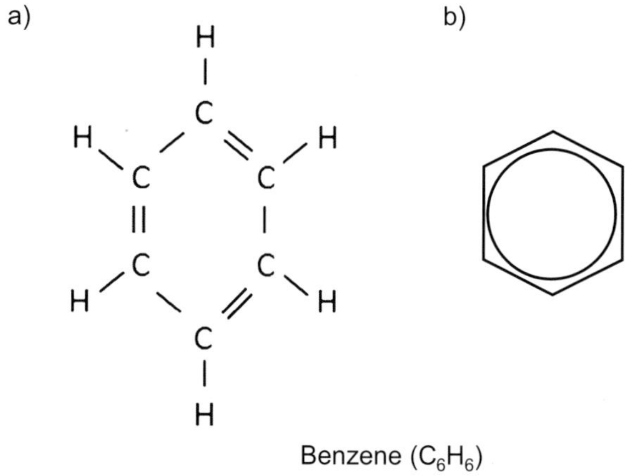

Figure 11. Benzene.

(thiols, mercaptans, disulfides). Crude oil samples containing sulfur cause corrosion and have a bad odor.

Some crude oils contain large amounts of wax in solution. Waxes belong to the straight-chain paraffin series of compounds and have high molecular weights. Wax content in crudes is usually removed on lowering the temperature when it solidifies and settles down.

Density of a crude oil is related to the overall molecular weight, and, for lighter oils, density correlates well to oil properties. Although density is defined as the mass per unit volume of the substance, for crude oil the same physical property is often expressed in terms of specific gravity, which is the ratio of the weights of equal volumes of the substance in question and pure water. As volume changes

Figure 12. (a) Molecular structure of crude Venezuelan asphaltene (Carbognani, 2009), and (b) 3D representation. Image courtesy of J. Murgich.

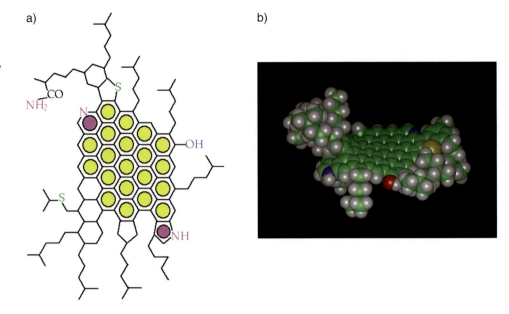

with temperature and pressure, these conditions should be specified. The American Petroleum Institute (API) has recommended the use of API gravity for crude oil, defined as the ratio of density of oil to the density of pure water both taken at 60°F and 1 atmosphere pressure.

$$°API = \frac{141.5}{Specific\ gravity\ at\ 60°F} - 131.5 \quad (1)$$

Classification of crude oil

Although API gravity has no units, it is expressed in degrees. API gravity is graduated on a special hydrometer designed for measuring specific gravities of petroleum liquids so that most values fall between 10° and 70°. This has become the standard used for comparing crude oil samples from different basins and countries.

The higher the API, the higher its commercial value. Interestingly, crude oil with API gravity greater than 10° floats in water; lower than 10°, it sinks. On the basis of its API gravity, crude oil is graded into light (>31.1°), medium (22.3°–31.1°), heavy (<22.3°), and extra heavy or bitumen (<10°).

This grading, recommended by the U. S. Department of Energy, is followed as a standard. Sometimes, heavy oil is so called because it is possible to recover it in its natural state by conventional production methods. However, some heavy oil less than 22.3° API may flow through wells very slowly with some form of stimulation in terms of heat or dilution. Oil that does not flow at all or cannot be pumped without some form of stimulation is called "bitumen." Bitumen mined from oil sands in the Athabasca area of Alberta, Canada, is approximately 8° API. It is upgraded to a higher API gravity (31°–33°), which is known as synthetic oil. Figure 13 shows a sample of heavy oil, dark in color, and slow to flow.

Figure 13. In its raw state, bitumen is a black, asphalt-like oil. Image courtesy of Syncrude Canada, Ltd.

Oil sands are naturally occurring mixtures of sand, clay, water, and bitumen. Bitumen and synthetic oil extracted from oil sands are often referred to as "unconventional" to distinguish them from the free-flowing crude oil recovered from oil wells. Oil sands have recently been incorporated to the world's oil reserves because the available technology can help in recovering oil and other useable products that are economically viable in current market conditions.

"Tar sands" was the term used for bituminous oil sands in the 19th and mid-20th century. However, tar also refers to the sticky viscous substance produced by destructive distillation of coal to pave roads. Petroleum product asphalt has come to replace tar and because naturally occurring bitumen is chemically more similar to asphalt than coal tar, the term "oil sands" is deemed more appropriate by the Alberta Energy Board. Figure 14 shows a sample of oil sands as they are mined in the Athabasca area looking like a crumpled mass of bitumen and sand together with some minerals and metals.

Oil sands and heavy oil are found in many countries including the United States, Mexico, Russia, China, and some in the Middle East (Figure 15). However, the largest deposits of oil sands are found in Canada and Venezuela and their combined reserves equal the world's total reserves of conventional oil. In Canada, oil sands are found in the Athabasca, Peace River, and Cold Lake regions of Alberta, covering an area of nearly 141,000 km^2 (Figure 16). Heavy-oil deposits (8°–19° API) are also found in the Alberta/Saskatchewan border in the area of Lloydminster.

The Athabasca deposit is the only one in the world where oil sands are present shallow enough they can be mined on the surface. Approximately 10% of the Athabasca oil sands are covered by less than 75 m of overburden. Close to 3400 km^2 of mineable area lies to the north of Fort McMurray. The overburden consists of a very thin (<3 m) water-logged muskeg layer that overlies a 75-m column of clay and sand. The oil sands below are typically 40- to 60-m thick and reside on top of a limestone formation.

Figure 14. Oil sands are bitumen and sand blended with some minerals and metals. Image courtesy of Suncor Canada, Ltd.

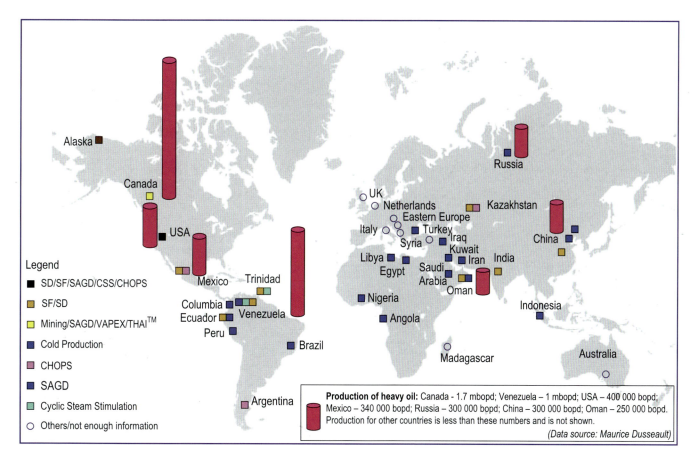

Figure 15. Production of heavy oil worldwide in barrels of oil per day.

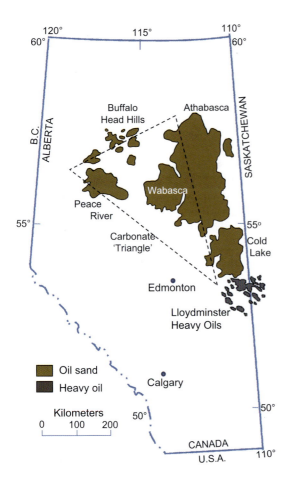

Figure 16. Alberta oil sand, carbonate triangle, and heavy-oil areas. Modified from Proctor et al., 1984.

Mechanisms for the Formation of Heavy Oil

An interesting characteristic of reservoir oil observed worldwide is that the specific gravity of reservoir oil decreases with depth; that is, API gravity increases with depth. The maturation of oil takes place in source and reservoir rocks, kerogen producing lighter oil with depth in the latter (Hunt, 1979). As a result, the average trend observed worldwide is that higher API gravity oil is found at increasing depths, although less oil is produced with depth. It must be pointed out that these are general trends, but there are exceptions.

There are several processes that alter the original oil that occur during migration of oil and its subsequent accumulation. These processes include biodegradation, water washing, oxidation, deasphalting/evaporation, and preferential migration of lighter components (Deroo et al., 1977).

Biodegradation of crude oil over geologic timescales can change its composition and physical properties. Various microorganisms are present in sediments bearing petroleum reservoirs (Bastin, 1926) and utilize the hydrocarbons as a source of carbon for their metabolic processes. The process can be aerobic or anaerobic. Typically, the hydrocarbons are oxidized to alcohols and acids. Simple straight chains are preferred, but as biodegradation continues, more complex molecules are progressively consumed. Long-chain paraffins are oxidized to yield di-acids. Similarly, naphthene and aromatic rings are oxidized to di-alcohols. This biodegradation results in a loss of the saturated and aromatic hydrocarbon content, accumulating resins and asphaltenes, and leads to a decrease of API gravity (increase in density). It has been found that biodegradation can occur if reservoir temperatures do not exceed 80° – 82°C (Hunt, 1979; Head et al., 2003).

Figure 17 shows gas chromatograms exhibiting the effect of biodegradation on a series of oil samples from the central part of the Western Canadian Sedimentary Basin (WCSB) toward the Athabasca oil sands. On the vertical axis is the relative amount of material exiting the "column" or filter material. On the horizontal axis is the time each compound takes to travel through the filter. Individual compounds, such as decane (C_{10}) will appear as sharp peaks. Other peaks are present, including biomarker isoprenoids (e.g., pristine and phytane). Figure 17a shows a conventional crude oil sample from Bellshill Lake pool. Numerous spikes are visible, primarily saturate or alkane chains. A similar analysis shown for the Edgerton sample in Figure 17b shows the depletion of normal alkanes and a relative enrichment of the pristine and phytane. At Flat Lake (Figure 17c), the normal alkanes have almost disappeared and pristine and phytane are quite prominent, although their absolute quantities have decreased. At Pelican (Figure 17d), which is adjacent to the McMurray area, even the isoprenoids have disappeared. Thus, this biodegradation results in an enrichment of the aromatic hydrocarbons, nonhydrocarbon resins, and asphaltenes. Hence, the vast Alberta heavy oils may represent only a small fraction of the original volume of lighter oil that migrated into the sands.

Water washing, oxidation, and deasphalting migration are processes that change crude oil within a reservoir. It is difficult to assess the impact of each of these effects separately.

Deasphalting and subsequent migration of lighter components depletes the residue in lighter hydrocarbons. In some situations, biodegradation can occur if oxygen, inorganic trace nutrients, and water are present; for instance, where there is an oil seep or if the oil accumulation is deeper but adjacent to an aquifer.

Asphalt seals can form at seep outcrops. Asphalt mats have also been formed at the oil-water interface of pools in contact with meteoric water (Hunt, 1996). All of these result from a combination of the above-mentioned processes and microbial alteration. In addition to the formation of asphalt seals, these processes can also cause a change in gravity of the trapped oil (Figure 18). In Lagunillas Field, in north-central Lake Maracaibo, circulating fresh water allows oil biodegradation, and this 5 to 10 km wide belt stretches from the outcrop to the shallowest oil with

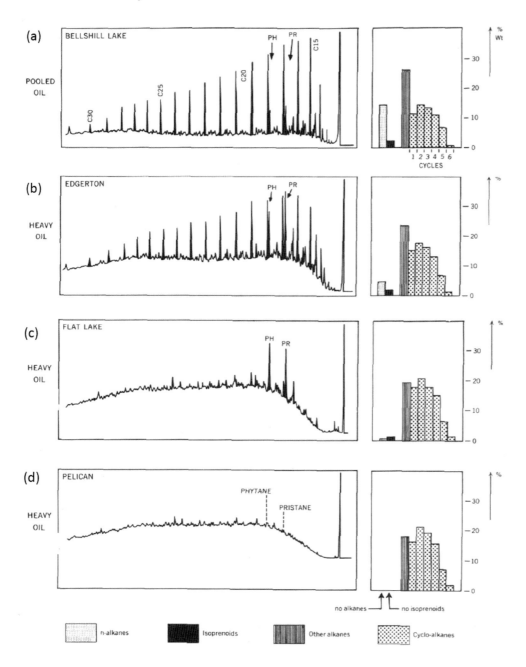

Figure 17. Effect of biodegradation on composition of saturated hydrocarbon samples. (a) Conventional oil from Bellshill Lake; (b) heavy oil from Edgerton, Mannville Formation; (c) heavy oil from Flat Lake area, Colony Formation; and (d) heavy oil from Pelican area, Wabiscaw formation. Adapted from Deroo et al., 1977.

12° API. The oil becomes lighter downdip (20° API) at 1500 m and still lighter (36° API) at deeper levels.

Another mechanism that can lead to heavier oils is precipitation of asphaltene components. If a crude oil is saturated with heavy components, a change in pressure, temperature, or oil-type mix can cause the high-molecular-weight components to drop out of solution.

General Phase Behavior of Hydrocarbons

The existence of a substance in a solid, liquid, or gas state is determined by the pressure and temperature acting on the substance. Just as lowering the temperature can change steam into water, and water into ice, by lowering the temperature further, hydrocarbon compounds individually or in mixtures also change their state or phase. This can happen by not only changing temperature but also pressure. The change of state that is brought about is known as "phase behavior." What type of fluid will be produced at the surface depends on the phase diagram of the reservoir fluid and the reservoir temperature and pressure. Hydrocarbon fluid-phase behavior is important in reservoir simulation, reserves evaluation, and forecasting. In addition, seismic velocity is very sensitive to even small amounts of gas that might develop as a phase boundary is crossed. Because heavy oil and bitumen in deep reservoirs cannot be produced naturally or in conventional ways, it is important to know the thermodynamic changes that these working fluids undergo. The phase behavior of fluids is expressed on phase diagrams, which are graphs showing how fluids behave under different conditions.

Single-component system

To understand the phase behavior of hydrocarbons, let us first look at a simplistic case — a single pure component. Crude oil is a mixture of single hydrocarbons and phase behavior is strongly controlled by composition. For a single hydrocarbon component (e.g., propane or butane), the phase diagram is shown in Figure 19.

The first observation is, as expected, that the vapor pressure of a liquid increases as temperature increases. With the increase in temperature, more liquid molecules escape into the vapor phase, thus increasing the vapor pressure. This is called the vapor pressure curve or boiling point curve. This segment also represents what is known as the bubble point curve or the dew point curve, overlaying one another, representing the transition between the vapor and liquid states. The bubble point refers to the pressure and temperature condition at which the system is all liquid and in equilibrium with a bubble or a very small quantity of gas. Similarly, dew point is the pressure and temperature condition at which a droplet or very small quantity of liquid is in equilibrium with vapor. For a single-component system, a single curve represents all three conditions: vapor pressure, dew point, and bubble point.

When a liquid system is cooled, we would expect the solid state to be formed. The green segment in Figure 19 represents the liquid-solid transition or the melting curve (or solidification). When pressure is low, it is also possible to go from the solid to the vapor phase (sublimation) without going to the liquid phase (the red segment curve in Figure 19). Notice also the critical point, which represents the point at which the fluid is in a supercritical state, and the triple point, which represents that point at which all three phases (solid, liquid, and gas) coexist in equilibrium. These two points bound the boiling point curve. For single-component substances, at the critical point the liquid and vapor phase are indistinguishable. Above this temperature (T_c), the fluid is supercritical and no separate liquid and vapor phases exist.

"Multicomponent systems" are the norm in real reservoir situations (i.e., mixtures of different compounds) from chemically simple to chemically complex.

For a mixture of two compounds, there is no single line segment on the pressure-temperature phase diagram, where bubble and dew point curves overlap (Figure 20). Instead, the bubble point curve shifts to the upper left (higher pressures) and the dew point curve shifts to the lower right (lower pressures), both joining at the critical point. Interestingly, the critical point is not necessarily at the apex of the two-phase region; a supercritical fluid exists at $T > T_c$ and $P > P_c$. Also, if this were a mixture of two components with one more volatile than the other, then at a given temperature the pressure at which this mixture is condensed to total liquid is lower than the pressure at which the lighter of the two components in the mixture would condense individually. Similarly, the pressure at which the mixture of the two components is vaporized to total gas is higher than the pressure at which the heavy component would vaporize individually. In other words, for a mixture, the critical point

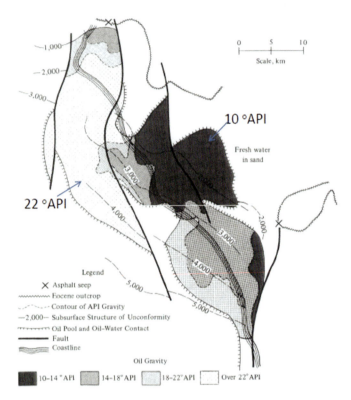

Figure 18. API gravities in the Lagunillas Field, Venezuela. Near the surface, the oil is heavily biodegraded, much denser, and more viscous. From Hunt, 1979.

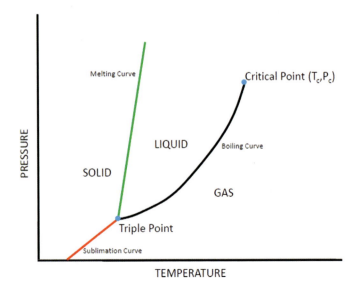

Figure 19. Schematic phase diagram for a single pure compound.

neither represents the maximum pressure nor maximum temperature for vapor-liquid coexistence.

Hydrocarbon Mixtures, Terminology, and Phase Diagrams

Crude oils typically contain gas in solution. As pressure is lowered and the bubble point line is crossed, gas will begin to come out of solution as a separate phase. During production, when crude oil is separated from the accompanying gas at the surface, the volume of gas evolved as oil reaches atmospheric pressure relative to the volume of remaining oil produced is the gas-oil ratio (GOR). It is usually measured in standard cubic feet (or meters) of gas per barrel (or cubic meter) of stock tank, or atmospheric oil (1 scf/bsto = 0.1781 m^3/m^3).

If the volume of gas evolved is low (2000 scf/bsto), the oil is said to be low-shrinkage or black oil (because of its dark color). When GOR is greater than 2000 but less than 3300 scf/bsto, the oil is said to be high-shrinkage or volatile oil (usually brownish). For GOR greater than 3300 and below 50,000, the oil is usually called condensate reservoir gas. Above 50,000 GOR, reservoir gas condenses as a liquid at surface conditions and is called "wet gas." The term "dry gas" refers to natural gas, primarily methane (70–98%) and small quantities of ethane, propane, and butane. Such reservoirs produce no condensate. GOR values associated with dry gas exceed 10,000 scf/bsto. It is called "dry" because no liquid condenses as the gas travels from the reservoir to the surface.

The shape and position of the curve on the phase (P-T) diagram is determined by the chemical composition and the amount of each constituent present. Different types of reservoir fluids have unique phase diagrams and it is interesting to note how they vary. Let us examine the behavior of low-shrinkage crude oil on a phase diagram. A given volume of oil in the reservoir exists in a saturated state with gas. As it makes its way in a horizontal direction toward the well bore, the pressure drops and the temperature remains the same. Figure 21 shows the phase diagram for low-shrinkage oil. At point A (bubble point), the gas starts coming out of solution and the volume of oil shrinks. As fluids travel upward through the well bore, the shrinking continues at the same temperature but reduced pressure. The shrinkage in Figure 21 is 75% of its original volume at point B. Other quality lines indicated for lower pressures are closely spaced along the dew curve. When the fluids reach the separator, both temperature and pressure would decrease and this is indicated with a dotted line to the left. Oil usually accounts for approximately 85% at the condition of the separator.

The phase diagram for high-shrinkage or volatile oil is similar to that of black oil except the quality lines, if drawn, would be close together along the bubble point curve and widely spaced at lower pressures. Oil usually accounts for approximately 65% at the condition of the separator. The above two cases have been discussed for oil reservoirs.

For gas reservoirs, we will consider cases for retrograde gas condensate, wet gas, and dry gas.

Retrograde gas condensate

Some reservoirs may contain gaseous hydrocarbons (initial condition) that exist above their critical temperature and pressure. As these fluids move closer to the wellbore, they encounter reduced pressure and instead of expanding

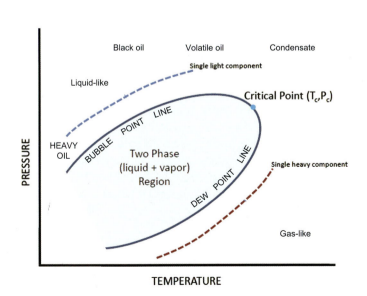

Figure 20. Schematic phase diagram for hydrocarbon mixtures.

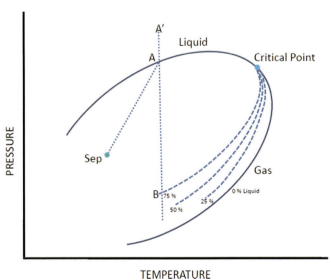

Figure 21. Schematic phase behavior for a low-shrinkage hydrocarbon system.

(for gas), they condense. In this sense, the term "retrograde" is used because normally gases liquefy when pressure is increased and not when pressure is decreased. So during gas production at the surface, condensate drops out in the separator. In such reservoirs, the fluids in place could be just gaseous or gaseous and liquid, depending on the pressure and temperature of the reservoir.

Wet gas

The temperature is above the critical condensing temperature of the gas mixture. At constant temperature, when there is a reduction of pressure, there is no condensation. However, when the gas from the reservoir reaches the separator, the temperature is lower and so liquid formation results.

Dry gas

The temperature is again above critical condensing temperature of the gas mixture and just as the wet gas, it does not condense with the reduction of pressure (at constant temperature). Also, when the gas reaches the separator, there is no condensation.

Properties of Heavy Oil

The bulk modulus (K) is a characteristic of pore fluids that strongly controls seismic properties. It is the inverse of the common engineering term *compressibility* (C). K relates the change in volume (V) for a given change in pressure (P).

$$K = -\frac{1}{V}\frac{dV}{dP} = \frac{1}{C} \qquad (2)$$

For a mixture of two separate phases, A and B each with their own bulk modulus K_A and K_B, the overall mixture modulus K_{mix} usually is described by Wood's average.

$$1/K_{mix} = V_A/K_A + V_B/K_B = V_A/K_A + (1 - V_A)/K_B, \qquad (3)$$

where V_A and V_B represent the volume fraction of phase A and B, respectively. Because of the inverse dependence on component modulus, a very low modulus of soft fluid will dominate the mixture modulus. Such is the case when even insignificant amounts of very low modulus (very compressible) gases are present.

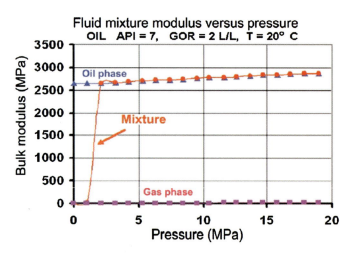

Figure 22. Calculated bulk modulus (mixture for a heavy oil sample) plotted as a function of pressure. With a decrease in pressure, the oil crosses the bubble point, lowering the modulus appreciably. From Batzle et al., 2006.

As shown in Figure 21, during the recovery process, pore pressure can cross the bubble point in either direction as the pressure or temperature changes. The calculated fluid bulk modulus (from Han and Batzle relations) of 7° API oil is plotted against pressure in Figure 22, in which sample the bubble point can be crossed at approximately 2 mPa. About this value, the bulk modulus of the homogeneous mixture is very high, 2600–2800 mPa, which is above the bulk modulus of water. However, once the bubble point is crossed, gas comes out of the solution and the modulus drops to near zero.

Similarly, the phase boundary can be crossed by changing temperature. As seen in Figure 23, the calculated bulk modulus is plotted as a function of temperature. Bubble point crossing occurs at approximately 120° C and the drop in bulk modulus value is of the same order as the pressure. Such appreciable changes in the

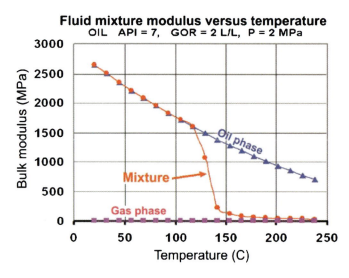

Figure 23. Calculated bulk modulus (mixture for a heavy-oil sample) plotted as a function of temperature. An increase in temperature causes the oil to cross the bubble point, appreciably lowering the bulk modulus. From Batzle et al., 2006.

bulk modulus suggest that the heavy-oil properties and their phase behavior need to be well understood for seismic monitoring.

Density correlates to many other oil properties. At standard conditions, density is used to define API gravity. However, density is not constant with changing pressure or temperature. Figure 24 shows the variation for density with temperature for an 8.5° API bitumen sample from Athabasca. Density decreases from 1.01 to 0.935 g/cm^3. Similar density correlation is seen in Figure 25 between Cold Lake bitumen and temperature, where measurements were taken over a range of 20°–250°C and pressure range of 3.5–7 mPa.

Viscosity is one of the defining attributes of heavy oils. Generally it increases with a decrease in temperature and in API. As the temperature decreases, heavy oil changes its phase from the low-viscosity liquid phase to a drastically higher quasi-solid phase and eventually to the glass phase, where the viscosity is over the glass point (Han et al., 2006), defined as that temperature at which viscosity is equal to 10^{13} poise.

Because heavy oil consists of complex heavy compounds, the simple empirical trends developed for estimating light oil fluid properties such as viscosities, densities, GORs, and bubble points seldom apply. Although at higher temperatures some of these empirical trends may be obeyed, at lower temperatures the viscosity of heavy oils is high, exhibiting different properties that necessitate special consideration. The viscosity of heavy oils is especially important because production methods exploit this property.

High API, light oil viscosity depends on temperature and the amount of gas dissolved in it; that is, the higher the temperature or the more gas content in solution, the lower the viscosity. Also, viscosity changes slightly with pressure.

Higher temperatures increase the molecular agitation, which in the absence of confining pressure increases the volume and so the intermolecular distances. This reduces molecular attraction and friction caused by colliding molecules.

The viscosity of heavy oils is mainly dependent on temperature and oil gravity. Figure 26 shows the variation of viscosity with temperature for 10.3° API Cold Lake bitumen. The curve exhibits a double logarithmic relationship between viscosity and temperature.

Beggs and Robinson (1975) developed the following empirical relationship between the viscosity and temperature as well as density.

$$\text{Log}_{10}(\eta^T + 1) = 0.505_y(17.8 + T)^{-1.163}, \qquad (4)$$

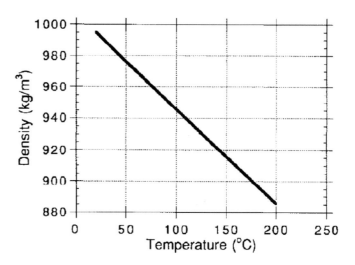

Figure 25. Density of Cold Lake bitumen as a function of temperature on the basis of laboratory measurements. From Eastwood, 1993.

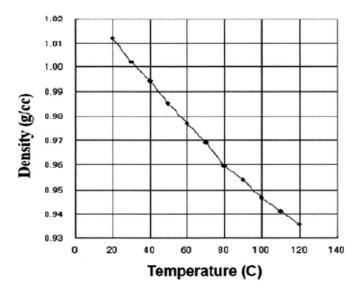

Figure 24. Temperature dependence of density for Athabasca bitumen. From Mochinaga et al., 2006.

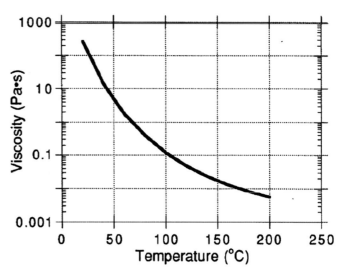

Figure 26. Viscosity of Cold Lake bitumen as a function of temperature on the basis of laboratory measurements. From Eastwood, 1993.

where

$$\mathrm{Log}_{10}(y) = 0.5693 - 2.863/\rho_o \quad (5)$$

and η is the viscosity in centipoises (cp), T is the temperature in °C, and ρ_o is the density at standard temperature and pressure (STP). The variation of viscosity with temperature is plotted in Figure 27 on the basis of the above equation. Note that the low temperature limit is fixed by the value 17.8, and so the equation is questionable for temperatures below 0°C.

The physical properties of heavy oil/bitumen must be understood to anticipate production performance and calculate reserves. These properties are determined from laboratory experiments on samples collected from formations of interest or from the surface. Empirical correlations derived from such experiments are applicable in a well-defined range of reservoir fluid characteristics; thus, when the laboratory pressure-volume-temperature (PVT) data become available, the required information can be derived from the empirical correlations.

Velocity is a crucial piece of information derived from seismic data. Velocities themselves depend on moduli of the material through which the seismic wave propagates. For example, velocity (V_P) for compressional waves is

$$V_P = [(K + 4/3G)/\rho]^{1/2}, \quad (6)$$

where K is the bulk modulus, G is the shear modulus, and ρ is density. For most fluids, the shear modulus is zero, and the compression velocity reduces to

$$V_{\mathrm{Pfluid}} = [K/\rho]^{1/2}. \quad (7)$$

However, as we shall see, heavy oils do not act like most fluids.

Nur et al. (1984) studied the effect of temperature and pressure on P- and S-velocities and amplitude by making ultrasonic measurements on reservoir samples of heavy oil and tar sands from Kern River, California; Maracaibo, Venezuela; and Athabasca, Alberta. They found that in fully oil-saturated sands, velocities showed a 40% decline as temperature increased from 25° to 150°C at constant pressure. This strong dependence is reversed when brine replaces oil in the samples, with velocities showing a strong dependence on differential pressure but little dependence on temperature. Similarly, amplitude decreases significantly with increasing temperatures in samples with oil as compared with brine. This behavior is characteristic of heavy-oil and tar samples, and neither brine nor gas-saturated samples display it. This raises the possibility of using seismic methods to detect temperature anomalies in reservoirs. Tosaya et al. (1987) and Wang and Nur (1988) performed laboratory experiments to demonstrate the velocity and attenuation dependence on temperature for several heavy-oil/tar-sand samples.

Wang et al. (1990) measured acoustic velocities at ultrasonic frequency (800 kHz) in oil samples of different API as a function of temperature and pressure. They found that velocities in oils increase with increasing pressure and decrease with increasing temperature. Figure 28 shows the measured velocities plotted as a function of pressure for different temperatures in 5°, 7°, and 12° API heavy-oil samples. Apparently, the velocities vary linearly with increasing pressure. At higher temperatures, the velocities are slightly more sensitive to pressure changes. Also, velocities decrease nonlinearly as temperature increases and decreases faster at lower (20°C–45°C) temperature ranges. This is because solid or semisolid (asphaltene and wax) components of heavy oil, which could cause a decrease in velocity while melting. Once these materials have melted, velocities decrease linearly with the increase in temperature, which is typical behavior of light oils.

Eastwood (1993) performed several experiments to measure the effect of temperature on acoustic velocities in oil-sands samples from Cold Lake, Alberta, Canada. Figure 29 shows the temperature dependence of compressional velocities in two samples of bitumen (fluid only) over the temperature range 22°–127°C and at a constant pressure of 0.1 mPa. The velocity is seen to decrease by

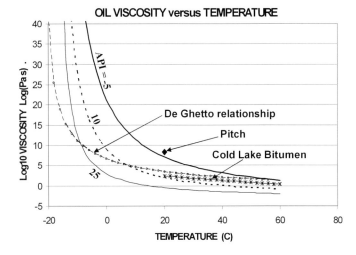

Figure 27. Variation of viscosity with temperature as computed from the empirical relationship given by Beggs and Robinson (1975), which produces a singularity at low temperatures. The data from Eastwood (1993) and Edgeworth et al. (1984) are also plotted. The heavy oil relationship from De Ghetto et al. (1995) is also indicated. From Batzle et al., 2006.

approximately 30% relative to the velocity at 22°C, and this dependence exhibits a linear trend for temperatures 60°C and higher and a departure from this linear trend for temperatures lower than 60°C. This observation is similar to the conclusion by Wang et al. (1990).

As we have seen, velocity and density decrease when temperature for heavy oil, oil sands, or bitumen increases, with the magnitude of the decrease depending on the temperature and differential pressure, which is the difference between the confining pressure and pore pressure. This decrease in velocity and density results in a corresponding change in the amplitude, which can be determined from two or more successive seismic surveys acquired for monitoring changes in the reservoir. Thus, these property changes suggest the usefulness of time-lapse seismic surveys to monitor steam fronts during steam flooding or SAGD operations.

Shear properties

As the viscosity of heavy oil becomes high, it effectively has a nonnegligible shear modulus. This transition can be tested in the laboratory by propagating a shear wave through the fluid sample. Batzle et al. (2006) noticed that for a very heavy oil sample (−5° API), at low temperatures (−12.5°C), a sharp shear-wave arrival is detected. At this temperature, the oil is almost a solid and so it has a shear modulus. As the temperature is increased, the shear velocity decreases and this also reduces the shear-wave amplitude (Figure 30). At such a stage, the oil is only marginally solid. Both the compressional and shear moduli decrease almost linearly with temperature, but the shear modulus approaches zero at approximately 80°C.

Because heavy oils act as a viscoelastic (semisolid) material, there is also frequency dependence. At room temperature, heavy oil supports a shear wave, but as the temperature increases, its shear modulus decreases rapidly, which in turn leads to a rapid drop in the shear modulus of the heavy-oil saturated rock (Figure 31a). At all temperatures, heavy oils would have a nonzero bulk modulus and the percentage change in bulk modulus would be smaller than the change in shear properties. This shear information can be more diagnostic than bulk modulus properties. Therefore, multicomponent seismic

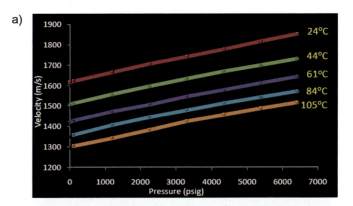
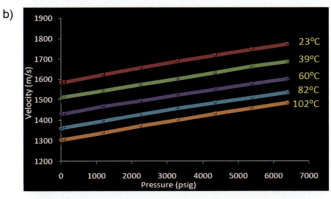
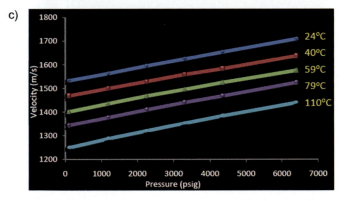

Figure 28. Measured acoustic velocities in heavy-oil samples with (a) 5° API, (b) 7° API, and (c) 12° API. From Wang et al., 1990.

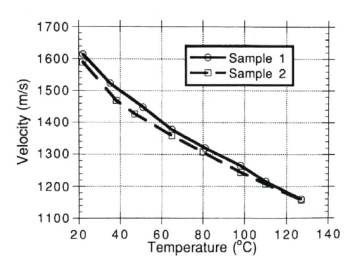

Figure 29. Experimentally measured compressional velocities (ultrasonic) in two Cold Lake bitumen (fluid only) samples as a function of temperature. The experiment was conducted at a pressure of 0.1 mPa. Velocities decrease almost 30% relative to the velocity at 22°C and exhibit a linear trend for temperatures above 60°C but not below. From Eastwood, 1993.

data should be able to help with the determination of V_P/V_S and the derivation of bulk modulus properties.

However, notice that the modulus increases with increasing frequency. This frequency dependence, also referred to as "dispersion," is directly related to the attenuation in the material. The typical measure of attenuation is the quality factor, and in Figure 31b, we show the quality factor plotted against the modulus for a heavy-oil saturated rock. At lower temperatures, the quality factor increases with frequency and initially decreases with temperature. However, at higher temperatures, Q and the shear modulus increase with increasing temperatures. This behavior is most likely due to a loss of lighter components at elevated temperature.

Rocks with Heavy Oil

Many various rock types are saturated with heavy oils. Because most heavy-oil reservoirs are near the surface, the rocks are often poorly consolidated. Figure 32a shows an example of Athabasca heavy-oil sand. Most of the grains are oil coated and dark. These sands are fluvial and migrated into the pores early after deposition, preventing significant cementation. In fact, the heavy oil itself acts like cement, and removing it usually causes the matrix to collapse. Thus, one of the primary assumptions in a Gassmann substitution is violated — that the matrix remains unchanged under different saturation conditions. One issue that needs further investigation is the matter of grain surface wettability. Under normal conditions, the Athabasca sands are presumed to be water-wet. This may not be the case for all rocks.

Many heavy-oil reservoirs are not in clastics. The Grosmont formation contains a substantial quantity of Alberta's heavy oil and, like most Middle East deposits, this formation is primarily composed of carbonates. Figure 32b shows a heavy-oil saturated carbonate from Texas. It is a quarry sample above the water table; therefore, in situ, this rock may be oil-wet. The matrix is made up of porous oolite grains. Note that the heavy oil is present only in the larger pore spaces.

In measurements of heavy-oil sands, Amos Nur and his coworkers noticed a dramatic temperature dependence of compressional velocity (Tosaya et al., 1987). This effect is most pronounced when the only pore fluid present is heavy oil (Figure 33). This observation led these researchers to first suggest that time-lapse seismic measurements would prove to be a useful reservoir monitoring tool.

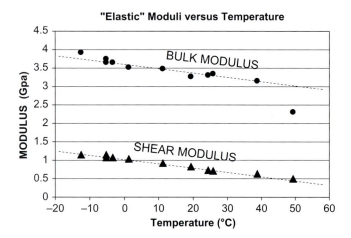

Figure 30. Measured (ultrasonic) bulk and shear moduli in a heavy-oil sample of $-5°$ API. From Batzle et al., 2006.

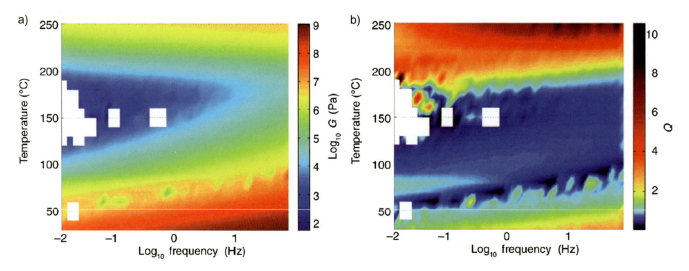

Figure 31. (a) Shear storage modulus (the real part of the complex modulus), and (b) quality factor Q of Uvalde heavy-oil rock. Measurements at temperature increments of $10°C$ and frequency increments of 0.1 on the $\log^{10}$ scale. Missing data points correspond to erroneous results (e.g., negative moduli) arising from noise and/or experimental errors and/or measurements lying outside of the sensitivity limit of the rheometer. From Behura et al., 2007.

Figure 32. (a) Reflected light view of a heavy-oil sand (from Han et al., 2008). (b) Scanning electron microscope image of the Uvalde carbonate saturated with heavy oil (from Batzle et al., 2006).

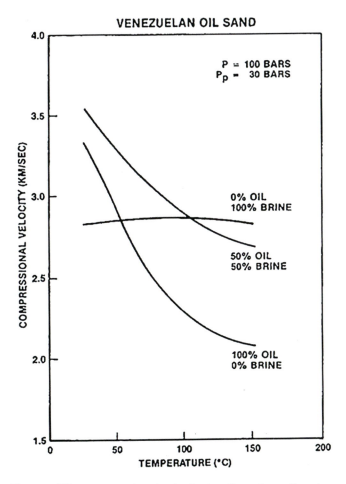

Figure 33. Compressional velocity in oil sand as a function of temperature and saturation. From Tosaya et al., 1987.

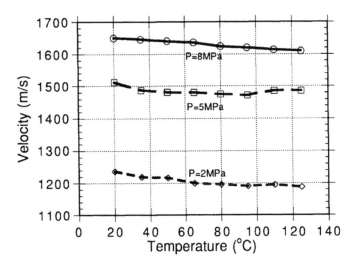

Figure 34. Experimentally measured compressional velocities (ultrasonic) in a Cold Lake oil-sands sample devoid of heavy oil as a function of temperature and effective stress. Notice the weak dependence of the solid rock matrix on temperature when compared with the properties of bitumen. From Eastwood, 1993.

The question arises, "How much of the temperature dependence is due to the rock frame, and how much is due to the oil?" Eastwood (1993) measured the effect of temperature on acoustic velocities in samples of the cleaned oil sands from Cold Lake. Figure 34 shows the temperature dependence of measured compressional velocity for a Cold Lake dry rock sample (no bitumen present) at different pressure values. The change in velocity with pressure is rather small, indicating that the dry rock matrix contributes little to the wave velocity for the temperature range at which measurements are made. Eastwood also found moderate temperature dependence for S-wave velocity for this sample. In contrast, Figure 35 shows velocity measurements made on a preserved Cold Lake oil-sands sample at different pressures. Notice the velocities decrease by approximately 12% relative to the initial velocities at 22°C and exhibit pressure dependence.

Another example of nonlinear dependence of velocity of heavy oil on temperature is shown in Figure 36. As the temperature is increased to 150°C, the velocity decreases by approximately 25%. Above 90°C, the velocity drops linearly with increasing temperature as is common with lighter oils. Below 70°C, the velocity departs from linearity, and in this particular case, the oil viscosity increases, approaching its glass point and beginning to act as a solid. When this happens, the change in the velocity is not only due to an increase in the bulk modulus, but also the appearance of a shear component, which is negligible when it is a liquid.

Dispersion in rocks (i.e., the variation of velocity with frequency) is important to be able to correlate laboratory acquired data with seismic or log data. Wang and Nur (1988) suggested that the velocity dispersion in rocks saturated with heavy oil is much larger than in the same rocks saturated with lighter fluids. Their results indicate that for heavy-oil saturated Berea sandstone, velocity dispersion could be as large as 10%. The authors try to explain the observed velocity dispersion in terms of the "local flow" mechanism (O'Connell and Budiansky, 1977; Mavko and Nur, 1979). This mechanism is probably not valid because of the high fluid viscosity.

Heavy oils themselves show strong frequency dependence (Figure 31). This fluid property is then directly imparted to the oil-saturated rock. Measurements on the Texas carbonate (Figure 32b) as a function of frequency are shown in Figure 37. Note the strong frequency dependence on compressional and shear waves. Dispersion becomes more pronounced as temperatures increase. Direct measurements indicate that observations made in the seismic band of 10–100 Hz often do not agree with standard acoustic log data (10,000 Hz) nor with ultrasonic (MHz) data.

Behura et al. (2007) measured in the laboratory the complex shear modulus and attenuation of a heavy-oil saturated rock sample at elevated temperatures within the seismic frequency band. As seen in Figure 38a, the modulus and quality factor (Q) of the heavy-oil saturated rock show a moderate dependence on frequency but are strongly influenced by temperature. These dependences are consistent with the measured properties of the extracted heavy oil as discussed before.

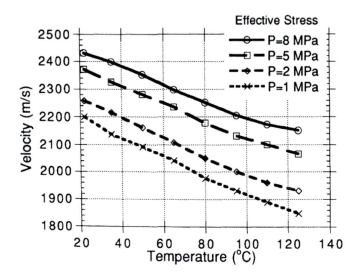

Figure 35. Experimentally measured compressional velocities in a saturated Cold Lake heavy-oil sand as a function of temperature. The ultrasonic experiment was conducted at a pressure of 0.1 mPa and measurements were performed in a drained configuration (i.e., as temperature increases, bitumen is allowed to flow out of the sample until equilibrium is reached). Velocities decrease almost 12% relative to the velocity at 22°C. From Eastwood, 1993.

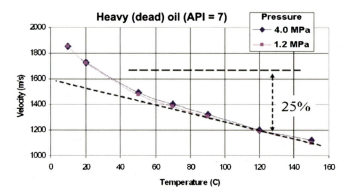

Figure 36. Measured P-wave velocity for a heavy-oil sample (7° API) at two different pressure values and at increasing temperature. Velocity variation is linear above 90°C, but below 70°C it becomes nonlinear. Pressure seems to have a minimal effect. From Batzle et al., 2006.

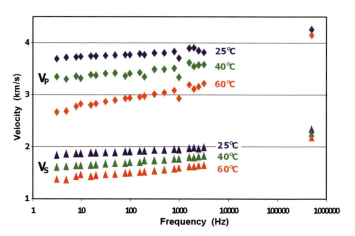

Figure 37. Frequency dependence of Texas heavy-oil saturated sample. Velocity decreases with increasing temperature. Also, dispersion is significant within the seismic band as temperature increases. From Kumar, 2003, and Batzle et al., 2006.

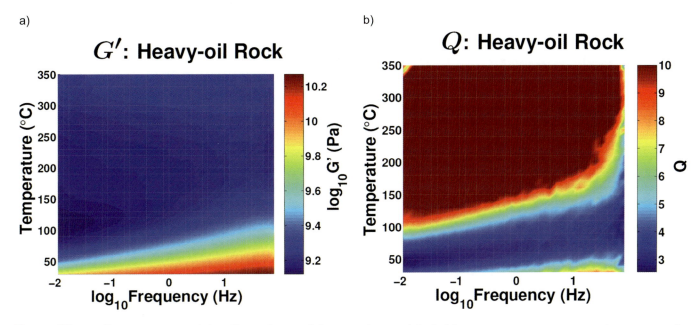

Figure 38. (a) Shear storage modulus (the real part of the complex modulus). Measurements at temperature increments of 10°C and frequency increments of 0.1 on the $\log_{10}$ scale. (b) Quality factor Q of Uvalde heavy-oil rock under the same measurement conditions. From Behura et al., 2007.

Schmitt (1999) also observed a significant difference between acoustic log and vertical seismic profile (VSP)-derived interval velocities in a heavy-oil sand reservoir (Figure 39). Thus, the viscoelastic (semisolid) behavior of the heavy-oil sands must be considered when applying data collected in one frequency band (logging or laboratory) with those collected in another band (seismic). Such discrepancies would result in differences in synthetic modeling in reservoirs because the reflection of heavy-oil sands based on acoustic logs would be different from the low-frequency seismic response.

Geology of Two Major Heavy-oil/Oil-sands Areas

Three-quarters of the world's heavy-oil reserves are found in two areas: the Orinoco Belt in Venezuela and the Athabasca region in the northern Alberta and Saskatchewan provinces of Canada.

Venezuela

To the north of the Orinoco River, there is an extensive 54,000 km² (600 × 90 km) area, the Orinoco Heavy-Oil Belt, which contains an estimated 270 billion barrels of recoverable oil (Talwani, 2002) — the equivalent to the oil reserves of Saudi Arabia. This elongated region is divided into four sections: Machete, Zuata, Hamaca, and Cerro Negro (Figure 40). The Orinoco Belt contains an estimated 1.2 trillion barrels of heavy oil.

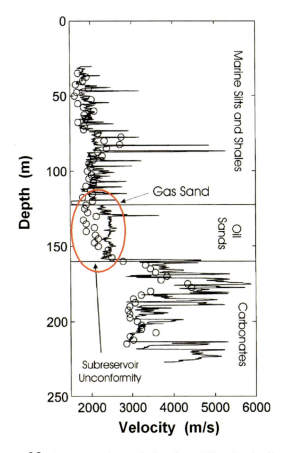

Figure 39. Segment of a sonic log from Pikes Peak, Canada. Interval velocity values from VSP (small circles) are seen overlaid on this curve. Significant difference between the two measurements (red circle) has been interpreted to be due to dispersion. From Schmitt, 1999, and Batzle et al., 2006.

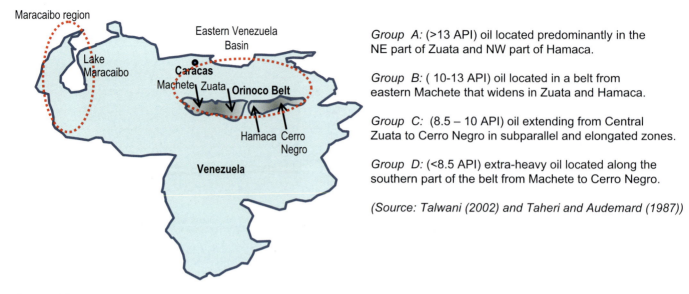

Figure 40. Orinoco Oil Belt in Venezuela comprises four production zones, each with distinct heavy-oil characteristics in terms of API gravity, viscosity, and metal and asphaltene content. The quality of the crude has been rated from highest A to lowest D. From Talwani, 2002.

These giant deposits of extra-heavy crudes (<10° API), the largest in the world, come from Tertiary/Cretaceous reservoirs and contain large amounts of vanadium, nickel, and sulfur (Talwani, 2002). Because extra-heavy oil cannot be processed in conventional refineries, it is blended with lighter Venezuelan crudes and upgraded into synthetic oil, which is exported to the tune of 600,000 b/d (Christ, 2007).

During the Mesozoic era, the supercontinent Pangea started drifting and gradually split into a northern continent, Laurasia, and a southern continent, Gondwana. Eventually, Laurasia split into North America and Eurasia, and Gondwana split into South America, Africa, Australia, Antarctica, and the Indian subcontinent, which later collided with the Asian plate during the Cenozoic, leading to the formation of the Himalayas.

With this breakup, the northern passive margin (transition between basaltic oceanic plate and granitic continental plates) of South America was created. With subsequent deposition, this margin started subsiding. Sedimentation during the Middle Cretaceous was rich in organic content, leading to the source rock in the La Luna formation in the Maracaibo region and the equivalent Querequal and San Antonio formations in the Eastern Venezuela Basin (Figure 40). These source rocks have produced large quantities of oil and gas in Venezuela.

The passive margin sequence terminated during the Oligocene when the Caribbean Plate collided against the South American Plate. The passive margin changed to a foreland basin and this oblique collision gradually migrated eastward from the late Oligocene to early Miocene, causing the formation of thrust belts, which got uplifted and eventually eroded. The foredeeps located south of the thrust belt received abundant sedimentation (shale) that created the Eastern Venezuelan Basin. South of this, flexural and isostatic uplift led to the elevation and subsequent erosion of the ancient Guyana Shield. The eroded sediments were transported by the north-flowing rivers into the Eastern Venezuela Basin. The formation of sandstone deposits took place during the Oligocene-Miocene and are represented by the Oficina formation and its equivalents. These formations contain the bulk of the oil in the Orinoco Belt.

Gradual and continuous loading of sediments continued in the Eastern Venezuela Basin and the eventual compaction subjected the organic matter to increasingly higher temperatures with greater depths of burial. Thermal degradation (also called "cooking") of organic-rich sediments took place from north to south, which is why the oil being formed migrated a few hundred kilometers updip to reach the southern part of the basin during the late Middle Miocene. During migration, the lighter oil fractions escaped or were acted upon by microbial activity, ending up as heavy-oil deposits.

Subsequent tectonic activity led to faulting, which cut off the supply of oil to the Orinoco Belt and so disrupted the migration pathways.

Canada

The bulk of Canadian oil sands are concentrated in a 142,000 km^2 region encompassing the Athabasca-Wabiskaw, Cold Lake, and Peace River areas (Figure 16), which contain at least 175 billion barrels of recoverable crude bitumen. Athabasca oil sands are Canada's largest deposits, holding 80% of the total.

In some areas with little overburden, deposits are so close to the surface that strip mining is the most efficient method of oil recovery. Athabsaca oil sands are typically 40–60 m deep, lying above an almost flat limestone rock. The overburden consists of muskeg, a 1 to 3 m thick water-soaked layer of decaying plant matter over a layer of clay sand and gravel. Because of these conditions, Athabasca is the only region in the world where strip mining oil extraction is possible over roughly 3450 km^2 of its total territory. This represents 2.5% of the total oil sands in Alberta.

The regional geology of the Western Canadian Sedimentary Basin (Figure 41) is shown in cross section AA' of Figure 42. The Alberta craton essentially consists of a simple monocline. A major unconformity separates the Paleozoic sediments from the gradually dipping Mesozoic strata (Deroo et al., 1977). The margins defining the basin are the Rocky Mountain Thrust Belt to the west and the boundary along which the sediments lap onto the Precambrian Shield to the northeast. To the southeast, the basin continues into Saskatchewan. The relatively deep sediments adjacent to the thrust belt thin out to the north and northeast because of depositional thinning and erosion.

During the period between Upper Devonian and Jurassic, the Alberta craton was dominated by deposition of marine carbonate and clastic sedimentary rock during a passive margin period. Marine sedimentation took place on a stable shelf including the deposition of Devonian reefs and associated off-reef facies. At the end of this period, orogenic movement to the west resulted in deposition of thick continental strata in the Alberta Foothills. This period ended in Tertiary time with the uplift of the land in the Canadian Cordillera, so that marine sedimentation made way for continental sedimentation. The continental sediments were derived and transported from the western uplift associated with the formation of the Western Canadian Cordillera to be deposited into the Inner Cretaceous Seaway to the east. This later progressed from the southwest and eventually from south to north. The uplift of land during this period created a deep, broad foreland basin. During the early Cretaceous, a major drop in the base level set forth erosion of the sediments deposited during late Jurassic through Devonian. This created an unconformity in the east-west direction truncating the formations (Figure 42).

With the subsequent uplift of the Canadian Cordillera, deposition of thick sediments from late Cretaceous into the Tertiary overlaid the unconformity with a succession of sandstones, shales, and conglomerate filled channels. The permeable sediments are thought to have provided a conduit for updip migration of the hydrocarbons and basinal fluids from the source to the host rocks where bitumen is found today. As the uplift of the Canadian Cordillera continued, downward bending of the adjacent region referred to as the Western Plains was taking place. This formed the Alberta Syncline, which received large quantities of clastic sedimentation through erosion of the Rockies. The uplifting Canadian Cordillera continued through the Neogene, which resulted in the erosion of the Tertiary sediments (Figure 42).

The widespread Devonian limestones, dolomites of the Grosmont formation, and associated reefs cover a large part of east-central Alberta. These are overlain by the Upper Devonian Winterburn and Wabamun formations or their stratigraphic equivalents. The eroded updip edge of the Grosmont formation, which is approximately 130 km southwest of the oil sands found at Fort McMurray, is overlain by the stratigraphic equivalent formations of the upper beds of the McMurray formation. The subcropping carbonates mostly fall in the carbonate triangle marked in Figure 16. The oil in these carbonates has the same chemical characteristics of the oil from the oil sands above, which suggests that the carbonates were sourced from above.

The commonly accepted theory for the origin of the oil in sands is that the shales associated with the Lower Cretaceous in an oil migration pattern from west to east

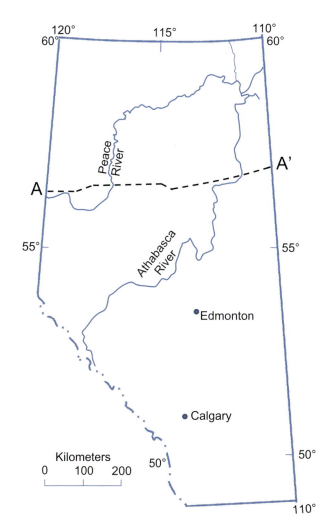

Figure 41. Location of line cross section AA' in Alberta. Modified from Deroo et al., 1977.

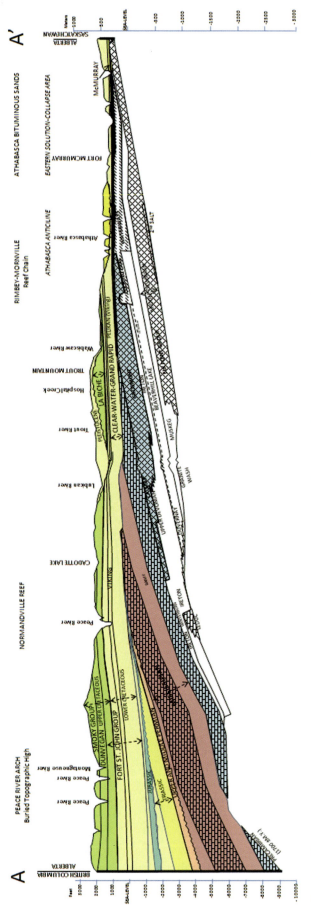

Figure 42. Regional cross section of Alberta Syncline showing structural and stratigraphic relationships. Modified from Deroo et al., 1977.

(Demaison, 1977). Cretaceous shales are organically rich (Deroo et al., 1977) and thermally mature downdip of the Peace River accumulation (Hacquebard, 1975), suggestive of a long-distance migration (approximately 80 km for Peace River and approximately 380 km for Athabasca) for oil. In addition to Lower Cretaceous shales as source rocks, shales of other ages have also been postulated as contributing to oil accumulation (Riediger et al., 2000). The origin of such a colossal volume of hydrocarbons and its timing are still controversial. What is generally accepted is that whatever the source, these hydrocarbons were generated deep in the basin, migrated long distances updip, and accumulated in shallow stratistructural traps near the basin's eastern edge. Over time, under the action of water and bacteria, the light crude was transformed into bitumen. Degradation of the trapped oil has been found to be more severe at the edge of the basin (8° API at Athabasca), and this severity decreases basinward in steps from one deposit to another (26° API at Bellshill Lake, 250 miles southwest) (Deroo et al., 1974; Jardine, 1974).

Most of the exploitable bitumen resides in unconsolidated Lower Cretaceous sandstones in the Athabasca, Peace River, and Cold Lake areas. The Devonian and Mississippian carbonate reservoirs unconformably overlain by the Athabasca and Peace River oil sands also host bitumen but have not been exploited commercially. Figure 43 shows the initial volumes of crude bitumen by formation and area. The Athabasca with the in-place and mineable Wabiskaw McMurray formations followed by the Grosmont formation (Hein and Marsh, 2008) is by far the most significant deposit. Other formations like Grand Rapids, Clearwater, and Bluesky/Gething hold significant in-place volumes of bitumen.

Figure 44 is the bitumen-pay thickness map for the Athabasca area on the basis of a lower cutoff of 6% porosity and 1.5 m thickness. The general shape of the contours indicates that most of the bitumen resource occurs as a north-south trend along the eastern margin of the Athabasca oil-sands area (Hein and Cotterill, 2006).

Heavy-oil Recovery

Techniques in heavy-oil recovery can be divided into surface mining and in situ recovery (Figure 45).

Surface mining

Two tons of oil sand mass render approximately one barrel of oil. Oil sands are mined by means of massive power shovels holding as much as 100 t of load (Figures 46 and 47). Dump trucks with a capacity of 400 t convey the mined oil sands to crushers where the bigger lumps are broken and other hard rocks are removed. The remaining sand is mixed with water and air at about 50°C, agitated in a cyclofeeder, and the slurry is then pipelined to the processing unit. The use of caustic soda and hot water has been discontinued to allow water recycling, thus reducing the volume of water that ends up in tailings ponds.

The slurry passes through vibrating screens that separate larger materials before it proceeds to the primary floatation separator. There, much of the bitumen content attaches to air bubbles, forming a froth layer while the coarse minerals and sand settle in the vessel. The middlings

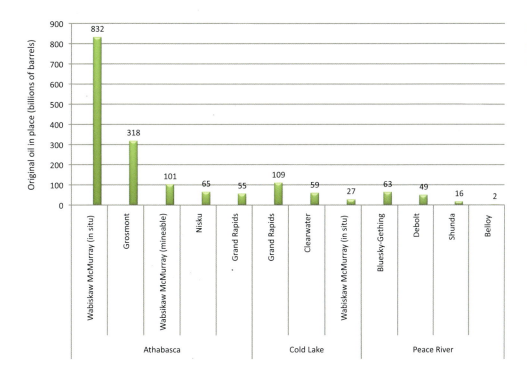

Figure 43. Original oil in place, Alberta bitumen. From Hein and Marsh, 2008.

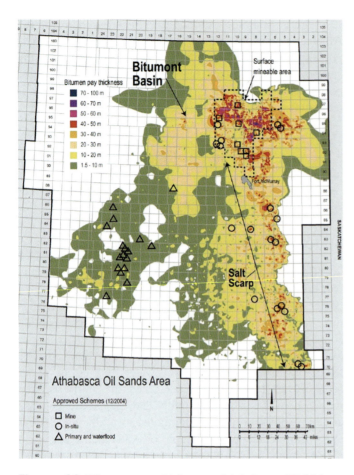

Figure 44. Bitumen pay thickness of Athabasca Wabiskaw McMurray deposit for areas with >1.5-m bitumen. Contour interval is 10 m. Annotations showing surface mineable area (dashed line) and approved schemes including surface mining (squares), in situ thermal (circles), and primary and waterflood (triangles). Modified from Alberta Energy and Utilities Board, 2008; adapted from Hein and Cotterill, 2006.

part is removed from the vessel and passed into an aerated secondary floatation vessel. The resulting froth is combined with the primary froth. The water from this step is recycled, the sand is stripped of oil, and the polluted water goes to the tailings ponds. The bitumen is directed to storage tanks from where it is transported for upgrading.

Tailings ponds are usually built in old mine pits and contain a mixture of water, clay, sand, some residual bitumen, and some toxic chemicals. Figure 48 shows an open-air tailings pond where clay and sand particles are allowed to settle down — a slow process that can take up to a decade. After the settling process is complete, water is siphoned off into another area of operation and the pond is worked over for reclamation, ensuring proper drainage. Although no land has been posted for reclamation at the time of this writing, the plan is to return lands to the province under reclamation certification; this will involve topsoil restitution and planting of native flora.

Figure 46. Truck and power shovel in a strip mining operation. Image courtesy of Syncrude Canada Ltd.

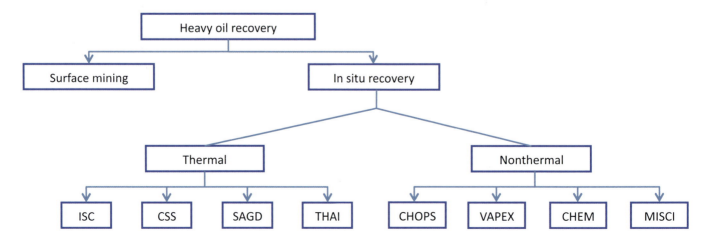

Figure 45. Block diagram showing the classification of different heavy-oil recovery methods.

Figure 47. Syncrude mine from the air. Layered mining 100 m into the earth. Image courtesy of C. Evans, The Pembina Institute.

Figure 48. Large tailings pond at the Syncrude plant. Image courtesy of D. Dodge, The Pembina Institute.

Because it is thicker than traditional oil, the combined bitumen froth is usually mixed with naphtha before going to a centrifugal separator where coarse and denser solids are removed. The separated bitumen is pipelined for upgrading into synthetic crude oil. New technology is being assimilated into the performance of tailings oil recovery (TOR) units, the diluent recovery units that recover naphtha, gas oils, and water from the froth and the inclined plate setters (IPS), all of which help recover close to 90% of the bitumen.

A process different from dilution centrifuging can separate heavy minerals (zirconium- and titanium-based compounds — rutile, ilmenite, and zircon) present in the bitumen during primary and secondary froth production (Kaminsky et al., 1979).

In situ recovery

Because only an estimated 20% of Alberta oil sands are close enough to the surface to be strip mined, heavy oils from most of Canada and all of Venezuela, the United States, Indonesia, and other countries are extracted mainly by means of in situ recovery techniques. These fall into two broad classes: thermal and nonthermal.

Thermal methods are most commonly applied because they supply heat to lower the viscosity of heavy oil and make it mobile so that it can be pumped to the surface.

In situ combustion (ISC) consists of burning a small amount of the oil in place to displace the rest closer to the producing wells. For this purpose, oxygen or an oxygen-rich gas mixture such as air (which may be preheated or otherwise) is injected into the reservoir, where the heavy crude oil is ignited. As combustion starts and the temperature rises, the lighter fractions of oil vaporize and, combined with the steam produced by vaporization of connate water, move forward in the reservoir matrix. As the surging combustion front comes in contact with cooler portions of the reservoir, the gases and vapors condense, transferring heat to the heavy oil and making it mobile to migrate. This process continues while the heavy residual coke left behind continues to burn, which produces hot gases provided that sufficient air supply is maintained. The different zones formed in the combustion process are shown in Figure 49.

Because air has a poor heat-carrying capacity, only 20% of the generated heat is carried forward ahead of the fire front. The rest of it is lost to the rock mass above and below. One of the efforts to utilize this lost heat is

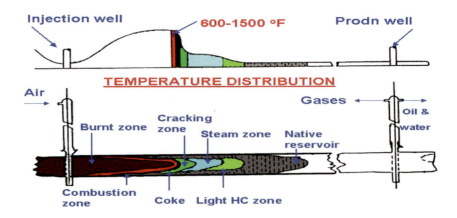

Figure 49. ISC zones forming the fire front as seen along a cross section of a formation. The temperature distribution along the formation ranges from 600°F to 1500°F. From Chattpoadhyay et al., 2005.

reverse combustion, in which once the fire is ignited, the initial injector well is made the producer and a different well is used for air injection. By reversing the direction of the air injection, oil is forced to move through the heated front to the producer, and in the process utilizes the heat in lowering its viscosity.

In another proposed technique, water is injected with air, which not only would reduce the air requirements but also help distribute the heat uniformly. As water reaches the combustion zone, it is converted into superheated steam, which passes through the flame and reaches the reservoir oil.

ISC is best suited for reservoirs that have a thickness greater than 3–4 m, contain 10–20° API (as heavier oils may deposit too much fuel), and permeability greater than 100 md to allow the flow of oil. ISC has been used for reservoirs 100–1200 m deep, but for greater depths compression costs are excessive, and ISC has lower sweep efficiency compared with other methods. Romania has the biggest ISC operation in the world, followed by India, and the United States (Louisiana).

Cyclic steam stimulation (CSS), also called steam soak or "huff-and-puff" method, uses steam injection to recover heavy oil. Steam is first injected into the well at high pressure and 300°C for several weeks, thereby heating the reservoir rock and the fluid and lowering the bitumen viscosity. After this, the well is shut in to allow heat to soak into the formation around the vertical well. Within the heavy-oil formation, there could be interfingering into high-permeability zones, but for simplicity one may assume that steam heats up the formation to a uniform temperature. Finally the well is put on production to pump out the mobile oil until the production rate declines several weeks later. The well is then put through another cycle of injection, soak, and production. This process may be repeated 20 or more times, as long as output justifies the cost of steam injection. For optimal results, this method requires a reservoir thickness of 15 m or more, a high well density (typically 1.5 hectare, or 15,000 square meters, per well), and the absence of base water aquifers that would vent away the injected heat. CSS has an average recovery factor of a little more than 20%.

High viscosity oil or bitumen (e.g., Cold Lake Field) may require larger quantities of steam than less viscous heavy oils (e.g., California and Venezuela). Also, for very deep targets, the effect of steam stimulation may be compromised by heat losses in the well bore and problems arising due to high temperatures. One performance index for this process is the oil-to-steam ratio (OSR), which is defined as the volume of oil produced for unit volume of water injected as steam at standard conditions (Ali, 1994). The majority of the oil sands industry uses the reciprocal performance index, called steam-to-oil ratio (SOR) = 1/OSR. For CSS, while companies generally look to generate production at an SOR of less than 5, a typical break-even SOR is between 6 and 7 m^3/m^3, implying an OSR cutoff of roughly 0.14 to about 0.17 m^3/m^3. Historically, a value of 0.15 m^3/m^3 has been taken as a typical OSR cutoff value. The cutoff usually depends on the price of natural gas (used as a heat source) and bitumen and so can change with time.

SAGD has a much higher recovery factor, potentially more than 70% of the original oil in place, because it overcomes the inherent limitation of CSS — insufficient lateral drive available to move the hot oil to the producer well. Because vertical wells have limited contact with the reservoir, the lateral or radial flow requires considerable pressure, which is not there. The SAGD pioneered by Butler (1985) and Butler and Stephens (1981) makes use of two long horizontal wells and gravity drainage to move the oil (its viscosity temporarily altered) to the production well.

High-pressure steam injection in vertical wells has been used for some time. In steam flooding, as it is called, injection is usually carried out in a pattern, the most common being a five-point pattern, with the steam injector well at the center surrounded by producers. The steam injected at the center produces an expanding heat front into the formation. As it advances laterally, it forms a hot waterflood zone just ahead of the steam zone, which in turn tends to cool down to formation

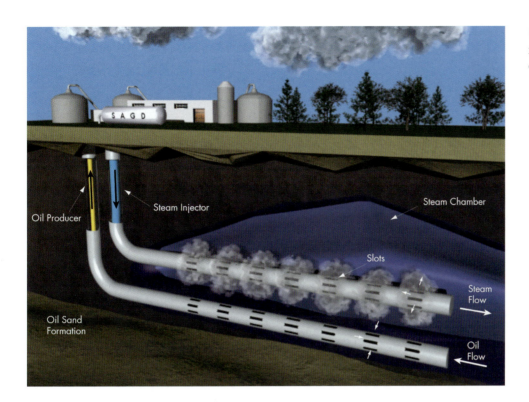

Figure 50. SAGD heavy-oil recovery process. Image courtesy of The Pembina Institute.

temperature. The gas drive effect that steam exerts outward from the injection point pushes the mobilized oil in the direction of the producers. This method works for heavy-oil formations but not for bitumen, which is difficult to push to start any adequate flow. Steam flooding has typical recovery factors greater than 50%.

SAGD improves on steam flooding principles. Two parallel horizontal wells are drilled into the formation in the same vertical plane (Figure 50). The upper well is used as a steam injector and the lower well as a producer. As steam is injected into the upper well, it rises to the top of the formation and sideways and forms a steam-saturated zone called a "chamber," which has the almost uniform temperature and pressure of steam. This heat is conducted to the bitumen, reducing its viscosity and making it mobile. As the steam chamber expands with injection, it also condenses at the periphery of the chamber. The bitumen and the condensate drain under gravity to be collected by the producer. For this to happen, the vertical permeability in the reservoir needs to be high. Consequently, the placement of the horizontal wells has to be such that neither shale stringers nor vertical barriers interfere between them. Figure 51 shows a schematic of the steam chamber, and its creation and growth are necessary for oil production to occur. Notice that the energy transport within the steam chamber is by convection (indicated in Figure 51 with arrows) and hence its shape is convex.

Used most effectively in the Alberta oil sands, SAGD is very suitable for bitumen reservoirs that are too deep to mine but shallow enough to permit high steam pressures. The

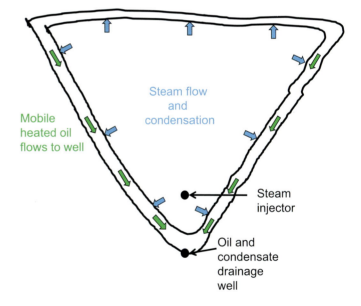

Figure 51. Steam injection and drainage in a SAGD operation.

efficiency of the process increases at higher temperatures and higher steam pressures, although it depends on the viscosity of the bitumen or heavy oil and on the properties of the reservoir zone being drained. Usually SAGD wells are drilled in groups off central pads and their lateral reach in terms of horizontal sections is very large. The distance between two SAGD wells is dependent on the thickness of the reservoir zone, but a 5-m separation is common. For thinner zones, the distance between wells is much shorter, 1 m or less for a 20-m pay.

Generally, SAGD is recommended for reservoir zones that have a thickness of at least 30–40 m. Several variations of the process aim to increase efficiency, surmount limitations, and improve economics. One which is being developed at this writing, steam and gas push (SAGP), calls for a noncondensable gas such as nitrogen or natural gas to be injected with steam to reduce its consumption and hence improve the economics. Another variation of SAGD is the nonthermal miscible technique, which will be discussed with nonthermal methods. In a process known as JAGD (J-well and gravity drainage) (Gates et al., 2008) the producing well has a J-shape, which is reported to have the advantage that production is less susceptible to the problems posed by permeability barriers such as shale layers.

Toe-to-Heel Air Injection™ (THAI) utilizes horizontal production wells paired with vertical air injection wells to recover heavy oil/bitumen (Figure 52). It represents a new approach to ISC (Greaves et al., 2004; Greaves and Xia, 2004). Traditional ISC operating between vertical wells has limited success in immobile bitumen reservoirs. THAI resolves some of the issues, which include long length of oil displacement as a consequence of using vertical producer wells; unpredictable burning surface and potential for spontaneous combustion at the surface, which is a direct result of air and oxygen injection at a significant distance from the producer wells; low air/oxygen flux; low-temperature oxidation; and difficult emulsions and gas override.

THAI is the first ISC process applicable to immobile bitumen reservoirs. Because it is less impacted by geologic variables found in oil sands than current steam-based technologies, it can be applied over a broader range of reservoirs (e.g., under 10-m thick, low pressure, previously steamed or depleted gas-over-bitumen, top or bottom water, and deeper reservoirs).

In the THAI process, horizontal production wells are drilled near the base of the reservoir with vertical air injection wells drilled near the toe of the production wells. A near-well steam preheat is conducted to establish communication between the injectors and the producers. Once the heavy oil/bitumen reaches the ignition temperature and mobility, air is injected, ignition occurs, and a combustion front develops. As air is injected into the formation, the combustion front moves from the toe to the heel of the horizontal well along its axis. The high temperatures within the combustion zone promote hydrocracking and result in partial upgrading of the in situ heavy oil/bitumen. The partially upgraded THAI oil along with vaporized water from the reservoir nitrogen and gases that form during combustion (primarily carbon dioxide) flow into the horizontal production well and are produced to surface facilities where the oil is treated and sent to market. An 8° API bitumen is produced as 12° API oil with a 1000-fold decrease in viscosity, greatly reducing diluent requirements to meet pipeline specifications.

The CAPRI technology uses a new horizontal well liner design. A catalyst is packed between an inner and outer slotted liner in the horizontal production well. THAI fluids flow from the reservoir into the outer liner and through the catalyst, further upgrading the oil before its entry into the inner liner of the well.

THAI and CAPRI processes require minimal surface facilities compared with other steam-based processes because of negligible natural gas consumption, steam generation, and water processing. This results in lower

Figure 52. THAI process. Image courtesy of Petrobank Energy.

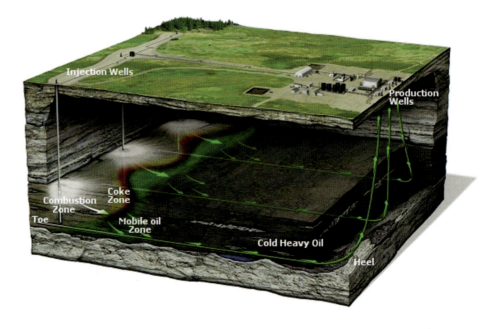

greenhouse gas emissions and a smaller surface footprint. The resultant upgraded product and lower operating costs are expected to lead to higher netbacks. Both processes are patented and, at this writing, being field tested. A chapter on the development of this project is included in the book in a later section.

Nonthermal methods are typically suitable for light and moderately viscous (less than 200 cp) oils. They are also suited for thin formations of 10 m or less and for deeper formations at 1000 m and more. The two main objectives in nonthermal methods are to lower viscosity and interfacial tension. We have included the cold heavy-oil production with sand (CHOPS) method under this class although it does not require the use of any agent for production, unlike other methods. The vapor extraction (VAPEX) method has also been included in this class because it is a nonthermal counterpart of the SAGD method.

CHOPS enables oil production from unconsolidated oil sands by encouraging their influx in production wells and then maintaining it. This nonthermal method differs from other well completion and lifting practices in that there are no filter screens or gravel packs that prevent sand from entering the wells. Historically, sand production in wells is minimized to save on workover costs that arise when rod pumps cannot handle sand inflow. However, in heavy-oil reservoirs, keeping the sand out lowers oil production to uneconomic levels. CHOPS wells allow the unconsolidated oil sands to enter unimpeded, which are then lifted to the surface with progressive cavity pumps designed to manage material that consists of sand up to half its bulk. A cavity pump basically consists of a twisted steel pipe inside of a cylinder and attached to a motor. As the sandy material enters the well, the turning twisted pipe brings it to the surface. Simultaneous extraction of oil and sand generates high-porosity channels, or "wormholes," that radiate away from the borehole (Figure 53b). Because wormholes have a permeability effect in heavy-oil reservoirs, the reservoir pressure falls below the bubble point, causing the dissolved gas to come out of solution to form foamy oil (Figure 53a). This gives rise to a partially saturated reservoir around the vertical borehole in the heavy-oil formation. The development of wormholes in the changing sand matrix also increases the porosity in the reservoir, lowering the seismic velocity in these zones.

Most cold-produced heavy-oil reservoirs do not have bottom water aquifers or top gas because these fluids could have detrimental effect on the production life of the wells (Mayo, 1996). CHOPS is a straightforward, cost-effective extraction method for a certain range of viscosities. It was pioneered in Canada where, in the mid-1990s, it became the primary heavy-oil production method (Wang et al., 2007). The method is also extensively used in Venezuela. The primary heavy-oil recovery factor with CHOPS is 5–10%.

The high volume of sand produced by CHOPS when a well is first put on production gradually decreases after some weeks. Excess sand is reinjected into the deep depleted formations by means of slurry fracture injection (SFI), which was developed in Canada. This environmentally friendly technology reduces the impact of waste on the surface.

VAPEX is similar to SAGD but uses light hydrocarbons instead of steam to reduce the viscosity of bitumen. Hydrocarbon gases such as ethane, propane, butane, or mixtures are injected in the upper horizontal well. These

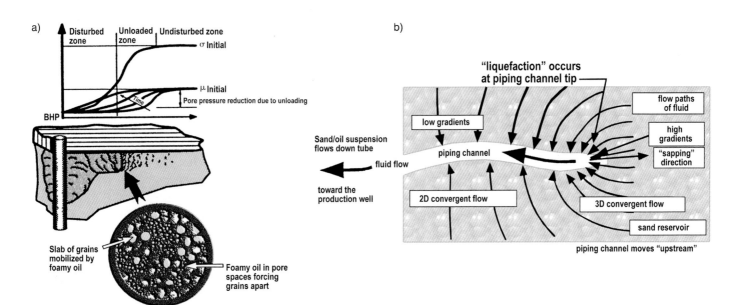

Figure 53. (a) Sand production and stress distribution around the borehole, and (b) single propagating wormhole in a sandstone formation as a result of cold heavy-oil production with sand. From Mayo, 1996.

solvents dilute the bitumen around the wellbore, allowing it to flow into the lower production well. A solvent vapor chamber is formed around the upper horizontal well, which propagates away from the borehole. Because viscosity reduction takes place in cold conditions, VAPEX takes longer than thermal methods. To enhance its efficiency, one variation involves establishing communication between the two horizontal wells by circulating steam between them to reduce the viscosity of the bitumen around the boreholes; this quickens the process when the solvent is injected in the upper well. Similarly, mixing a small fraction of steam with the solvent has been suggested in the interest of enhancing efficiency.

Chemical flooding, although unsuitable for heavy-oil/bitumen recovery, is briefly discussed for the sake of completeness. In polymer flooding, water-soluble polymers are injected as a waterflood to help improve the sweep efficiency. Recovery is generally higher than just water flooding. Surfactant flooding lowers interfacial tension between oil and water, enhancing oil displacement efficiency, but the flooding becomes ineffective after a short distance because of surfactant adsorption and/or reaction with rock minerals. Alkaline flooding consists of an aqueous alkaline solution of sodium hydroxide or some other alkali. When injected into the light oil formation, the alkaline components react with the acidic components in the oil to form surfactants in situ, which reduce interfacial tension. Formation of emulsions also takes place and these help reduce water mobility and improve volumetric sweep efficiency. Emulsion flooding and carbon dioxide flooding have also been suggested.

Injected fluids are miscible or immiscible upon first contact with the resident oil. Ethane, propane, or butane are some of the fluids injected, sometimes followed by injection of natural gas or an inert gas such as nitrogen. Carbon dioxide has also been used as a fluid for displacement purposes.

In addition to the methods discussed above, there are evolving hybrid and proprietary technologies that hold promise. Imperial Oil has used proprietary liquid-assisted steam-enhanced recovery (LASER) in the Cold Lake area of Alberta. Essentially it adds a low concentration of diluents to the steam in mid-life steaming cycles, and these lead to improved recovery uplift.

Hybrid approaches combining production methods are evolving in the interest of higher outputs. Some of these are being tried and others are ideas to be experimented with. For instance, the SAGD-VAPEX approach, a mixture of steam and miscible solvents, should lead to reduced steam-to-oil ratios (SORs). Simultaneous CHOPS-SAGD — high permeability zones created by CHOPS in the horizontal well — followed by breakthrough of steam, should help SAGD recovery. Single or horizontal wells can be operated as moderate CSS wells in combination with a wide steam chamber, which will reduce the SOR by approximately 20%.

Pressure pulse technology (PPT) in combination with CHOPS and other methods is an economical and promising recovery approach. PPT is based on the premise that a low-frequency, high-amplitude pressure pulse acting on fluid-saturated porous media can enhance the flow of fluids. The mechanism involved is that the impinging wave generates a pore-scale dilation and contraction in the porous media through an elastic response so that the fluids flow in and out of the pores, overcoming capillary barriers, suppressing viscous fingering, and reducing pore throat blockage.

Geophysical Characterization of Heavy-oil Formations

Soon after Pickett (1963) showed on the basis of laboratory data that V_P/V_S can be a lithology indicator and assigned values of 1.9 to limestone, 1.8 to dolomite, and 1.6–1.7 to sandstones, Tatham (1982) and Domenico (1984) followed up on this work and suggested correlations between the V_P/V_S ratio, porosity, and lithology.

Tatham and Stoffa (1976), and McCormack et al. (1984) demonstrated the combined use of V_P and V_S from seismic data for identification of hydrocarbon anomalies and stratigraphic interpretation. This was followed by development of empirical relationships between V_P and V_S in clastics (Castagna et al., 1985; Han et al., 1986) and carbonates (Rafavich et al., 1984). As such applications were reported from time to time, it was also realized that a lowering of the V_P/V_S ratio in response to one or more specific reservoir properties may not be unique or unambiguous (McCormack et al., 1985). This is because for a given rock type, there may be several parameters affecting the V_P/V_S ratio, some of which McCormack et al. (1985) have listed as the number and distribution of mineral grain types that comprise the rock matrix, type of cement and degree of cementation, rock density, type of pore fluid and its density, effective stress, formation temperature, depth of burial, geological age of the formation, and porosity and pore aspect ratio, among others. Thus, for low-porosity rocks, the pore space may not have a significant effect and the V_P/V_S ratio could be the determining factor; for rocks with higher porosity, the pore aspect ratio is generally important. The V_P/V_S ratio can be used to distinguish sands, carbonates, and dolomites or for identification of sand-prone versus shale-prone environments (Brown et al., 1989). However, if the pore aspect ratios/porosities for different lithologies have an overlap in V_P/V_S, then such a discrimination could be questionable.

The relationships between seismic data and lithology can also be determined at well control points by multivariate analysis or neural networks, and then the lithology between the wells can be predicted from these determined relationships using linear or nonlinear methods. Such approaches are often resorted to if the correlation between lithology and chosen seismic attributes is weak.

Alternatively, rock physics analysis can help understand the relationship between lithology and related rock parameters and select those lithology-sensitive rock parameters that can be seismically derived. Once this is done, the chosen parameters can be derived from the available seismic data. Such elastic parameters can be estimated from conventional seismic data using amplitude variation with offset (AVO) analysis or from multicomponent seismic data.

For AVO analysis, the requirements include true amplitude processing, high signal-to-noise ratio, and long-offset data (for three-term AVO inversion), which require accurate moveout corrections and stretch compensation, among other things. The main advantage is that the available P-wave data can be used for this analysis.

Elastic parameters from AVO analysis

Because the V_P/V_S ratio is a good indicator of lithology, attempts are made to derive this ratio by different methods. Dumitrescu and Lines (2007) compare this ratio derived by AVO analysis of conventional P-wave data and the ratio derived from multicomponent data. The seismic data are from a heavy-oil field (oil sands of the Devonian-Mississippian Bakken formation) near Plover Lake, Saskatchewan, Canada. Simultaneous inversion of prestack time-migrated PP gathers was performed to derive P- and S-impedance and V_P/V_S volumes. This provides a significant improvement over performing separate inversions on P- and S-reflectivities. The result of this analysis is shown in Figure 54a in the form of a horizon slice at Bakken $+2$ ms from the V_P/V_S volume.

Next, the joint inversion of PP and PS (registered in PP time) poststack data was performed using the method of Hirsche et al. (2005) described later. The V_P/V_S ratio derived using this approach as equivalent to Figure 54a is shown in Figure 54b. Both of these displays compare well with a similar display (Figure 55a) derived using traveltimes. However, the V_P/V_S volumes based on simultaneous and joint inversion produce values with a vertical resolution of a 2-ms sample interval, whereas the previous results from traveltimes were averaged over an interval of 60 ms (Figure 55). Thus inversion results offer sharper details for identification of sand or shale.

Xu and Chopra (2009) demonstrate a practical workflow for mapping heterogeneity facies within the bitumen-bearing

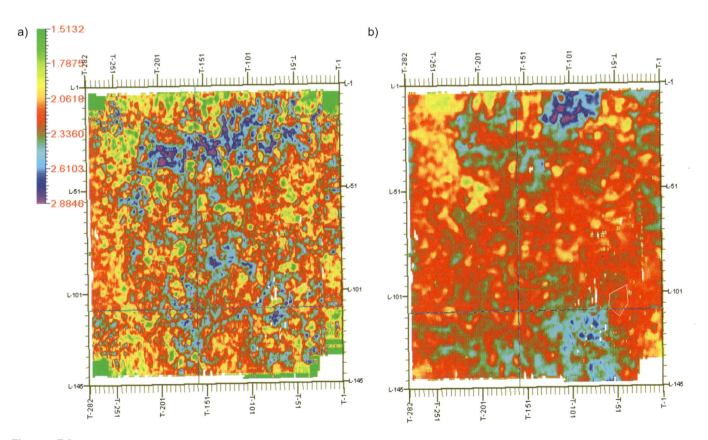

Figure 54. Horizon slice at Bakken $+2$ms on V_P/V_S volume obtained from (a) simultaneous inversion on prestack PP data and from (b) joint inversion of poststack PP and PS data. From Dumitrescu et al., 2007.

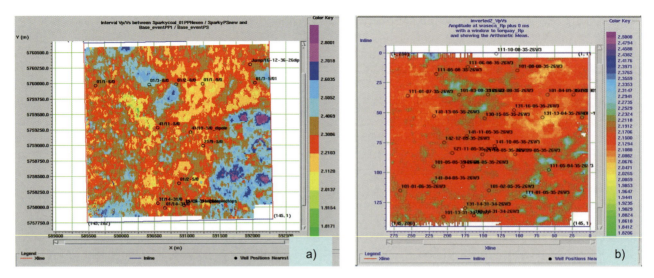

Figure 55. (a) Average V_P/V_S for interval Sparky-Torquay as derived from traveltimes (from Lines et al., 2005), and (b) average V_P/V_S for interval Waseca-Torquay from simultaneous inversion on prestack PP data. From Dumitrescu et al., 2007.

McMurray reservoir in northeast Alberta, Canada. As stated earlier, determining density from three-term AVO analysis usually requires long-offset data. In this approach, the authors describe an improved inversion that relaxes the requirement for large angles and yields a more reliable output. As the first step, rock physics analysis is carried out by crossplotting different pairs of parameters for the McMurray formation reservoir, which is at a depth of approximately 100 m. One or more pairs of parameters that show good correlation are then selected with the objective to derive these from the seismic data. Density is seen to correlate with gamma ray, and so density is determined from the three-term improved AVO analysis. This density attribute can then be used to determine lithology. Figure 56a shows the density reflectivity extracted from seismic data; the equivalent inverted relative and absolute density sections are shown in Figure 56b and c. Notice the lateral variation of density.

Determination of density attribute from AVO analysis of P-wave seismic data requires long offsets. For seismic data recorded with less than 30° offset angle, determination of density is not possible by conventional AVO analysis. In such a situation, Dumitrescu et al. (2005) show the use of neural network analysis to determine density in the Long Lake oil-sands area near Fort McMurray. This area produces approximately 70,000 b/day of raw bitumen via SAGD from the McMurray formation. Three-dimensional three-component data were acquired over this 42-km^2 area in 2002–2003. Bin size was 10 × 10 m. The area has 38 wells with dipole sonics uniformly distributed, which were used for this analysis. After carrying out AVO analysis for the data, P- and S-reflectivity attributes were generated. This was followed by poststack inversion, which yielded P- and S-impedance. Probabilistic neural network analysis was conducted using migrated stack and the above AVO attributes. Figure 57 shows the derived density attribute along a line from the 3D volume, which provides useful variation of the attribute in the McMurray reservoir.

Tonn (2002) demonstrated an interesting application of neural network analysis for discrimination of shale heterogeneities in a sand-dominated heavy-oil reservoir. To achieve this objective, seismic attribute volumes were used to predict gamma-ray value distribution and so produce an equivalent gamma-ray volume.

This particular exercise began by establishing a relationship between gamma-ray and seismic attributes and using log data and the attribute volumes. Making use of simple options such as single attribute regression and then going to multiattribute correlation and neural networks forms a crucial testing phase in this exercise. It was found that neural network analysis using the migrated seismic; P- and S-impedance volumes as input volumes; and the energy, median frequency, and median gradient gave the best results. This exercise involves validation steps and blind well tests during the analysis. After this, the neural network was applied to the full 3D seismic volume and the 3D gamma-ray cube computed.

This exercise helped support the decision with regard to an optimal location for the horizontal SAGD wells for which clean sand bodies had to be delineated. Figure 58 shows two sections extracted from the predicted gamma-ray volume along the horizontal well pairs. Gamma-ray curves from the horizontal wells (logged later) and vertical wells are overlain. Good correlation between the predicted and measured gamma-ray values suggested that chances to encounter clean sand were high.

On drilling and subsequent logging, the horizontal wells encountered clean sand with a value less than 60° API. In one

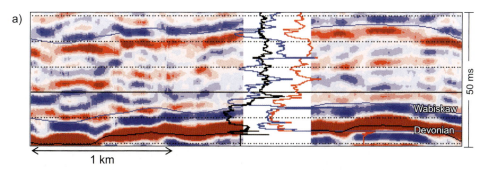
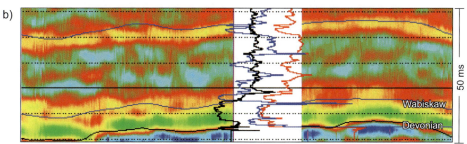
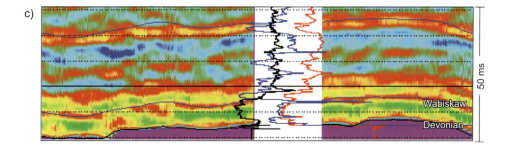

Figure 56. (a) Density reflectivity extracted from AVO analysis, (b) relative density, and (c) density derived after inversion of data shown in (a). From Xu and Chopra, 2009.

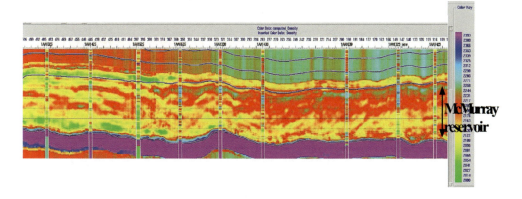

Figure 57. Estimated density along an arbitrary line through the 3D volume derived using multiattribute analysis. Only the second density log from the right was used for the training. All of the other ties shown lie within a 50-m projection distance. From Dumitrescu et al., 2005.

well (Figure 58b), shale was encountered in a small 30-m portion (indicated with a blue curve) that was not predicted. Over this distance, measured gamma-ray values were 60° API whereas predicted values were 50° API. However, over the same interval just 5 m lower, the production well encountered clean sand with a gamma-ray value less than 30° API. Therefore, the upper injection well may have encountered a thin layer that was below seismic resolution. The exercise demonstrates that prediction of gamma-ray attributes from seismic data can be useful in delineating sand facies.

Bellman (2008) discusses a proprietary seismic transformation and classification (STAC) process, which involves using AVO, inversion, and multiattribute analysis to determine various rock physics attributes (Lambda*Rho, Mu*Rho, and density) from seismic data. These attributes are classified based on derived facies and fluid rock physics relationships from well logs and cores and translated back to the seismic volume to produce the top image shown in Figure 59. The bottom image is a conventional migrated stack display.

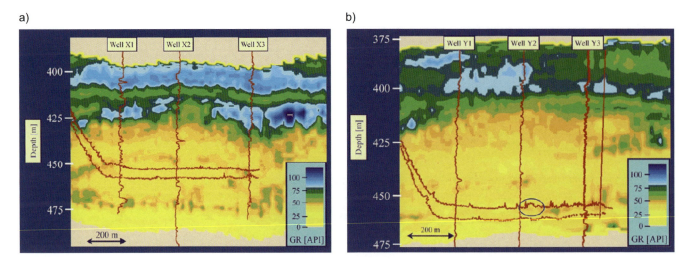

Figure 58. Tracks of two different horizontal wells and projection lines with predicted gamma-ray values. Gamma-ray curves of upper injector well, lower producer, and three vertical test wells are included. From Tonn, 2002.

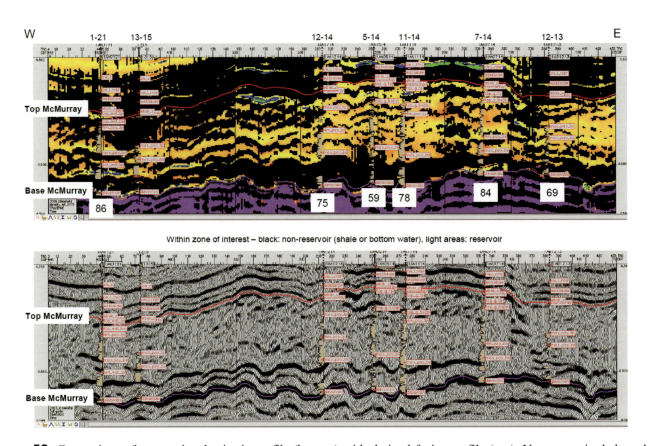

Figure 59. Comparison of conventional seismic profile (bottom) with derived facies profile (top). Nonreservoir shale or bottom water (black), bitumen reservoir (yellow), wet reservoir (blue), gas reservoir (green), and Devonian (purple). Gamma-ray logs with 0° to 70° (at baseline) API range are displayed on the profiles. Only well 13-15 was used in the derivation of facies shown above, the rest were drilled after the facies volume was completed. Numbers posted on the facies section represent the percentage of the well (within the McMurray interval) correctly predicted by the seismic facies and fluid estimates. All wells with a number below them were drilled after the facies volume was produced. From Bellman, 2008.

Multicomponent data for characterization of heavy-oil formations

Multicomponent data do not require long offsets because P- and S-wave modes are available at short offsets. Also, the P- and S-reflectivities are involved in better balanced conditions, allowing somewhat lower signal-to-noise ratios than for single wave mode (Garotta et al., 2000). Inversion outputs from multicomponent data (e.g., V_P/V_S ratio) are more reliable than those derived from a single-wave mode.

Another advantage, although not strictly in the context of heavy-oil applications, is that P-wave AVO is sensitive to V_P/V_S, but these ratios are almost the same for low- and high-gas-saturated sands. P- to S-converted-wave amplitudes depend on shear-wave velocity and density changes. Of these, density is sensitive to low and high gas saturations and rock porosity variation. Consequently, converted-wave data are helpful. Also, because P-waves are attenuated in partially gas-saturated structures, converted waves in multicomponent data are used for imaging such structures.

Multicomponent data require the accurate association of P- and S-propagation times corresponding to the same time. In addition, the upfront cost of acquiring multicomponent data could be a deterrent. Kendall et al. (2005) propose a general workflow for using multicomponent data and other derived attributes for heavy-oil reservoir characterization (Figure 60).

The acquired multicomponent seismic data are processed in terms of PP and PS data volumes. Well log correlation is done on PP stack volume followed by horizon interpretation. AVO analysis on the prestack PP data is carried out. At the same time, PS data processing continues using the velocities and statics obtained from PP processing. Once the PS stack volume is processed, well log correlation is first done in PS time to pick similar geologic events as those picked on PP stack. Horizons so picked are used for registering the stacked PS data to PP time (Kendall et al., 2002) and used with the PP velocity model to perform PP-AVO and PS-AVO. Following the workflow in Figure 60, the 18 attributes listed in Table 1 were generated for a typical heavy-oil case study (Kendall et al., 2005).

Multiple attributes can be overwhelming for an interpreter and so it is advisable to integrate them into a few meaningful attributes to obtain an accurate reservoir model. To this end, various types of neural networks (Hampson et al., 2001; Leiphart and Hart, 2001) have been proposed. On the basis of the available well control, neural networks are used to derive relationships between the chosen attributes and the target log property. The choice of the probable attributes to be used in the analysis is done at what is known as the training phase. A probabilistic neural approach is typically preferred to other available techniques that include back-propagation and multilayer feed-forward neural networks. In the discussed case study, the goal is to characterize the bitumen-bearing McMurray formation. The petrophysical analysis indicated that density had the best correlation with the interpreted volume of shale, V-shale, or V_{sh} estimate. V_{sh} is a term frequently used by petrophysicists to establish a gamma-ray cutoff value, which can help to determine a shale from a clean sandstone in a clastic depositional setting. Consequently, a probabilistic neural network was used to first estimate density and this in turn was used to estimate V_{sh}. Figures 61 and 62 show these estimates along an arbitrary seismic line passing through four wells. Notice the

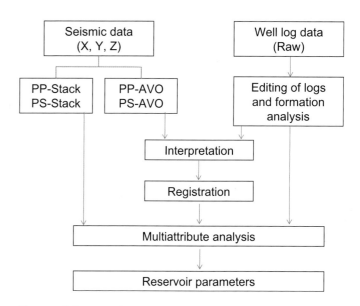

Figure 60. Workflow for multicomponent seismic reservoir characterization. From Kendall et al., 2005.

Table 1. Attributes generated for a typical heavy-oil case study.

PP-data	PS-data
a. Stack	a. Common conversion point stack
b. P-wave reflectivity	
c. Shear-wave reflectivity	b. S-wave reflectivity
d. Density reflectivity	c. Density reflectivity
e. Fluid factor	d. Pseudo-shear impedance
f. Acoustic impedance	e. S-wave impedance
g. P-wave impedance	f. Density (inversion)
h. S-wave impedance	g. V_P/V_S ratio from event registration
i. Density (inversion)	
j. Lambda*Rho (incompressibility)	
k. Mu*Rho (rigidity)	

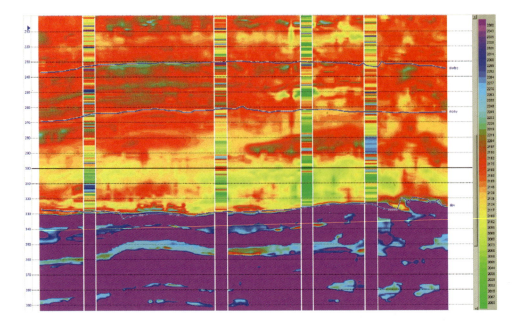

Figure 61. Estimated density section along an arbitrary line through PP and PS volumes. From Kendall et al., 2005.

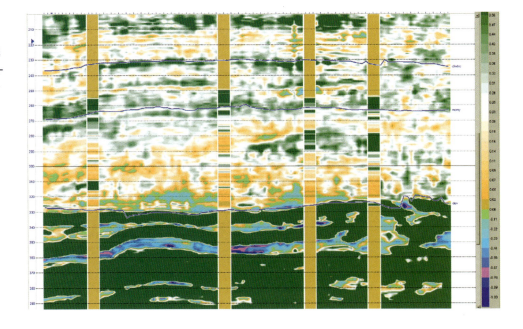

Figure 62. Estimated shale volume section along an arbitrary line through the PP and PS volumes derived using multiattribute analysis. Of the nine attributes used for the analysis, five were from the PP volume and four were from the PS volume. From Kendall et al., 2005.

variation seen in these estimates in the formation of interest. These estimates have correlated well with drilling results after completion of this work. This type of analysis has been incorporated in the workflow to guide future SAGD well pairs to avoid drilling risks associated with shale plugs within the McMurray zone.

Peron (2004) describes a similar multicomponent data workflow, but instead of using neural networks on derived attributes, the determination of V_P/V_S ratio was demonstrated by comparing P-wave and converted-wave 3D seismic data. Following the workflow in Figure 60 up to the point where log correlation is done, horizons are picked for top and bottom of the reservoir on the PP and PS sections. This is followed by trace-by-trace horizon matching of the events on the PP and PS sections. At well locations, horizons would match because they represent the same depth horizon. However, away from the well locations, differences would exist because of variations in the V_P/V_S ratios. Thus, horizons are snapped together, which introduces time shifts and consequently their time-depth curves need to be corrected. In Figure 63, corrections required to align horizons were calculated at horizon times T_0 and T_4 from differences between P-wave horizon and PS-wave horizon times in the same PP time domain. Taking the corrections to be 0 ms at the surface, the corrections are then proportionally distributed between

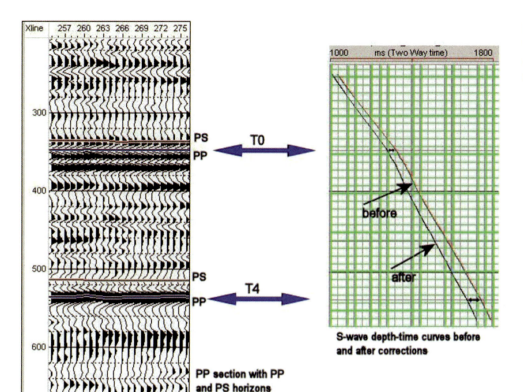

Figure 63. Correction to S-wave depth-time curve resulting from horizon matching. From Peron, 2004.

the surface and the top horizon and then between the horizons. For simplicity, S-wave depth-time curves are recalculated separately for each trace, and this allows the determinations of the spatial variation of V_P/V_S.

Once the horizons for specific geologic interfaces are picked, the interval V_P/V_S ratio is calculated from horizon times and displayed as a map. In the example for heavy-oil sands in northern Alberta (Figure 64), the map shows the distribution of V_P/V_S in the interval between T_0 and T_4. Lower V_P/V_S values corresponding to cleaner sands dominate the southern part of the survey, and higher V_P/V_S ratios correspond to shaly areas in the northern part.

Hirsche et al. (2005) developed a new method for simultaneously inverting PP and PS seismic volumes. This approach honors the physical relationship between P- and S-wave velocities and provides a significant improvement over independent inversions of the two data volumes, especially for the V_P/V_S ratio estimates. The method requires pre- or poststack PP and PS seismic traces as input along with an initial estimate of the background trend relationship between P- and S-impedance, obtained by crossplotting well log information. Also, a trend relationship between P-wave velocity and density is required. Simultaneous inversion is based on an extended conjugate gradient technique, which starts from an initial low-frequency model of P- and S-wave velocity, the fit between the recorded seismic traces, and the model-based synthetic traces are improved by locally modifying the

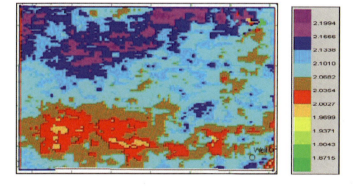

Figure 64. V_P/V_S ratio map for reservoir interval between T_0 and T_4 horizons. From Peron, 2004.

P-impedance model with local deviations of the relationship between P-impedance, S-impedance, and density.

Results from simultaneous inversion demonstrate that V_P/V_S ratio estimates are significantly better than those independently obtained by PP and PS data sets and were subsequently used in a multivariate statistical technique to predict gamma-ray log over the entire 3D volume. The predicted gamma-ray log curves in Figure 65 show a very good fit to the recorded gamma-ray log curves. This exercise provides a significant insight into the distribution of sand and shale in the reservoir interval, which is otherwise difficult because of their similar impedances.

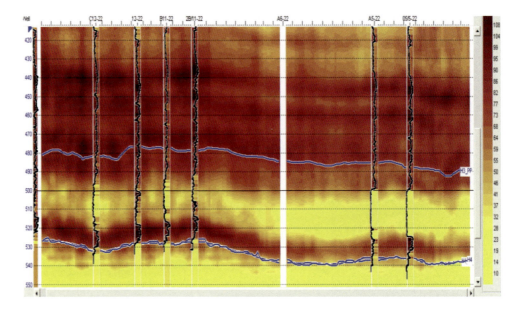

Figure 65. An arbitrary line from predicted gamma-ray volume with actual logs inserted with same color spectrum. From Hirsche et al., 2005.

Lines et al. (2005) discuss a straightforward and robust method for creating V_P/V_S maps from multicomponent data. Watson (2002) and Pengelly (2005) have also described this method.

Flat events on vertical stacks are predominantly PP reflections but on radial stacks are mostly due to PS conversions. If the thickness between any two reflectors is Δd, then with reference to Figure 66 this thickness is given by

$$\Delta d = \frac{V_P \, \Delta t_{PP}}{2} = \frac{V_S \, \Delta t_{SS}}{2}, \qquad (8)$$

where Δt_{ss} is the two-way S-wave traveltime. The converted-wave traveltime can be expressed as

$$\Delta t_{PS} = \frac{\Delta t_{PP}}{2} + \frac{\Delta t_{SS}}{2}. \qquad (9)$$

Solving equations 8 and 9 we get

$$\frac{V_P}{V_S} = \frac{2\Delta t_{PS} - \Delta t_{PP}}{\Delta t_{PP}}. \qquad (10)$$

So, traveltimes for reflection events above and below the interval of interest are picked on the PP and PS sections by first identifying the markers seeking guidance from the available logs or generated synthetic seismograms. Once the traveltimes are obtained, equation 10 above is used to compute V_P/V_S for the entire data set.

This method has been successfully used for mapping oil sands in heavy-oil reservoirs. Figure 67 is a V_P/V_S map for the producing sands at Bakken level in Plover Lake Field (Lines et al., 2005). The low V_P/V_S values in the middle of the map indicate a thickening Bakken,

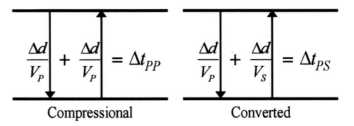

Figure 66. PP and PS interval traveltimes between the top and bottom of a horizontal layer of thickness Δd. From Lines et al., 2005.

while higher V_P/V_S values to the lower right correspond to the eroded Bakken and thicker Lodgepole Formation. The V_P/V_S map was an effective indicator of lithology.

Michelena et al. (2001) demonstrate the use of multicomponent seismic data for characterization of a heavy-oil formation in the Orinoco Oil Belt. The reservoir consists of closed interbedded heavy-oil bodies at very shallow depths. In this area there is little correlation between gamma-ray and acoustic impedance. There is poor correlation of velocities with lithology, although the S-wave correlation is better than the P-wave correlation. V_P/V_S and gamma-ray correlation is also poor. However, the correlation between gamma-ray and density is high, which is suggestive of its use to help differentiate lithology. When the two parameters that show the highest correlation with gamma ray (S-wave and density) are crossplotted and the cluster points are colored with gamma-ray values (Figure 68), the sands and shales appear to be separated. This observation suggests that these elastic parameters can be estimated from seismic data and used to differentiate shales from sands.

An arbitrary line passing through three wells (with log curves available) was extracted from the PP and PS seismic volumes. These two lines were inverted for impedance using commercial software and the results combined as per the method of Valenciano and Michelena (2000) to yield density of the medium. This method combines the PP and PS impedances into a single expression for density, which depends, in addition to the input impedances, on the product and ratio of the P- and S-wave velocities.

Figure 69 shows a segment of the density section with the overlaid density log. The target level is below 0.44 s and is seen in this figure as low density. In Figure 70, the lithology distribution along the arbitrary line is extracted from the 3D seismic volumes. Using the log curves for density and velocity available at the middle well and calibrating it with the derived elastic parameters at that location, a neural-net-based classification algorithm was used on the derived elastic parameters to yield distribution of lithology. Sands are seen in blue, shales in red, and transition zones are shown in yellow.

Within the formations, abrupt changes of clay content and porosity, in a lateral sense, give rise to reservoir heterogeneity. Similarly shale barriers interfere with the continuity of the reservoir sands and pose a problem for exploitation of the resource. It follows that an important exploration objective is to determine areas with clean sand within a given sequence.

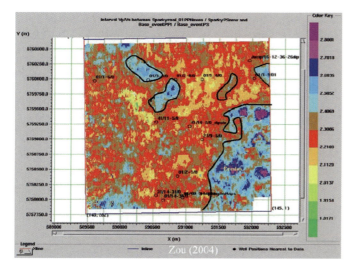

Figure 67. V_P/V_S map for a 2.75- by 2.75-km area of the Plover Lake Field. Low V_P/V_S values in the middle of the map correspond to thicker Bakken and Lodgepole, whereas higher V_P/V_S values to the lower right of the map correspond to a zone where the Bakken sand and Lodgepole formation have been eroded, suggesting that V_P/V_S is a good lithology discriminator. From Lines et al., 2005.

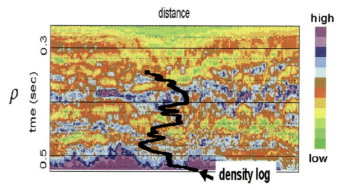

Figure 69. Density section estimated after combining PP and PS inversions. There is good agreement between estimated and expected densities except between 0.2 and 0.4 s, where the seismic-derived density shows lateral variations around the well. From Michelena et al., 2001.

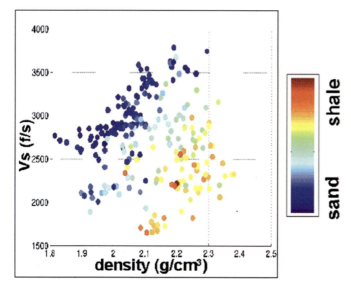

Figure 68. Crossplot for density versus S-wave velocity that is color-coded with gamma-ray values. Sands are seen clearly separated from shales. From Michelena et al., 2001.

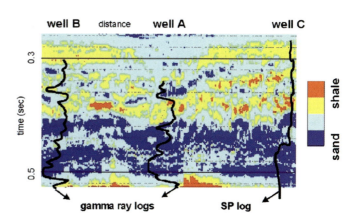

Figure 70. Lithology distribution after classifying the estimated elastic parameters using a neural-net-based algorithm. There is good agreement between seismic-based predictions and well logs, in particular below 0.44 s, where the target is located. The initial model was built using only information from well A. From Michelena et al., 2001.

Enhanced Oil Recovery, Time-lapse Monitoring, and Crosswell Imaging

Enhanced oil recovery (EOR) processes are methods adopted to extract oil from reservoirs after primary production or from reservoirs that contain heavy oil or bitumen that cannot be produced by conventional means. In other words, EOR methods are utilized for improving the efficiency of reservoir production. There are three different types of EOR processes used for different reservoirs: chemical flooding (which includes alkali and polymer flooding), miscible displacement (which includes CO_2 or hydrocarbon injection), and thermal recovery. In the context of heavy-oil production, thermal and chemical EOR processes are preferred.

Chemical methods involve injecting suitable solvents into heavy-oil formations to make oil less viscous and/or form an emulsifying solution that can be recovered more efficiently. These methods cut down on water and energy usage and pollution that some of the thermal methods entail.

As discussed in the section on heavy-oil recovery, thermal EOR processes comprise ISC (fireflood), SAGD, CSS, and THAI, all designed to reduce viscosity of the bitumen and induce petrophysical changes in the target reservoir formation in terms of velocity, density, conductivity, and other parameters. It is of interest to monitor such changes with geophysical methods.

High-resolution 3D surface seismic surveys for designing and monitoring EOR processes are carried out over specific time intervals, generating time-lapse seismic images of the reservoir and of the changes being induced in terms of seismic attributes that correlate with reservoir conditions and properties.

Thermal processes (ISC and steam injection) also lower the viscosity of heavy oil and bitumen to enable pumping the hydrocarbon to the surface. How fluids migrate away from the injection point will depend on permeability heterogeneities or other heterogeneities, or anisotropy induced by fracturing that exists or is induced in the reservoir during injection. Knowledge of some of these reservoir characteristics is helpful before any EOR process is started and, of course, determining the rate and movement of the thermal front, its shape, and overall areal extent of the heated zone provides valuable information for the management of the EOR project.

As steam is injected into the formation, it displaces the heavy oil from the pores of the medium. The rock frame and fluid compressibility increase, resulting in a decrease in the P-wave velocity and an increase in attenuation in the heated zone. Also, as the steamed zone reaches production stage, gas saturation increases. All of these changes result in time delays or pushdowns on seismic amplitudes of reflections from the steam-invaded zones. For example, in a typical sandstone reservoir sandwiched between shale configuration, the reflection from the top of the heated reservoir would have an increase or a brightening of the amplitude, whereas the reflection from the base of the heated reservoir would have a time-delayed reflection and a decrease or dimming of the amplitude reflection. Britton et al. (1983) used a conventional seismic survey over a steam-flooded area to demonstrate that traveltime delays could be seen around the steam injection well. Similarly, Macrides et al. (1988) concluded that seismic waves traveling through the steam-flooded zone get delayed in time and also undergo changes in amplitude.

Thus, time-lapse seismic images generated during EOR processes may exhibit amplitude anomalies corresponding to the changing conditions in the reservoir over time in terms of gas saturation, mobility, phase, temperature, and distribution of the reservoir fluids (Eastwood et al., 1994; Kalantzis, 1994; Kalantzis et al., 1996a, b).

In one of the earliest applications, Mummery (1985) demonstrated the use of seismic inversion to measure or detect the movement of the steam stimulation front within the Clearwater formation at Marguerite Lake, Cold Lake area. Oil sands in these deposits occur in four members of the Manniville Group (Mummery, 1985).

The McMurray formation rests unconformably upon Devonian Beaverhill Lake carbonates. It is composed mainly of nonmarine, well-sorted, mature quartose sandstones interpreted to be channel fill point bar deposits. The Clearwater formation was deposited during a transgressive period and is composed of marine salt-and-pepper sandstone containing some glauconite. Thick marine bar sands and interbar sands, silts, and shales make up the Clearwater formation in this region. The Lower Grand Rapids formation represents a period of northward propagation after the Clearwater deposition. Clastics in this formation represent a prograding deltaic sequence composed of delta plain and channel deposits. The Upper Grand Rapids formation represents a gradual return to transgressive conditions and contains beach deposits, shallow marine bar sands, and occasional thick channel sands.

Figure 71 shows a typical set of log curves from the Cold Lake area. Three-dimensional seismic data were acquired in 1980 and 1981 with identical field acquisition parameters and then processed following the same sequence of steps with care being taken that the data had enough high-frequency content.

The Clearwater formation is 35 m thick at a depth of 440 m and is subdivided into sand reservoir units C1, C2, and C3; the latter is the thickest and contains the most oil.

Seislog sections were generated for three lines and the sonic and density log curves from the two wells were overlaid on these sections to allow calibration. Comparison of the pre- and poststimulation data and the Seislog sections indicated that steam stimulation lowered impedance in the C3 zone and that this application could be used to monitor the movement of the steam front (Figure 72).

Monitoring ISC fire front

Greaves and Fulp (1987) utilized time-lapse 3D seismic data to monitor the propagation of ISC fire front in a small portion of the Holt Field in north central Texas, which is not impregnated with heavy oil. During a one-year period, three sets of 3D surveys were acquired at pre-, mid-, and postburn times. The objective was to detect the change in seismic character that could be attributed to the movement of the fire front. Because the combustion process changes gas saturation levels in the reservoir, reflections from the top and bottom of the reservoir will be more pronounced so that amplitude anomalies, both bright and dim, should be clearly seen. In this application, reflection strength, or envelope amplitude seismic attribute, was useful to detect anomalies.

The three processed seismic data volumes were transformed into envelope amplitude volumes. Difference volumes corresponding to midburn-preburn and postburn-preburn were then generated.

Figure 73 shows the comparison of a profile from pre-, mid-, and postburn attribute volumes. The reflection from the top of the reservoir is identified as a trough and on the preburn profile appears as one of enhanced amplitude, as indicated with arrows (Figure 73a). At the midburn display, this amplitude level increases in lateral extent and intensity (Figure 73b). At the postburn stage, there is further increase in intensity and lateral extent of these amplitudes (Figure 73c).

Time slices at the top of the reservoir through the difference in envelope amplitude for the midburn-preburn and postburn-preburn are shown in Figure 74. The bright amplitude is larger, covering most of the area in the lower half for the postburn difference volume.

This application demonstrated the use of seismic envelope attribute for detecting anomalous seismic responses attributable to reservoir processes.

Zadeh et al. (2007) discuss the seismic monitoring of ISC in the Balol heavy-oil field in Cambay Basin, India. The objective of this exercise was to test the ISC process to improve secondary recovery from a heavy-oil sandstone reservoir. The oil is 15° API at 72°C and a fluid pressure of 104 kg/cm^2. Because the primary recovery rate of the viscous heavy oil is only 10%–12%, it was decided to enhance production by means of ISC. The

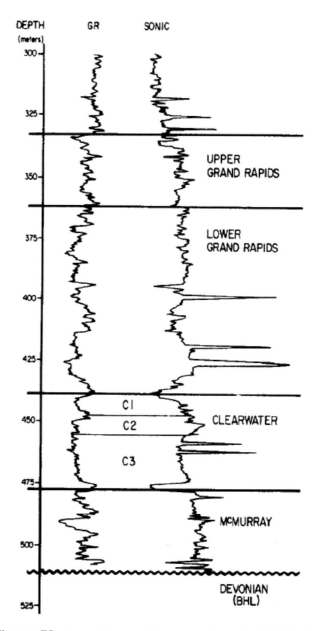

Figure 71. A typical set of log curves from the Cold Lake area. From Mummery, 1985.

combustion front was expected to move by approximately 50 m a year, and acquiring time-lapse seismic data before and after ISC would monitor its advance.

With the increase in temperature in the reservoir, P-wave velocity and density decrease. Monitor seismic surveys will therefore exhibit pushdowns because of longer traveltimes. Also, reflection amplitudes in the zones that are affected by heating will exhibit lower impedance. Both of these changes can be determined by mapping changes in the seismic response before and after heating, which may be due to combustion or steam heating.

Synthetic seismic carried out as part of this exercise showed that P-wave velocity initially decreases rapidly with increasing gas saturation and, after reaching a

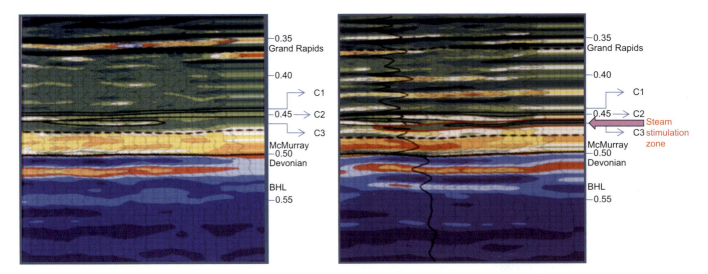

Figure 72. Seislog sections generated before (left) and after (right) steam injection indicated that steam stimulation lowered impedance in the C3 zone, and that this application could be used to monitor the movement of the steam front. From Mummery, 1985.

Figure 73. A profile from (a) preburn, (b) midburn, and (c) postburn 3D seismic data volumes. The reflection traces are overlain by a color scale of the computed envelope amplitude. A bright spot is created at the top of the reservoir by the midburn stage and this increases in lateral extent and intensity by postburn stage. From Greaves and Fulp, 1987.

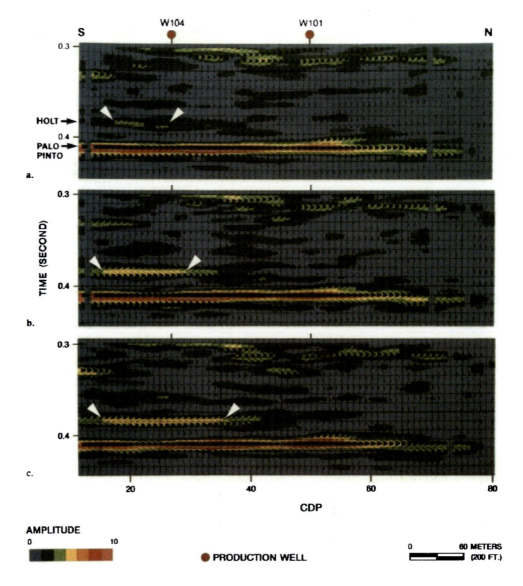

minimum, it increases again (Figure 75). However, density decreases monotonically with an increase in gas saturation (Figure 75b). Fluid substitution effect from oil/water to gas is observed as a dimming of the top reservoir event and a brightening of the base reservoir event. This was actually seen on the monitor survey.

Two 3D seismic surveys were acquired for the purpose, the baseline data in 2003, and the monitor data in 2005. Because the monitor data had some repeatability issues, it was found that migrated near-offset (30–700 m) stack data were best suited for this analysis. After application of time-shift correction for overburden, the root mean square (RMS) of the difference in the baseline and monitor data sets was taken, and it showed good anomalies at the location of the injectors (Figure 76).

Steam flood monitoring

De Buyl (1989) demonstrated the monitoring of steam flooding in the Athabasca oil sands of Alberta. He showed that examining time differences and differences

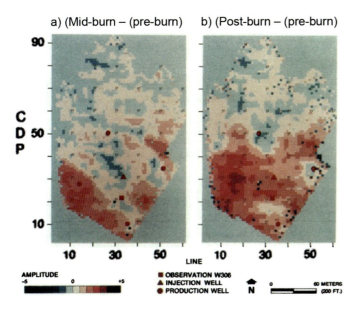

Figure 74. Time slices from the difference in envelope amplitude volumes at a level corresponding to the top of the reservoir. Bright amplitudes seen to the left have increased in their areal extent to the right. From Greaves and Fulp, 1987.

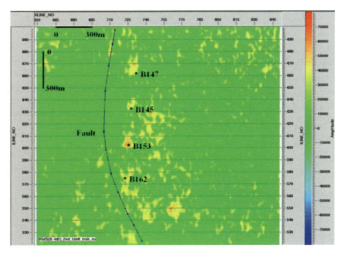

Figure 76. The RMS of the difference between the base and monitor surveys in the reservoir window after time shift correction. From Zadeh et al., 2007.

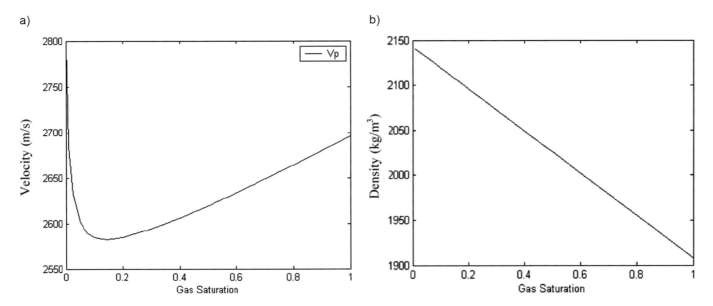

Figure 75. Variation of (a) P-velocity (m/s) and (b) density (kg/m^3) with gas saturation. From Zadeh et al., 2007.

in pseudovelocity volumes between a baseline survey and a monitor survey can reveal the preferential direction of steam movement, and that this information can be used to optimize the EOR process deployed and maximize reservoir depletion. A 3D baseline survey was recorded before injecting steam into the 50-m oil-sands formation. This was done with a dense geometry wherein a 4-m common midpoint (CMP) geometry produced a 1-m depth resolution after processing. After this baseline survey, three monitor surveys were recorded at intervals.

In this analysis, time difference changes were mapped across the steamed reservoir. Seismic amplitudes were then inverted to pseudovelocity profiles. The inverted 3D profiles for the baseline survey were subtracted from similar profiles from subsequent monitor surveys, and vertical and horizontal displays were studied.

Figure 77 illustrates the enhanced spatial extent of the heat zone in the period between monitor 1 and monitor 2 surveys. Similarly, in Figure 78, the inverted velocity profile differences between the base survey and a particular monitor survey illustrate the complex pattern of steam propagation across the heterogeneous reservoir. Again, as part of this interpretation, the vertical and horizontal displays through the pseudovelocity difference volumes are shown in Figures 79 and 80. The increase in the size of these patterns indicates the rate of progress of steam injection during the period over which the steam was injected.

A pioneering seismic monitoring experiment was conducted from 1985 to 1987 at the Gregoire Lake In Situ Steam Pilot (GLISP) in northeastern Alberta, approximately 40 km south of Fort McMurray (Pullin et al., 1986; Pullin et al., 1987 a, b), which demonstrated the

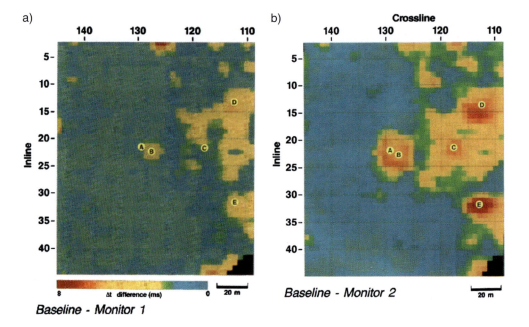

Figure 77. Time difference maps obtained by subtracting the time values for (a) monitor 1 and (b) monitor 2 from the baseline survey. Notice the growing spatial extent of the heat zone in the period between monitors. From De Buyl, 1989.

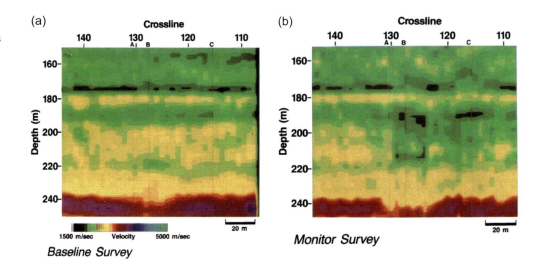

Figure 78. Inverted velocity profiles passing through wells A, B, and C for (a) the baseline survey and (b) the monitor survey. In (b), the pattern of the lower velocity zones (green and brown) between 190 and 215 m indicates the irregular vertical distribution of heat in the reservoir. From De Buyl, 1989.

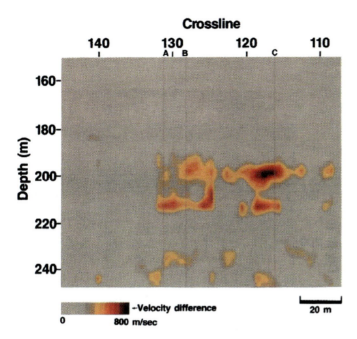

Figure 79. Difference of the inverted velocity profiles shown in Figure 78. Two horizontal anomalies correspond to high permeability thief zones. They are connected by a vertical pathway that may correspond to a fracture zone affecting encased lower permeability oil sands. From De Buyl, 1989.

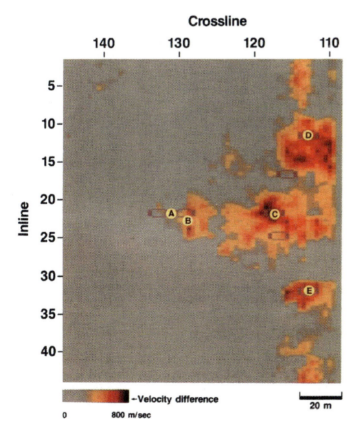

Figure 80. Horizon slice at a depth of 200 m through the velocity difference profiles shown in Figure 79. From De Buyl, 1989.

use of high-resolution 3D seismic for delineating steam fronts in oil sands. At the time this site was selected by Amoco Canada along with partners Alberta Oil Sands Technology Research Authority (AOSTRA) and Petro-Canada, the first well drilled, H-3, had indicated the presence of a 50-m thick, relatively homogeneous McMurray formation, having approximately 30% porosity and 85% oil saturation (Hirsche et al., 2002). The McMurray oil sands are resting on a Devonian unconformity approximately 240 m below the surface. H-3 had also encountered a 4-m sand acquifer directly above the erosional surface. It was assumed that the basal wet zone shown in Figure 81 would be crucial to the success of the project in that it would allow the steam to be injected at a significant rate below the oil sand to mobilize the oil formation above. The second well, HO-7, indicated that the acquifer was absent. Thus, the proposed seismic survey was desirable for accurately mapping the area to help with the in situ design. Thorough planning of the seismic surveys began with sampling of cores, measuring velocities under in situ conditions as a function of temperature, and carrying out field tests for establishing seismic acquisition parameters (i.e., burying the seismometers, determining charge size for highest frequency content, offset, etc.). An initial 3D survey was recorded as a baseline seismic response for the unheated reservoir and the surrounding lithologies. Three more monitoring surveys followed over the next two years (Figures 82–85).

As part of the seismic field tests, Figure 82 shows a shot-record comparison between the case in which surface seismometers were used for recording the data and later and when buried seismometers were used for the purpose. Also shown are the synthetic seismograms for

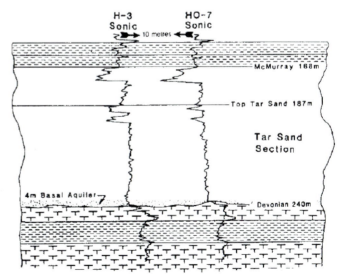

Figure 81. GLISP geologic cross section showing the assumed distribution of the basal acquifer. From Pullin et al., 1987.

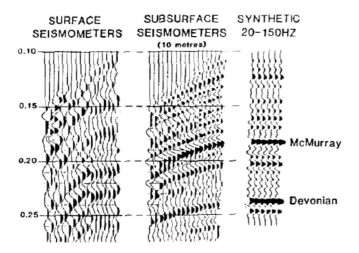

Figure 82. Comparison of shot records recorded with surface seismometers, subsurface seismometers, and their correlation with a synthetic seismogram. From Pullin et al., 1987.

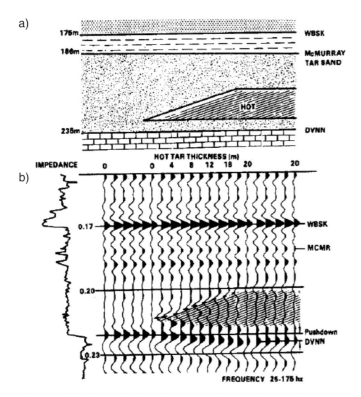

Figure 83. A geologic oil-sands wedge model and its seismic response. From Pullin et al., 1987.

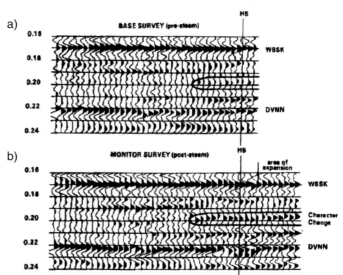

Figure 84. Comparison of base and monitor seismic sections. From Pullin et al., 1987.

reflection could help in estimating the areal extent of the heated zones and their thickness. Figure 84 shows a comparison of two seismic profiles, one from the baseline survey and the other from a follow-up monitor survey. Notice the pushdown of the Devonian reflection caused by the lowering of the velocity of heated oil sand and the character change observed between the two surveys at approximately 0.2 s on the monitor survey. Figure 85 shows the time delays for the three monitor surveys (Hirsche, 2006) and confirms that seismic data could track the progress of the steam flood in the reservoir.

Kalantzis et al. (1996a) discuss 3D seismic monitoring over Imperial Oil's D3 pad undergoing CSS in the Cold Lake area of Alberta. Steam has been injected since 1985 into the 450-m deep Clearwater oil-sands formation at 310°C, 10 mPa, and a high rate of 240 m^3/day for several weeks. After a desired volume of steam has been injected, the well is shut in for a soak period of approximately two weeks and the production begins from the same wellbore. The first survey was carried out in 1990 during the sixth production cycle, followed by another one in 1992 during the eighth steam injection cycle. Both of these surveys were centered on well D3-8 and included 15 of the 20 injector/producer wells on the pad and five observation wells (Figure 86). The well spacing in the east-west direction is 96 m and 167 m in the north-south direction.

Field acquisition geometry for both surveys was maintained identical. The energy source was dynamite and sample rate was 1 ms. There was no baseline survey available because the D3 pad has been undergoing CSS since 1985. Figure 87 shows inline 42 (depth profile) from the production (1990) and steam (1992) depth-migrated volumes. The Devonian marker (Figure 87b) is pushed down during the steam cycle as the velocity gets

correlation with the shot records. Figure 83a shows a wedge model generated from computer modeling results done using sonic log and rock property data. The oil sand wedge model of 75°C increases in thickness from 0 to 20 m. Its seismic response (Figure 83b) shows that the boundaries between hot and cold oil sands produce reflections, and a 5-m thick oil sand layer can be detected with 175-Hz frequency seismic data. In addition, a pushdown directly below the heated oil sand on the Devonian

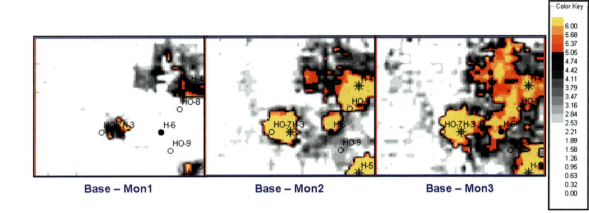

Figure 85. Time delay maps from the GLISP surveys. From Hirsche, 2006.

lowered in the steam-heated area above, between 450 and 480 m in the Clearwater formation. For observing lateral changes in the steam zone and to track the steam front, depth slices were summed over 5-m intervals (five samples per trace) so that the sum of amplitudes for the 470- to 475-m interval are shown in Figure 88. A small anomaly corresponding to each of the wells is seen in Figure 88a. For the steam injection volume, Figure 88b shows significant anomalies between the upper two rows of wells, indicating that this area of the reservoir was affected appreciably by the injected steam. Measurements made at OB1 and OB6 wells had shown significant increase in temperature, consistent with the large anomaly at OB1. The anomalies seen to the north of the first row are in agreement with the observation that there is communication between D3 and another pad to the north. Also seen on Figure 88b are the hyperbolic-shaped amplitude anomalies below wells D3-1 to D3-4. These are not due to processing artifacts but could be associated with movement of an expanding steam front. However, this hypothesis would need more field examples for validation.

Figure 89 shows the differenced isopach for the Devonian (below the reservoir) and Grand Rapids (above the reservoir) markers from production and steam versions, which represent the depth difference for the Devonian. It clearly shows the main depth pushdown to be located between the row of wells D3-1 to D3-4 and D3-6 to D3-9, which is the area affected by steam injection. Figure 90 shows the amplitude difference and amplitude ratio between the 1990 (production) and 1992 (steam) Devonian horizon interpreted from the two volumes. A large anomaly on both maps between the D3-1 to D3-4 and D3-6 to D3-9 row of wells indicates that the Devonian horizon is dimmed in this area because of the large steam-heated zone in the reservoir above.

In a similar analysis, Kalantzis et al. (1996b) show examples from another study performed in a Cold Lake area where at the time Mobil Canada had a 23 vertical-well pilot using steam stimulation for EOR. The reservoirs in this area

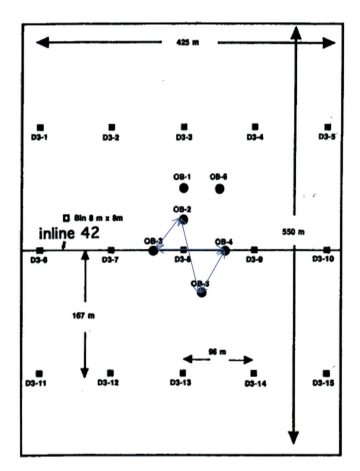

Figure 86. Site map of Imperial Oil's D3 pad (under cyclic steam stimulation) at Cold Lake showing location of producer/injection wells (D3-1 to D3-15) at reservoir depth level, observation wells (OB1 to OB6), and inline 42 of the seismic grid (8- by 8-m bins). From Kalantzis et al., 1996a.

are the Sparky and Waseca formations, which with the Colony, McLaren, General Petroleum, Clearwater, and McMurray formations constitute the Mannville Group. Sparky consists of Upper, Middle, and Lower high-viscosity (150,000 cp) bitumen-saturated sands separated by shales.

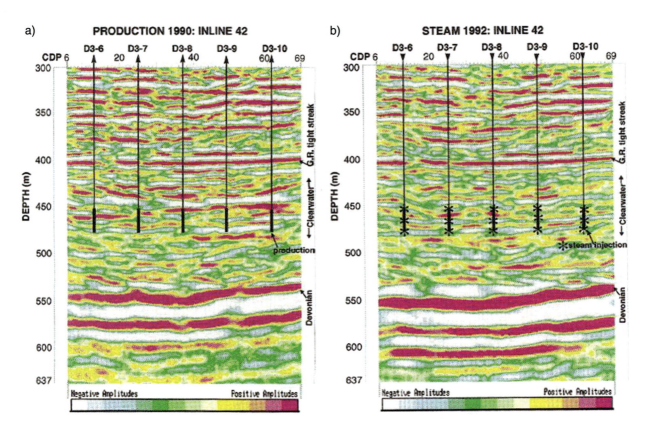

Figure 87. Inline 42 from (a) production (1990) and (b) steam (1992) depth-migrated data volumes. The seismic data are zero phase and negative polarity. From Kalantzis et al., 1996a.

Although the reservoir has good lateral continuity, it is interrupted by discontinuous shale barriers, tight cemented siltstones, and calcified tight streaks that could affect vertical conformance of the steam stimulation.

Steam was injected into the three Sparky sands at a depth of 360 m at a constant rate of 200 m³/d and wellhead pressures of 8–10 MPa (Kalantzis et al., 1996a,b).

A baseline 3D seismic survey was acquired in 1987 and a monitor survey in 1998, almost at the end of the second cycle of steam injection. After basic processing of the two volumes, the stacked data were examined. In Figure 91, an inline from the base and monitor volumes is shown. Bright amplitude anomalies associated with steam injection are seen clearly as indicated within the blue box. Furthermore, the reflectors below the Sparky sands (between 0.375 and 0.4 s) show time delays. The time delay (approximately 6 ms) is more obvious for the Devonian reflector as indicated with yellow arrows in Figure 91. Comparison of time slices from the two volumes at two different times (Figure 92) shows time delays at the Devonian level.

The two volumes are also processed through a one-pass 3D depth migration, and a comparison of the inline in Figure 91 is again compared in Figure 93. Notice that the amplitude of the reflections within the Sparky in the vicinity of the steaming wells has increased, and they are looking brighter. Also, the depth pushdown at the Devonian is evident. Isaac and Lawton (2006) revisited these same data arriving at similar conclusions.

Monitoring SAGD steam flooding

Li et al. (2001) discuss a case study from East Senlac, Saskatchewan, Canada in which the producing reservoir is unconsolidated fluvial channel sand of the Lower Cretaceous Dina formation. This sand directly overlies the erosional surface of Paleozoic carbonate, is 15-m thick, has an average porosity of 33%, a permeability of 5–10 darcies, and is highly saturated with viscous heavy oil. Oil is being produced using SAGD technology. Three pairs of horizontal wells A1–A3 were drilled first, followed by an infill horizontal well pair A4 to enhance production.

Three 3D surveys were acquired at different times to monitor the growth of the steam chamber, determine steam sweep efficiency, and examine how reservoir heterogeneity affects the heated reservoir zones. The first survey was shot in the winter of 1990 using vibrator as the source and yielded a bin size of 15 × 30 m. The second survey was shot in September 1997 after 18 months of continuous steam injection. This time the source was a truck-mounted weight drop to avoid damaging surface

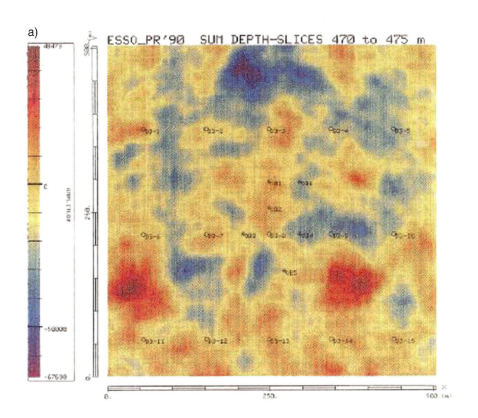

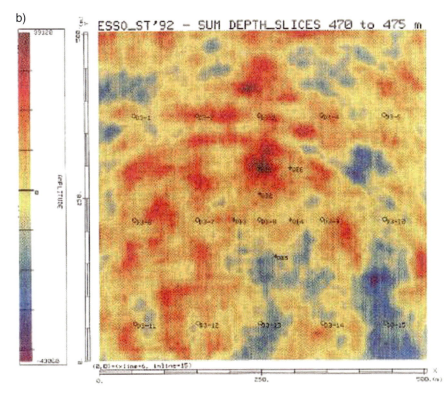

Figure 88. Depth slices summed over a 5-m interval, 470–475 m, from (a) production (1990) and (b) steam (1992) depth-migrated data volumes. From Kalantzis et al., 1996a.

facilities surrounding the steam plant and wells. This survey yielded a bin size of 10 × 20 m. The third survey was acquired in the spring of 1998, when steam had continuously been injected for years. For this survey, Mega-Bin geometry was used in the interest of recording wider-band seismic signals containing higher signal-to-noise ratio and better offset and azimuth distributions. The dynamite charges used in 6-m drill holes remained safe. The bin size was 20 × 40 m, and the data had much higher folds than the previous two surveys.

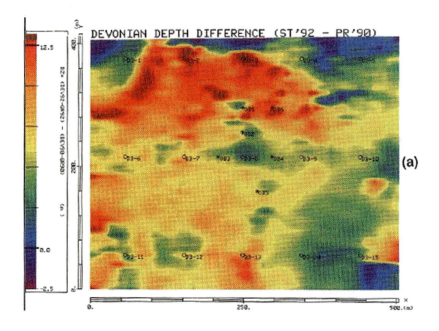

Figure 89. Devonian apparent depth pushdown between steam (1992) and production (1990) depth-migrated data. From Kalantzis et al., 1996a.

Evidently, the three surveys were not designed optimally for time-lapse monitoring, in which repeatability is the key so that differences between surveys can be attributed to changes in reservoir properties.

To extract meaningful information from the data sets, careful processing of the adopted sequence aimed at preserving relative amplitudes and intersurvey balancing is needed. Figure 94 shows a comparison of the three survey profiles with sections cut along the same line across the toes of the four SAGD horizontal wells. The steam chamber migration is clearly noticeable in Figure 94b and c. After two years of steam injection, the reflection from the top of the reservoir is strong and thickening compared with the other two profiles.

In a similar comparison (Figure 95), the profiles are taken along the horizontal trajectory of well A4. Again the significant influence of steam injection is clearly noticeable. The net changes in the reservoir over a period of time can be affected by taking different sections of these profiles. Figure 96 illustrates the difference of the profiles along horizontal trajectory A4, which demonstrates the length of the steam chamber around the wellbore and its vicinity. Production logging in this well confirmed the interpretation that after two years of steam injection, the full length of the horizontal well had been heated.

Theune et al. (2003) and Theune (2004) modeled the anticipated seismic responses due to steam injection into this reservoir that incorporated fluid substitution of steam and water for bitumen under expected in situ conditions. Because the Senlac reservoir is relatively thin (approximately 10 m), it was suggested that seismic monitoring of the reservoir would be difficult because traveltime attributes were not observable. However, in a subsequent analysis, Zhang (2006), while supporting Theune's observation, showed that steamed reservoirs could be detected using variations in seismic amplitudes. Similar studies by Bianco and Schmitt appear later in this book.

Frequency attenuation

During steam injection, because of high temperature and pressure conditions, gas in the form of steam can exist only in a small, restricted zone at the wellbore. Consequently, gas saturation is close to zero. However, during production, the volume of free gas (carbon dioxide, methane, and steam) is considerably larger than during injection. Dilay and Eastwood (1995) report that based on the neutron logs, well pressure, temperature data, and reservoir stimulation, gas saturation around the wellbore (radius 20–30 m) is greater than 5% during oil production.

Such conditions could lead to higher attenuation of high frequencies in the heated zones near the wellbore than in the unheated reservoir away from the wells. Dilay and Eastwood (1995) demonstrated this by an interesting application of spectral analysis to seismic monitoring of CSS. Three seismic time windows were chosen for spectral analysis: one of length 110 ms above the reservoir, a second of 80 ms on the reservoir, and a third of 120 ms below the reservoir. The choice of the windows' lengths was based on the criteria that they should ensure stable results, not intersect any seismic reflectors, and that neither the above nor the below window should sample the reservoir.

A signal power spectrum was estimated for each stacking bin of the three windows and for the injection and production cycles.

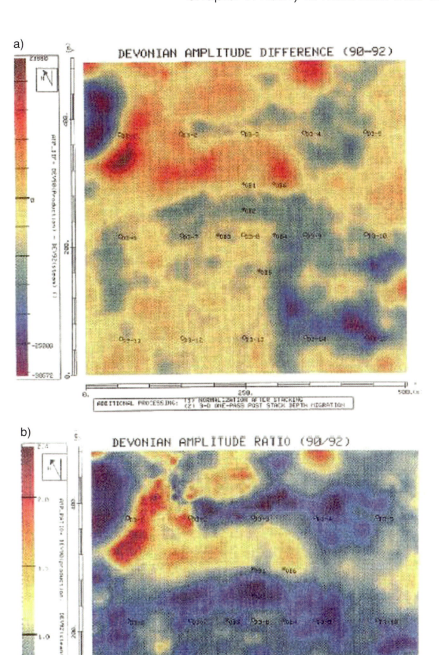

Figure 90. Devonian (a) amplitude difference and (b) ratio between production (1990) and steam (1992) depth-migrated data volumes. From Kalantzis et al., 1996a.

Figure 97 shows the results of the spectral analysis averaged over five traces about a well signal spectra (injection and production), which are similar (Figure 97a), and the spectra for the middle window, which is centered on the reservoir (Figure 97b). This time the two are quite different, particularly for higher frequencies, and the observed changes could be attributed to different phenomena such as velocity sag, changes in impedance contrasts, intrinsic attenuation of higher frequencies, geologic scattering, and degradation of stacking velocities. Figure 97c shows the spectra for the

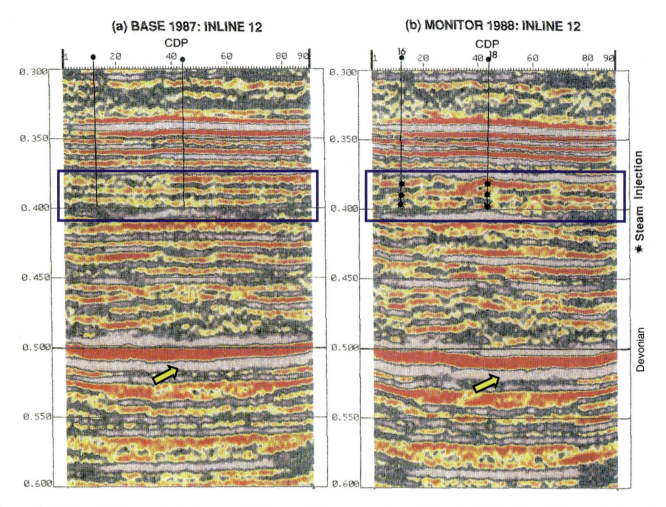

Figure 91. Inline 12 from stacked data sets. (a) Base and (b) steam monitor. Notice the brightening of the amplitude at the Sparky level around the steaming wells and the time pushdown at the Devonian. From Kalantzis et al., 1996a.

window below the reservoir and the change in this could be due to intrinsic attenuation.

Dilay and Eastwood (1995) have further analyzed the power spectra by partitioning them into user-defined energy segments. A useful quantile frequency is the 85% energy segment that produces a spectral energy surface for injection and production surveys. Figure 98 shows these spectra across the entire 3D survey. In Figure 98a and b, the spectral tracking for the injection and production surveys are similar. In Figure 98c and d, a similar set for below the reservoir is quite different, deviating by as much as 30 Hz at some points. These show strong attenuation during the production cycle, which is attributed to partial gas saturation.

In the foregoing examples, the use of 3D seismic data for delineating steam fronts in oil sands, time maps, and time-difference maps for reflection data was demonstrated. Lines et al. (1989) utilize traveltime modeling of steam injection into oil sands to conclude that velocity models computed from seismic traveltimes indicate zones of oil sand heating. Three steps are followed in their approach. First, the tar sand layer thickness $d(x,y)$ is estimated by computing depth maps before steam injection. This is done by converting the time-structure maps to depth maps by using the image-ray model described by Hubral (1977). Second, TOR successive monitor surveys after steam injection, differences in reflection traveltime $\Delta t(x,y)$ caused by injection of steam are computed. Third, the change in slowness of the oil sand layer can then be computed by using

$$\Delta s(x,y) = \frac{\Delta t(x,y)}{2d(x,y)}, \qquad (11)$$

where Δt is the difference in reflection time because of steam injection, $2d$ is the two-way distance of the reflected raypaths in the oil sands, and Δs is the change in the slowness caused by steam injection.

Using the same data as described by Pullin (1987b), Lines et al. (1989) showed the velocity model (Figure 99a) for the first monitor survey, in which the velocity near the

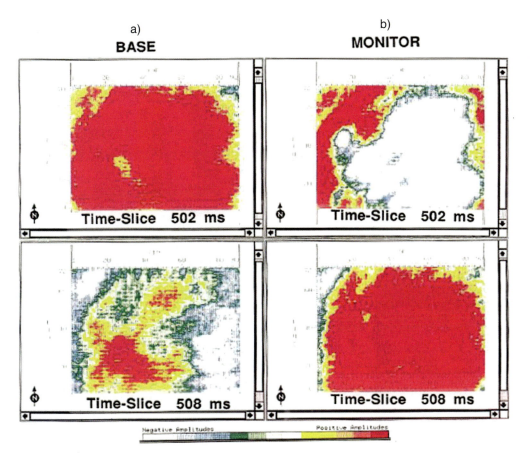

Figure 92. Time slices from the stacked data volume (a) base and (b) steam monitor showing the time delay at the Devonian marker. From Kalantzis et al., 1996a.

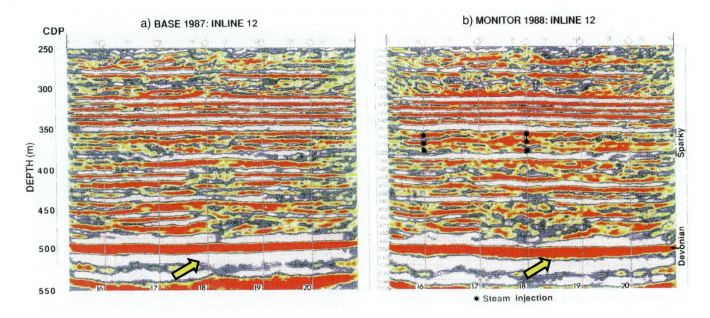

Figure 93. Inline 12 from depth-migrated data (a) base and (b) steam monitor surveys. Notice the brightening of the amplitude at the Sparky level around the steaming wells and the depth pushdown at the Devonian. From Kalantzis et al., 1996a.

steam injector wells (marked with dots) has been lowered by temperature increase.

Apart from using surface seismic data, seismic arrival times are also recorded by borehole seismometer at an observation well. When these values are included in the slowness calculation, the velocity model obtained (Figure 99b) is similar to the one obtained only from reflection data (Figure 99a).

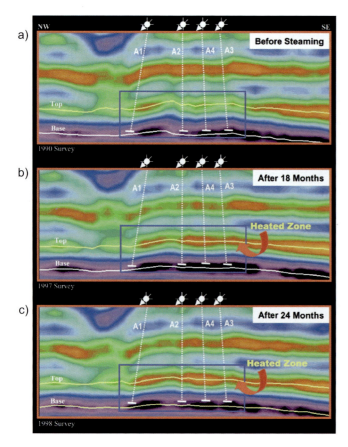

Figure 94. Steam chamber developing across the pattern of four SAGD horizontal wells. The steam migration is clearly noticeable in (b) and (c). From Li et al., 2001.

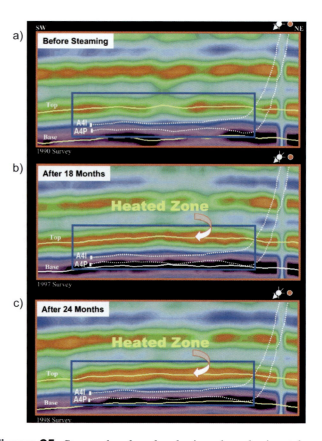

Figure 95. Steam chamber developing along horizontal well pair A4. The steam migration is clearly noticeable in (b) and (c). From Li et al., 2001.

Seismic tomography methods

Just as reflection seismic is used for monitoring hydrocarbon reservoirs, crosswell seismic tomography produces images that determine changes therein. In a tomographic survey, a source of seismic energy is lowered to a suitable depth to span the zone of interest in one borehole, and a string of receivers is lowered to cover the same zone in another borehole. Sources and receivers will occupy many positions in the boreholes and are sampled at regular intervals covering the zone of interest. For each shot in one borehole, recording is done in the other borehole, providing coverage of the zone between the two boreholes with thousands of traces. Justice et al. (1993) review some of the applications of crosswell tomography for EOR monitoring. There are basically two types of measurements derived from seismic data and used for tomographic image reconstruction. One is based on traveltimes, which allow seismic velocity field (P or S) to be reconstructed, and the other is based on attenuation, which allows the Q-factor to be reconstructed in the reservoir. Most applications are based on traveltime measurements and so they are more common.

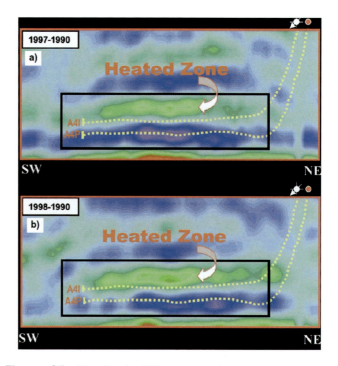

Figure 96. 4D seismic difference sections between (a) 1997 and 1990 surveys and (b) between 1998 and 1990 surveys. Highlighted sections show that the steam chamber has expanded along horizontal well pair A4. From Li et al., 2001.

Laine (1987) showed that crosswell methods could be used to construct tomographic images in a heavy-oil steam flood environment.

Bregman et al. (1989) discussed a crosswell seismic experiment for monitoring the fire front in a 17-m heavy-oil saturated sand between two wells 51 m apart at the target depth. The final velocity model obtained after an iterative procedure showed gross features that correlated with well logs. The application yielded encouraging results.

Paulsson et al. (1994) demonstrated successful application of crosswell tomography for monitoring steam injection in the McMurray formation at a location 60 km northeast of Fort McMurray, Alberta. The application defined geology and detected the movement of the thermal front. Figure 100 shows the P-wave velocity tomograms obtained by this method. The crosswell surveys were acquired in wells CH1–CH4 located at the corners of a square 75 m on a side. At the center of the square was an injection well IN1. CH1 and CH4 were the source wells, and CH2 and CH3 were the receiver wells. Two 80-level 3-C geophone arrays were cemented into CH2 and CH3 boreholes.

The velocity tomograms obtained after processing the data are shown in an unfolded form that traverses the four sides of the square. The upper set of tomograms are the pre-steam acquisition and the lower set for the survey acquisition done after 72 days of steam injection. Notice the effect of the steam injection is seen clearly in terms of the increase in red colors, signifying reduced velocity in the lower figure.

Eastwood et al. (1994) also report on the application of crosswell tomography in their CSS analysis at Cold Lake Alberta. The crosswell tomograms indicate the velocity structure in planes surrounding the central well D3-08 (marked with blue lines). The tomograms in Figure 101 show good correlation with sonic logs and good velocity correlation at common well locations. Data quality below 480 m (below the reservoir) is not reliable because of poor coverage at that level. In an attempt to test their hypothesis that calcified tight streaks were impeding vertical conformance of steam, the geologic model was overlain on the tomograms. Direct correlation between the location of the tight streaks and anomalous slow velocity in the reservoir were strong evidence that tight streaks were impeding vertical growth of the heated region.

Mathisen et al. (1995) describe the application of time-lapse crosswell seismic data processed as P- and S-wave tomograms for imaging heavy-oil lithofacies and changes as a result of steam injection. Baseline reservoir properties and conditions were established using P- and S-wave tomograms acquired before steam injection and then interpreted with timelapse and difference tomograms. Figure 102 shows a baseline S-wave tomogram that clearly differentiates the high-velocity channel facies in

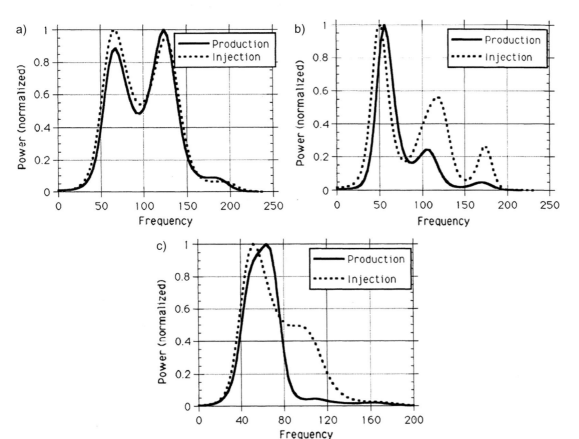

Figure 97. Signal power spectra at a well location (a) above the reservoir, (b) at the reservoir, and (c) below the reservoir. From Dilay and Eastwood, 1995.

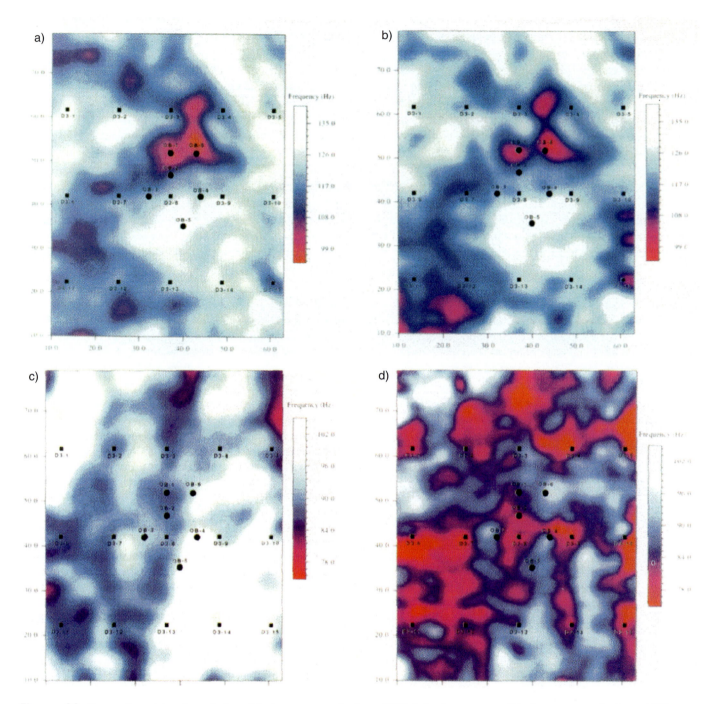

Figure 98. Spectral tracking for window (a) above reservoir from 1992 injection survey, (b) above reservoir from 1990 production survey, (c) below reservoir from 1992 injection survey, and (d) below reservoir from 1990 production survey. From Dilay and Eastwood, 1995.

yellow from the lower velocity, low-to-moderate flow levee facies (brown). Laterally continuous sequences of channels are seen extending in zone 3 and along the tops of zones 4b, 5, and 6. Velocity heterogeneity within each 300-ft/s (90 m/s) color-defined velocity unit is indicated in the display to the right with 150-ft/s (45 m/s) velocity contours. An anomalously high velocity unit (green-blue) is seen near the base.

The baseline P-wave tomogram exhibits the same reservoir lithology variations and structure as the S-wave tomogram (Figure 103). The low velocity zones seen are a result of the previous steam-heat injection and the formation of gas. A high velocity zone seen in green-blue near the base of the tomogram correlates with the S-wave tomogram.

Time-lapse tomograms were acquired and then processed to show the effects of additional steam injection

a) 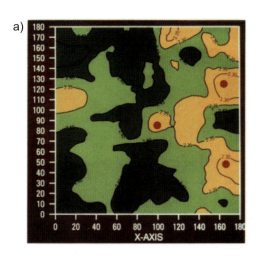 b)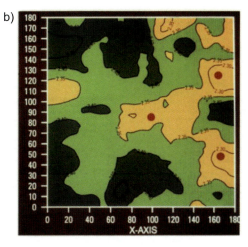

Figure 99. Map of oil-sands velocity model for (a) first monitor survey (red dots indicate three injector wells for initial steam injection), and (b) cooperative inversion of reflection and borehole traveltimes in first monitor survey. From Lines et al., 1989.

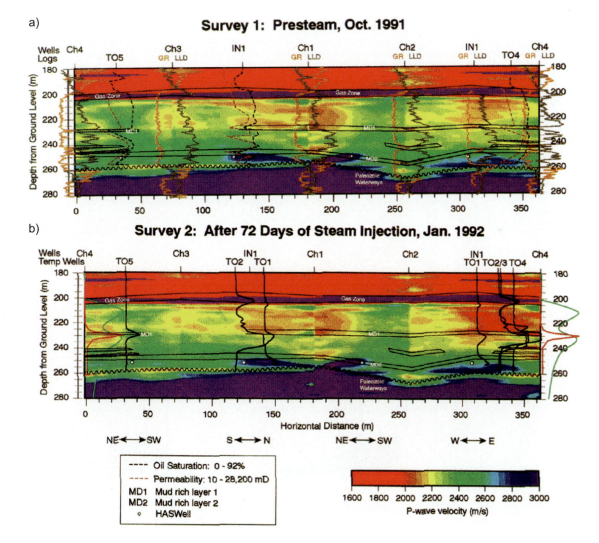

Figure 100. Four P-wave velocity tomograms. Overlain on (a) are gamma and resistivity logs from indicated wells; bitumen saturation and permeability from core measurements in T05, IN1, and T04; and the geologic model. In (b), the decrease in P-wave velocity indicated by reddening in section CH4CH2 shows the direction taken by the steam. The steam, injected near the center of the figure at approximately 250 m, appears to have risen approximately 20 m around the wellbore and then escaped to the east. Also shown are temperature logs from several wells. Two are shown for CH4. The one with the single spike was taken at the time of the second crosswell survey; the other, taken six weeks later, shows how the heat has expanded upward, which is also apparent on the tomogram. From Paulsson et al., 1994.

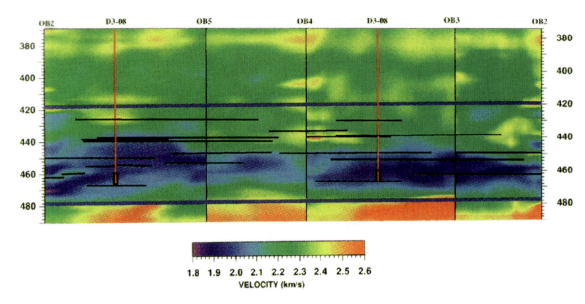

Figure 101. Composite plot of four crosswell tomogram planes (see Figure 86 for location). Overlain on this plot are the top and bottom of the reservoir (blue lines at 418 and 478 m), tight streaks within the reservoir (black lines), and the location of well D3-08. A low-velocity zone (P-wave velocity of 1900–2100 m/s) is predominantly in the lower portion of the reservoir. The tight streaks correlate with the top of the anomalous velocity zone. From Eastwood et al., 1994.

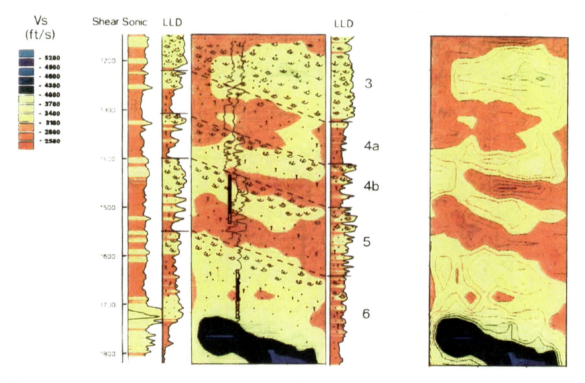

Figure 102. Baseline heavy-oil sand S-wave tomogram that shows the imaging of interwell structure, lithofacies, and porosity variations. Sands facies variations are documented by resistivity logs. Channel sands with excellent reservoir quality are imaged by higher velocities (yellow) than moderate reservoir quality bioturbated levee sands (brown). From Mathisen et al., 1995.

into zones with significant steam heat-gas content. Figure 104 shows the P-wave tomograms acquired after 8, 27, 52, and 109 d after the baseline tomogram. Notice the progressive reduction in the P-wave velocity as seen with the increase in the red velocity field.

Challenges for Heavy-oil Production

Heavy-oil reservoirs, mostly on land, have been produced for decades around the world, but they entail extra

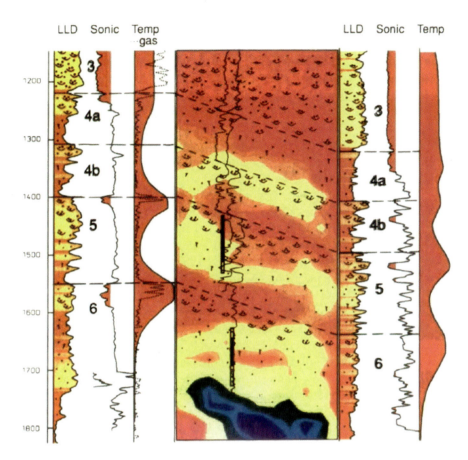

Figure 103. Baseline heavy-oil sand P-wave tomogram with low-velocity zones (red) because of the presence of gas. It images the same structure and lithofacies variations (yellow and brown) as the S-wave tomogram (Figure 13). Temperature logs (black curve) document portions of the reservoir being heated by steam, whereas sonic logs (red curve) document zones with low-velocity anomalies at the wells (red) that are due to gas. The fact that the tomographic low-velocity zones coincide with high-temperature steam zones and the presence of gas suggests that low tomographic velocities displayed in red are imaging areas where gas has formed as a result of the steam process. From Mathisen et al., 1995.

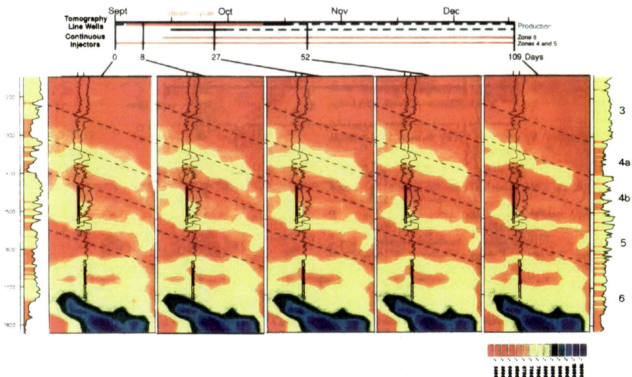

Figure 104. Time-lapse P-wave tomograms that document progressive reductions in P-wave velocity (increase in red velocity field) due to cyclic and continuous steam injection. The red low-velocity field increases more because of the shallow inline injector than because of the deeper injector, which is projected into the line. Continuous injectors outside of the survey area also cause velocity reductions and growth of the red low-velocity fields. From Mathisen et al., 1995.

effort, investment, and operating challenges. The single most important factor behind many of the challenges is keeping the impact that bitumen or heavy-oil production has on the environment to a minimum. Some of the main concerns are discussed below.

Land surface disturbance and reclamation

Surface or open mining of oil sands in Athabasca involves first clearing large areas of the boreal forest of trees and of the upper muskeg-laden layer that comprises the wet swampy vegetation. Approximately 3500 km^2 of Alberta's 381,000 km^2 total boreal forest are mineable, causing significant impact on the land surface and wildlife.

Mine tailings disposal

Mine tailings disposal is another major concern. Oil-sands production requires large quantities of hot water to separate bitumen from the sand and other material. Two to five barrels of water are used to produce a barrel of bitumen. Fresh water used for this purpose is currently being recycled 18 times, and the rest of it is released into the tailings ponds, which occupy 130 km^2 of boreal forest. Tailings are a sticky mass of clay and small quantities of sand, water, fine silts, and residual bitumen as well as some other contaminants. The sand and the heavier rock pieces settle down, but the fine clay remains in the water and takes more than 10 years to settle in the ponds.

Open tailings ponds not only emit foul odors but also are generally lethal to the birds and waterfowl they attract. This is particularly problematic in spring because migratory birds in northeastern Alberta find the warm water tailings ponds inviting for stopovers while other water bodies are still frozen. Inevitably, their plumes become oiled with the sticky bitumen, and they ingest polluted clay-laden water and die. Oil companies try to reduce bird killing using scarecrows and propane-fired canons, but birds eventually tend to become habituated to these deterrents. A more effective method relies on radar to detect approaching birds, simultaneously setting off in their path various deterrents such as propane canons, noise cannons, flashing lights, and robotic falcons powered by solar panels. This system is particularly effective at night, keeping ducks, geese, and shore birds, which are nocturnal and diurnal migrants, away from the lethal tailings ponds.

A mechanism patented in 1983 by Jan Kruyer (1988) recovers bitumen from oil-sands tailings, middlings, and sludge ponds. The process consists of passing the tailings mixture through an aperture cylindrical cage, which is in contact with an aperture oleophilic endless sieve. The sieve and the cage rotate continuously. As the tailings pass through the cage, the aqueous phase falls through the sieve apertures and is removed. The viscous oil phase is captured by the oleophilic sieve surface and is carried out of the separation zone into the recovery zone where it is removed.

Because the clay particles are not dispersed during the process, the resulting tailings consolidate easily. Kruyer's patent holds promise, but it has not been implemented so far because of the high cost of replacing existing processes (Smith, 2009a).

Another major concern is that the contaminated water from the tailings ponds, laced with carcinogenic polycyclic aromatic hydrocarbons and trace metals, could seep into aquifers or underground water flows and end up in rivers (Robinson and Eivemark, 1985). One way to control seeps is by means of a perimeter tailings beach and perimeter collector ditches of the free water pond. Plastic liners may also prevent contaminated water from escaping.

Recent research suggests that as dry tailings technology becomes commercially viable, sludge ponds may become a problem of the past. An article by Collison (2008) suggests that new bitumen mining techniques may be able to eliminate the sludge pond altogether.

Water consumption

In addition to contamination of fresh water resources, the large water demands represent an adverse effect of oil-sands mining. In Alberta, water availability fluctuates with climate changes and seasonal flows. More than 75% of the Athabasca River water allocation is being used by Suncor, Syncrude, and smaller players like Albian Sands and Canadian Natural Resources Limited, among others. The Alberta Energy and Utilities Board is still issuing more licenses for water usage as the oil-sands industry demands continue to grow. Environmentalists fear the potential irreversible cumulative effects this large increase in water consumption could have in the maintenance of a healthy ecosystem. Although the Canadian government is committed to not have oil-sands development adversely affect the environment, experts suggest that to maintain a healthy aquatic ecosystem a minimal water level must be maintained in the rivers, which can only be achieved by storing water during high flow times so that it is not depleted during low flow times. Encouraging water recycling is another suggestion that most, but not all, operators are already following.

Fuel consumption

Canadian oil-sands production entails the consumption of natural gas for heating treatment water, steam production, and for upgrading and refining heavy oil. Natural gas is already used for electricity generation, heating, and chemical processing, and additional demand for heavy-oil refining is stretching the available supply thin. It also drives the price of natural gas higher, which in turn adds to production costs, putting the economy of heavy-oil extraction in question.

Natural gas consumption also affects the environment through carbon dioxide emissions. Donnelly and Pendergast (1999) concluded that deployment of nuclear-powered oil-sands projects would reduce a substantial fraction of Canada's emissions. Predictably, any suggestion of nuclear power draws opposing arguments (Caldicott, 2007); for example, nuclear power plants release no carbon dioxide, but everything leading to their construction and operation does. The debate is sure to continue.

Upgrading of heavy oil

Canadian heavy-oil crude production in 2008 surpassed 1.2 million b/d, mostly from oil sands in Athabasca and extracted by Suncor and Syncrude. Overall production is likely to reach 2 million b/d by 2015. Bitumen production needs to be upgraded to synthetic oil of higher API gravity and reduced sulfur content to facilitate its transportation and to enhance their value to most refineries, which are designed to handle lighter crudes. Upgrading essentially involves the use of temperature, pressure, and catalysts to break long hydrocarbon chains into small ones. Because bitumen has more carbon and less hydrogen during upgrading, chemical processes that tend to add hydrogen and remove carbon are adopted.

The easiest way to reduce viscosity and increase mobility is to heat the bitumen so it can be pipelined to regional upgraders. Injection of steam in the subsurface is often used to lower the viscosity of the oil sufficiently to allow it to flow. An alternative is dilution (i.e., adding a light oil or natural gas condensate to further sufficiently reduce the viscosity of oil so that it can flow through pipelines with ease). But because large volumes of light oil are required, adequate capacity needs to be built in place for diluents to be recycled. To this end, a parallel pipeline must be laid to return the diluents to the upgrader.

Dilution of Athabasca bitumen in *n*-pentane or *n*-heptane yields a solid precipitate of asphaltenes, which consist of carbon, hydrogen, nitrogen, oxygen, and sulfur and have high molecular masses. Deasphaltation itself yields crude oil with higher API gravity.

Because bitumen consists primarily of highly condensed polycyclic aromatic hydrocarbon molecules, these do not distill over when bitumen is heated; therefore, distillation typically yields low levels of distillates. Consequently, heavy oils must be cracked to get smaller molecules, a process referred to as "primary upgrading." The cracked products are rich in sulfur and nitrogen, which are then reduced during secondary upgrading.

Coking is one of several cracking processes that has traditionally been used and is essentially the thermal cracking (above 45°C) of heavier fractions to produce gasoline and fuel gas. This process does not remove the metal content and carbon residue significantly; however, the coking liquids and residue have a high content of sulfur, olefins, and heavy aromatics. These require additional treatment before they can be used as transportation fuels or other fuels as such. An alternative to coking is visbreaking — a primary noncatalytic cracking process like coking but characterized by higher temperature and shorter residence time. What this means is that the cracking reactions are terminated before the coke is formed as residue. By doing so, although the sulfur, nitrogen, metal, or asphaltene content of the heavy oil is not changed significantly, the molecular weight is reduced, which results in a reduction in the boiling range and a lowering of viscosity. The presence of sulfur (a pollutant) and nitrogen causes problems in some downstream processes such as catalytic cracking.

There are some other variations of the cracking process (e.g., delayed and fluid coking) used for primary upgrading of heavy oils. In each case, the products are gases, distillate oils, and coke as residue. In delayed coking the bitumen is heated before being fed into the coking chamber to allow sufficient time to undergo cracking reactions. The distillates from delayed cracking are usually passed through a chamber with cobalt molybdate catalyst; this process is called "hydrotreating," in which the sulfur- and nitrogen-produced hydrogen sulfide and ammonia are removed.

Fluid coking allows preheated feed oil flow (bitumen) into the coking chamber in the form of a spray, where it gets thermally cracked above 500°C. The products are lower hydrocarbons, which are condensed with coke as deposit on fluidized coke particles. The yield from fluid coking is higher than that from delayed coking.

As stated above, coke has limited use as boiler fuel because of excessive emission of sulfur dioxide. However, if the bitumen is mixed with a small quantity of a reagent like calcium hydroxide and carbonized in a laboratory at 475°C, the resultant coke has significantly reduced sulfur dioxide emissions during combustion (George et al., 1982).

In "hydrocracking," hydrogen is added, resulting in better distillate quality and lower levels of sulfur dioxide. However, because vanadium and nickel are present in bitumen, the catalysts gradually lose their properties and need to be replaced periodically. The outputs from the hydrocracker and the delayed or fluid cokers are hydrotreated and then pumped through pipelines to refineries.

Bitumen can also be upgraded by reforming, which involves splitting hydrogen atoms to transform the bitumen into a synthetic lighter crude.

In Venezuela, the heavy oil is warm enough so that it can flow. However, diluents still need to be added to pipeline it to the upgrading facility. Venezuela has developed a process for marketing extra-heavy oil in the form of Orimulsion, which is essentially an emulsion of approximately 70% natural Cerro Negro bitumen (8.5° API) and 30% water with less than 1% alcohol-based emulsifiers to help bitumen droplets remain suspended in the emulsion. This emulsion serves as inexpensive feed for power stations. It has good combustion characteristics and lower carbon dioxide emissions than coal.

Aquaconversion is a thermal-catalytic steam conversion process that upgrades heavy or extra-heavy crudes (9° API) to 15° API syncrude that can be transported without the need for diluents and still be processed to final fuels in conventional refineries (Pereira et al., 2005). Aquaconversion does not produce any solid residue byproduct such as coke and does not require a hydrogen source or high-pressure equipment. The process can be set up in the production area, eliminating the need for external diluents and transport over long distances.

Heavy-oil upgrading is also achieved by means of sonic generators that make use of low-frequency sonic vibrations to generate resonant frequencies that create cavitation in the heavy-oil samples. When used effectively, cavitation (i.e., the formation of vapor bubbles in the flowing liquid) can accelerate physical and chemical processes. Sonic generators have been used to deasphalt heavy oil and upgrade it to 10° API or higher (Smith, 2009b).

Greenhouse gas emissions

The development of oil sands in Alberta, although generating significant revenue for the province, has also been the cause of controversy because of its effect on the environment, high rates of fossil energy use, and the associated greenhouse gas (GHG) emissions (Charpentier et al., 2009).

Canada produces 2% of the world's GHG emissions. Oil-sands exploitation accounts for only 4% of the country's GHG emissions, which is 8 times less than the Canadian emissions from transportation, 4.5 times less than electricity and heat generation, and less than half of the emissions from agriculture (Alberta Government Brochure, 2008).

It may be pointed out that although the GHG emissions produced per barrel have been gradually reduced (intensity of emissions per barrel of production reduced by 38% since 1990), overall GHG emissions are up as a result of enhanced production over the last few years. Therefore, the challenge is to reduce the overall GHG emissions to acceptable levels despite the projected increase in production.

Conscious attempts are being made by the Alberta government and by oil-sands operators to monitor GHG emissions because these could increase substantially with oil-sands production of syncrude, which is expected to reach 5 million b/d by 2030. The long-term plan by the government of Alberta is to reduce GHG emissions by 14% below the 2005 levels by 2050, and in the short-term, the emissions should be stabilized by 2020.

There are still many challenges ahead for the exploitation of heavy-oil sands, perhaps none more important than its environmental impact. Plans to come up with a solution to curb GHG emissions include carbon sequestration. It involves capturing carbon at the source and directing it into depleted reservoirs for storage or into producing reservoirs for enhanced recovery. This will come at a cost. A recent report (Hoberg and McCullough, 2009) by the Canadian Energy Research Institute (CERI) has suggested that it is possible to make oil sands environmentally sustainable. However, this will require the benchmark price for oil to be approximately $105/barrel(US$). Efforts are gradually under way to draw attention to the ways and means of reducing GHG emissions from oil sands below conventional oil so that technological advancements in this area help in producing "green" bitumen. Technology must be further improved to address critical thermal EOR processes such as CSS and SAGD, which are routinely used in deeper formations that cannot be mined. Two problems with these processes are cost and emissions. At the scale at which oil-sands operations are predicted to increase, they will use up the natural gas resources in Western Canada. Therefore, another cheap, clean, and efficient source of energy needs to be found.

There is a growing concern that in situ production can warm groundwater, thereby liberating arsenic and other heavy metals from deep sediments, posing risks to human health. More work is required to ascertain if there is truth in this and if in situ production needs to be avoided. On the other hand, in situ upgrading of heavy oil to obtain lower viscosity and higher API may also be a possibility in the future.

Ultimately, the main obstacle to oil-sands exploitation remains that the cost incurred to produce a barrel of synthetic oil is higher than the cost of producing conventional oil. Furthermore, only special refineries can handle synthetic oil with its various impurities. For this reason, synthetic oil generated from oil sands sells for a lower price than conventional crude, which is processed in regular refineries. During the industry's cyclic downturns, this disparity widens and the result is that some of the new projects become uneconomic.

One of the ways to keep heavy-oil or bitumen production costs low is by enhancing production itself. Various ways and means have been suggested. In reservoirs where fluids occupy strata that are almost horizontal, drilling horizontal wells has become the preferred method of oil and gas recovery. Over the last many years, the cost of drilling such wells has come down, so that it costs only marginally more today. However, production from these wells is 15–20 times higher, which makes them an attractive choice for oil and gas producers.

Multilateral wells also justify their existence by allowing multiple wells to be drilled from a single main wellbore, eliminating the need to drill vertical stems for individual wells and thereby saving costly rig days. This arrangement also has the ability to tap several zones from branches emanating from a single wellbore.

Future outlook

As stated above, production from oil sands is energy intensive and causes concerns about their effect on the environment. The term "energy intensive" refers to the net energy available after accounting for the energy spent in producing the resource in a useable form. For oil sands, this value is much less than conventional oil. Coupled with this are concerns about the use of natural gas needed for generating steam to recover the bitumen, upgrading it to synthetic crude oil, its impact on the environment by emission of carbon dioxide, not to mention the risk of pollutant seepage from tailings ponds into freshwater aquifers. As many of the oil companies operating in the oil sands and heavy-oil areas around the world confront these issues, the reality is that production from oil sands and heavy-oil reservoirs is expensive compared with conventional oil, making the economics of the business vulnerable to the varying price of the barrel and the energy costs. However, as much as we may not want to believe, the steady decline that is taking place in mature reservoirs and others that will fall in this category will need to be balanced out. Heavy-oil deposits with their worldwide distribution have large potential as major long-term oil sources. In recent years, the discoveries of heavy crude in deep waters of Brazil, in western Africa, and in the Gulf of Mexico have further encouraged efforts aimed at production of this unconventional resource.

Application of newer technologies in the last two decades has reduced operating costs for the production of heavy oil in different areas of the world. Going forward, the key to enhanced growth and improved economic viability lies in the development of new technologies to help sustain a continuous supply of environmentally responsible energy. As long as the price of conventional oil remains reasonable and the development of new technologies continues, the abundance of this resource and the stable political regimes in most areas around the world will encourage oil companies to decide on the high financial commitments required to become players in the future energy growth.

Advancements in geophysical characterization of heavy-oil reservoirs have a definite contribution to make. As seen over the last two decades, poststack seismic inversion has been used for detecting the movement of steam stimulation from within a heavy-oil reservoir. Seismic attributes have been used for differentiating heated from unheated parts of a reservoir and imaging the areas affected by steam injection. Prestack AVO attributes have been used for characterizing bitumen and heavy-oil formations in terms of their density. Time-lapse monitoring has been used to ascertain reservoir depletion. Multicomponent seismic offers the advantage of characterizing heavy-oil formations in terms of the V_P/V_S ratio in addition to being able to correlate the P and S sections. All of these efforts have been ably supported by 3D visualization technology advancements. It is expected that advancements in all these areas will grow. We will see the application of sophisticated and accurate techniques in each of the areas mentioned above. Multicomponent seismic and neural network applications are two technologies likely to be favored in the near future for heavy-oil reservoir characterization.

References

Alberta Government Brochure, 2008, Alberta's oil sands, resourceful, responsible, http://www.energy.gov.ab.ca/OilSands/pdfs/Oil_Sands_Opp_Balance.pdf, accessed 18 May 2009.

Ali, F., 1994, CSS — Canada's super strategy for oil sands: Journal of Canadian Petroleum Technology, **33**, 16–19.

Bastin, E., 1926, Microorganisms in oilfields: Science, **63**, 21–24.

Batzle, M., R. Hofmann, and D.-H. Han, 2006, Heavy oils — seismic properties: The Leading Edge, **25**, 750–756.

Beggs, H. D., and J. R. Robinson, 1975, Estimating the viscosity of crude oil systems: Journal of Petroleum Technology, **27**, 1140–1141.

Behura, J., M. L. Batzle, R. Hofmann, and J. Dorgan, 2007, Heavy oils — their shear story: Geophysics, **72**, no. 5, E175–E183.

Bellman, L. W., 2008, Oil sands reservoir characterization — an integrated approach: CSEG Recorder, **33**, 7–10.

BP Statistical Review of World Energy, 2005, BP plc, 6.

Bregman, N. D., P. A. Hurley, and G. F. West, 1989, Seismic tomography at a fire-flood site: Geophysics, **54**, 1082–1090.

Britton, M. W., W. L. Martin, R. J. Leibrecht, and R. A. Harmon, 1983, Street ranch pilot test of fracture-assisted steamflood technology: Journal of Petroleum Technology, **35**, 511–522.

Brown, R. L., R. N. Cook, and D. Lynn, 1989, How useful is Poisson's ratio for finding oil/gas?: World Oil, **231**, 99–106.

Butler, R. M., and D. J. Stephens, 1981, The gravity of drainage of steam-heated oil to parallel horizontal wells: Journal of Canadian Petroleum Technology, **20**, 90–96.

Butler, R. M., 1985, A new approach to the modeling of steam-assisted gravity drainage: Journal of Canadian Petroleum Technology, **24**, 42–50.

Caldicott, H., 2007, Nuclear CO_2 warming costs, http://www.helencaldicott.com/pdf/070521.pdf.

Campbell, C. J., and J. H. Laherrère, 1998, End of cheap oil: Scientific American, **278**, 80–86, accessed 10 April 2011.

Carbognani, L, 1992, Molecular structure of asphaltene proposed for 510c residue of Venezuelan crude, INTEVEP S.A. Tech. Report, in G.A. Mansoori, A unified perspective on the phase behaviour of petroleum fluids; International Journal of Oil, Gas, and Coal Technology, **2**, 141.

Castagna, J. P., M. L. Batzle, and R. L. Eastwood, 1985, Relationships between compressional-wave and shear-wave velocities in clastic silicate rocks: Geophysics, **50**, 571–581.

Charpentier, A. D., J. A. Bergerson, and H. L. MacLean, 2009, Understanding the Canadian oil sands industry's greenhouse gas emissions: Environmental Research Letter, http://www.iop.org/EJ/article/1748-9326/4/1/014005/erl9_1_014005.pdf?request-id=308453ac-0bbf-4836-b2ab-49ef929a9640, accessed 18 May 2009.

Chattopadhyay, S. K., B. Ram, R. N. Bhattacharya, and T. K. Das, 2005, Enhanced oil recovery by in-situ combustion process in Santhal Field of Cambay Basin, Mehsana, Gujarat, India–a case study: Presented at the 13th European symposium on improved oil recovery, Abstract D24, SPE Paper 89451.

Christ, S., 2007, Oil, Chavez, and the Orinoco Belt, http://www.energyandcapital.com/articles/chavez-orinoco-oil/355, accessed 12 February 2009.

Collison, M., 2008, The end of the sludge pond?: Oilsands Review, **3**, 45–46.

De Buyl, M., 1989, Optimum field development with seismic reflection data: The Leading Edge, **8**, 14–20.

De Ghetto, G., F. Paone, and M. Villa, 1995, Pressure-volume-temperature correlations for heavy oils and extra heavy oils: SPE Paper 30316.

Demaison, G. J., 1977, Tar sands and super giant oil fields: AAPG Bulletin, **61**, 1950–1961.

Deroo, G., B. Tissot, R. G. McCrossan, and F. Der, 1974, Geochemistry of the heavy oils of Alberta, in L. V. Hills, ed., Oil sands — fuels of the future: CSPG Memoir 3, 50–67.

Deroo, G., B. Powell, B. Tissot, and R. G. McCrossan, with contributions by P. A. Hacquebard, 1977, The origin and migration of petroleum in the Western Canadian Sedimentary Basin, Alberta — Geochemical and thermal maturation study: Geological Survey of Canada Bulletin, **262**, 136.

Dilay, A., and J. Eastwood, 1995, Spectral analysis applied to seismic monitoring of thermal recovery: The Leading Edge, **14**, 1117–1122.

Domenico, S. N., 1984, Rock lithology and porosity determinations from shear and compressional wave velocity: Geophysics, **49**, 1188–1195.

Donnelly, J. K., and D. R. Pendergast, 1999, Nuclear energy in industry: Application to oil production: Presented at the 20th Annual Conference of the Nuclear Society, http://www.computare.org/Support%20documents/Publications/Nuclear%20oil%20sand.htm, accessed 18 May 2009.

Dumitrescu, C., L. W. Bellman, and A. Williams, 2005, Delineating productive reservoir in the Canadian oil sands using neural network approach: CSEG Abstracts, 209–212.

Dumitrescu, C., and L. Lines, 2007, Heavy oil reservoir characterization using V_P/V_S ratios from multicomponent data: Presented at the EAGE Conference and Exhibition.

Eastwood, J., 1993, Temperature-dependent propagation of P- and S- waves in Cold Lake oil sands: Comparison of theory and experiment: Geophysics, **58**, 863–872.

Eastwood, J., P. Lebel, A. Dilay, and S. Blakeslee, 1994, Seismic monitoring of steam-based recovery of bitumen: The Leading Edge, **13**, 242–251.

Edgeworth, R., B. J. Dalton and T. Parnell, 1984, The pitch drop experiment: European Journal of Physics, **5**, 198.

Garotta, R., P. Granger, and H. Darvis, 2000, Elastic parameter derivations from multicomponent data: 70th Annual International Meeting, SEG, Expanded Abstracts, 154–157.

Gates, I., J. Adams, and S. Larter, 2008, The impact of oil viscosity heterogeneity on production from heavy oil and bitumen reservoirs: geotailoring recovery processes to compositionally graded reservoirs: CSEG Recorder, **33**, 42–49.

George, Z. M., L. G. Schneider, and M. A. Kessick, 1982, Coking of bitumen from Athabasca oil sands. 1. Effect of calcium hydroxide and other reagents: Fuel, **61**, 169–174.

Greaves, R. J., and T. J. Fulp, 1987, Three-dimensional seismic monitoring of an enhanced oil recovery process: Geophysics, **52**, 1175–1187.

Greaves, M., T. X. Xia, S. Imbus, and V. Nero, 2004, THAI-CAPRI processes: tracing downhole upgrading of heavy oil: Proceedings of the Canadian International Petroleum Conference.

Greaves, M., and T. X. Xia, 2004, Downhole catalytic process for upgrading heavy oil: produced oil properties and composition: Journal of Canadian Technology, **43**, 25–30.

Hacquebard, P. A., 1975, Correlation between coal rank paleotemperature and petroleum occurrence in Alberta, Canada: Geological Survey Paper, **75-1**, 5–8.

Hampson, D. P., J. S. Schulke, and J. A. Querein, 2001, Use of multiattribute transforms to predict log properties from seismic data: Geophysics, **66**, 220–236.

Han, D., J. Liu, and M. L. Batzle, 2006, Acoustic property of heavy oil — measured data: 76th Annual International Meeting, SEG, Expanded Abstracts, 1903–1906.

Han, D., A. Nur, and D. Morgan, 1986, Effects of porosity and clay content on wave velocities in sandstones: Geophysics, **51**, 2093–2107.

Head, I. M., D. Martin Jones, and S. R. Larter, 2003, Biological activity into the deep subsurface and the origin of heavy oil: Nature, **426**, 344–352.

Hein, F. J., and D. K. Cotterill, 2006, The Athabasca oil sands — a regional geological perspective, Fort McMurray area, Alberta, Canada: Natural Resources Research, **15**, 85–102.

Hein, F. J., and R. A. Marsh, 2008, Regional geologic framework, depositional models and resource estimates of oil sands of Alberta, Canada: Presented at the World Heavy Oil Congress.

Hirsche, K., 2006, A personal perspective on the past, present, and future of time-lapse seismic monitoring: CSEG Recorder, Special Issue, 136–139.

Hirsche, K., D. Hampson, J. Peron, and B. Russell, 2005, Simultaneous inversion of PP and PS seismic data — a case history from Western Canada: Presented at the EAGE/SEG Research Workshop.

Hirsche, K., N. Pullin, and L. Matthews, 2002, The GLISP 4D monitoring project — a lesson in how to avoid a successful failure: 64th EAGE Conference and Exhibition, Extended Abstracts, A-27.

Hoberg, G., and G. McCullough, 2009, Can the oil sands be made environmentally sustainable? Canadian Energy Research Institute report, http://greenpolicyprof.org/wordpress/?p=237, accessed 10 April 2010.

Hubral, P., 1977, Time migration — some ray theoretical aspects: Geophysical Prospecting, **25**, 738–745.

Hunt, J. M., 1979, Petroleum geochemistry and geology: W. H. Freeman & Co.

Hunt, J. M., 1996, Petroleum geochemistry and geology, 2nd ed; W. H. Freeman & Co.

Isaac, J. H., and D. C. Lawton, 2006, A case history of time-lapse 3D seismic surveys at Cold Lake, Alberta, Canada: Geophysics, **71**, no. 4, B93–B99.

Jardine, D., 1974, Cretaceous oil sands of Western Canada, in L. V. Hills, ed., Oil sands — fuels of the future: CSPG Memoir **3**, 50–67.

Justice, J. H., M. E. Mathisen, A. A. Vassiliou, I. Shiao, B. R. Alameddine, and N. J. Guinzy, 1993, Crosswell seismic tomography in improved oil recovery: First Break, **11**, 229–239.

Kalantzis, F., 1994, Imaging of reflection seismic and radar wavefields: monitoring of steam-heated oil reservoirs and characterization of nuclear waste repositories: Ph.D. thesis, University of Alberta.

Kalantzis, F., E. R. Kanasewich, A. Vafidis, and A. Kostykevich, 1996a, Seismic reflection modeling and imaging of a thermal enhanced oil recovery project at Cold Lake, Canada: First Break, **14**, 91–103.

Kalantzis, F., A. Vafidis, E. R. Kanasewich, and A. Kostykevich, 1996b, Seismic monitoring and modeling enhanced oil recovery project at Cold Lake, Canada: Canadian Journal of Exploration Geophysicists, **32**, 977–989.

Kaminsky, V., L. Trevoy, and A. Maskwa, 1979, Method for concentrating heavy minerals in the solids tailings from hot water extraction of tar sands: U. S. Patent 4138467.

Kendall, R. P., J. Pullushy, D. Karpluk, and C. Seibel, 2002, A proposed workflow for converted-wave event registration — the step between processing and interpretation: 64th EAGE Conference and Exhibition, Extended Abstracts.

Kendall, R., P. F. Anderson, L. Chabot, and D. Gray, 2005, A proposed workflow for heavy oil reservoir characterization using multicomponent seismic data, Presented at the EAGE/SEG Research Workshop.

Kumar, G., 2003, Fluid effects on attenuation and dispersion of elastic waves: M.S. thesis, Colorado School of Mines.

Kruyer, J., 1988, Separating oil phase from aqueous phase using an aperture oleophilic sieve in contact with an aperture cylindrical cage wall: U. S. Patent 4,740,311.

Laine, E. F., 1987, Remote monitoring of the steam-flood enhanced recovery process: Geophysics, **52**, 1457–1465.

Leiphart, D. J., and Hart, B. S., 2001, Comparison of linear regression and probabilistic neural network to predict porosity from 3D seismic attributes in Lower Brushy Canyon channeled sandstones, southeast New Mexico: Geophysics, **66**, 1349–1358.

Li, G., G. Purdue, S. Weber, and R. Couzens, 2001, Effective processing of nonrepeatable 4D seismic data to monitor heavy oil SAGD steam flood at East Senlac, Saskatchewan, Canada: The Leading Edge, **20**, 54–62.

Lines, L. R., R. Jackson, and J. D. Covey, 1989, Seismic velocity models for heat zones in Athabasca tar sands: 59th Annual International Meeting, SEG, Expanded Abstracts, 751–754.

Lines, L., Y. Zou, A. Zhang, K. Hall, J. Embleton, B. Palmiere, C. Rene, P. Bessette, P. Cary, and D. Secord, 2005, V_P/V_S characterization of a heavy-oil reservoir: The Leading Edge, **24**, 1134–1135.

Macrides, C. G., E. R. Kanasewich, and S. Bharatha, 1988, Multiborehole seismic imaging in steam injection heavy oil recovery projects: Geophysics, **53**, 65–75.

Mathisen, M. E., A. A. Vasiliou, P. Cunningham, J. Shaw, J. H. Justice, and N. J. Guinzy, 1995, Time-lapse crosswell seismic tomogram interpretation: Implication for heavy oil reservoir characterization, thermal recovery process monitoring, and tomographic imaging technology: Geophysics, **60**, 631–650.

Mavko, G., and A. M. Nur, 1979, Wave attenuation in partially saturated rocks: Geophysics, **44**, 161–178.

Mayo, L., 1996, Seismic monitoring of foamy heavy oil, Lloydminster, Western Canada: 66th Annual International Meeting, SEG, Expanded Abstracts, 2091–2094.

McCormack, M. D., J. A. Dunbar, and W. W. Sharp, 1984, A case study of stratigraphic interpretation using shear and compressional seismic data: Geophysics, **49**, 509–520.

McCormack, M.D., M. G. Justice, and W. W. Sharp, 1985, A stratigraphic interpretation of shear and compressional wave seismic data for the Pennsylvanian Morrow formation of southeastern New Mexico, in O. R. Berg and D. G. Woolverton, eds., Seismic stratigraphy II—an integrated approach: AAPG Memoir **39**, 224–239.

Meyer, R. F., and E. Attanasi, 2003, Heavy oil and bitumen – strategic petroleum resources: USGS Fact Sheets, Reston, VA, U. S. Geological Survey, 7.

Michelena, R. J., M. S. Donati, A. J. Valenciano, and C. D'Augusto, 2001, Using multicomponent seismic for reservoir characterization in Venezuela: The Leading Edge, **20**, 1036–1041.

Mochinaga, H., S. Onozuka, F. Kono, T. Ogawa, A. Takahashi, and T. Torigoe, 2006, Properties of oil sands and bitumen in Athabasca: 2006 CSPG-CSEG-CWLS Joint Convention, 39–44.

Mummery, R. C., 1985, Quantitative application of seismic data for heavy oil projects: 55th Annual International Meeting, SEG, Expanded Abstracts, 393–394.

Nur, A. M., C. Tosaya, and D. Vo-Thanh, 1984, Seismic monitoring of thermal enhanced oil recovery processes: 54th Annual International Meeting, SEG, Expanded Abstracts, 337–340.

O'Connell, R. J., and B. Budiansky, 1977, Viscoelastic properties of fluid-saturated cracked solids: Journal of Geophysical Research, **82**, 5719.

Paulsson, B. N. P., J. A. Meredith, Z. Wang, and J. W. Fairborn, 1994, The Steepband crosswell seismic project: reservoir definition and evaluation of steamflood technology in Alberta tar sands, The Leading Edge, **13**, 737–747.

Pengelly, K., 2005, Processing and interpretation of multicomponent seismic data from the Jackfish heavy oil field, Alberta: M.S. thesis, University of Calgary.

Pereira, P., R. Marzin, M. McGrath, and G. J. Thompson, 2005, How to extend existing heavy oil resources through aquaconversion technology, http://mail.worldenergy.org/tech_papers/17th_congress/2_1_05.asp, accessed 26 September 2009.

Peron, J., 2004, Multicomponent data interpretation (registration) in a heavy oil reservoir: 66th EAGE Conference and Exhibition, Extended Abstracts, D005.

Pickett, G. R., 1963, Acoustic character logs and their application in formation evaluation: Journal of Petroleum Technology, **15**, 659–667.

Proctor, R. M., G. C. Taylor, and J. A. Wade, 1984, Oil and natural gas resources of Canada 1983: Geological Survey of Canada, Paper 83-31.

Pullin, N., R. K. Jackson, L. W. Matthews, R. F. Thorburn, and K. Hirsche, 1987a, 3D seismic imaging of heat zones at an Athabasca tar sands thermal pilot: 57th Annual International Meeting, SEG, Expanded Abstracts, 391–394.

Pullin, N., L. W. Matthews, and K. Hirsche, 1987b, Techniques applied to obtain very high resolution 3D seismic imaging at an Athabasca tar sands thermal pilot: The Leading Edge, **6**, 10–15.

Pullin, N., L. Matthews, and K. Hirsche, 1986, Techniques applied to obtain very high resolution 3D seismic imaging at an Athabasca tar sands thermal pilot: 56th Annual International Meeting, SEG, Expanded Abstracts, 494–497.

Rafavich, F., C. H. St. C. Kendall, and T. P. Todd, 1984, The relationships between acoustic properties and the petrographic character of rocks: Geophysics, **49**, 1622–1636.

Riediger, C., S. Ness, M. Fowler, and T. Akpulat, 2000, Timing of oil migration, Paleozoic and Cretaceous bitumen, and heavy oil deposits, eastern Alberta: CSEG Annual Convention, Expanded Abstracts, http://www.cseg.ca/conferences/2000/2000abstracts/819.pdf, accessed May 2009.

Robinson, K. E., and M. M. Eivemark, 1985, Soil improvement using wick drains and preloading: 11th International Conference on Soil Mechanics and Foundation Engineering, Proceedings, 3, 1739–1744.

Schmitt, D. R., 1999, Seismic attributes for monitoring of a shallow heated heavy oil reservoir: a case study, Geophysics, **65**, 368–377.

Schmitt, D., 2005, Heavy and Bituminous oils: Can Alberta save the world?: Preview, 22–29.

Simmons, M. R., 2005, Twilight in the desert: the coming Saudi oil shock and the world economy: Wiley.

Smith, M., 2009a, Good vibrations: New Technology Magazine, **15**, 42–44.

Smith, M., 2009b, Barrier to entry: New Technology Magazine, **15**, 20–29.

Stell, J., 2008, Canadian oil imports: Oil and Gas Investor, **28**, no. 1, 119–121.

Stonehouse, D., 2008, The world gets heavier: Higher prices and growing demand put global focus on unconventional oil plays: Oilsands Review, http://www.oilsandsreview.com/osr-article.asp?id=7373, accessed 10 April 2010.

Taheri, M. and N. M. Audemard, 1987, Application of multivariate statistics in crude quality characterization and regional distribution in Orinoco Oil Belt, in R. F. Meyer, ed, Exploration for heavy crude oil and natural bitumen, AAPG Studies in Geology, no. 25, 175–181.

Talwani, M., 2002, The Orinoco heavy oil belt in Venezuela, http://www.rice.edu/energy/publications/docs/Talwani_OrinocoHeavyOilBeltVenezuela.pdf, accessed 12 February 2009.

Tatham, R. H., 1982, V_P/V_S and lithology: Geophysics, **47**, 336–344.

Tatham, R. H. and P. L. Stoffa, 1976, V_P/V_S — a potential hydrocarbon indicator: Geophysics, **41**, 837–849.

Theune, U., 2004, Seismic monitoring of heavy oil reservoirs: rock physics and finite element modeling, Ph.D. thesis, University of Alberta.

Theune, U., D. R. Schmitt, and C. D. Rokosh, 2003, Feasibility study of time-lapse seismic monitoring for heavy oil reservoir development — the rock-physical basis: 73rd Annual International Meeting, SEG, Expanded Abstracts, 1418–1422.

Tonn, R., 2002, Neural network seismic reservoir characterization in a heavy oil reservoir: The Leading Edge, **21**, 309–312.

Tosaya, C., A. Nur, D. Vo-Thanh, and G. Da Prat, 1987, Laboratory seismic methods for remote monitoring of thermal EOR: SPE Reservoir Engineering, **2**, 238–242.

Valenciano, A. A., and R. J. Michelena, 2000, Stratigraphic inversion of poststack PS converted waves data: 70th Annual International Meeting, SEG, Expanded Abstracts, 150–153.

Veazeay, M. V, 2006, Rigzone, http://www.rigzone.com/training/heavyoil/insight.asp?i_id=193 accessed 26 April 2010.

Wang, T., L. Lines, and J. Embleton, 2007, Seismic monitoring of cold heavy oil production: CSPG/CSEG Convention, Expanded Abstracts.

Wang, Z., and A. M. Nur, 1988, Velocity dispersion and the 'local flow' mechanism in rocks: 58th Annual International Meeting, SEG, Expanded Abstracts, 548–550.

Wang, Z., A. M. Nur, and M. L. Batzle, 1990, Acoustic velocities in petroleum oils: Journal of Petroleum Technology, **42**, 192–201.

Watson, I., K. Brittle, and L. R. Lines, 2002, Heavy-oil reservoir characterization using elastic wave properties: The Leading Edge, **21**, 736–739.

Xu, Y., and S. Chopra, 2009, Mapping lithology heterogeneity in Athabasca oil sands reservoirs using surface seismic data: case history: CSEG Convention Abstracts, 547–551.

Zadeh, H. M., R. P. Srivastava, N. Vedanti, and M. Landro, 2007, Seismic monitoring of in-situ combustion in the Balol heavy oil field: 77th Annual International Meeting, SEG, Expanded Abstracts, 2878–2882.

Zhang, Y., 2006, A case study: seismic monitoring of a thin heavy oil reservoir: Ph.D. thesis, University of Alberta.

Section 1

Rock Physics Aspects

Chapter 2

Seismic Properties of Heavy Oils — Measured Data

De-hua Han,[1] Jiajin Liu,[1] and Michael Batzle[2]

Introduction

With a high demand of hydrocarbon worldwide, conventional oil production is quickly approaching its peak. Inevitably, heavy oil and bitumen (ultraheavy oil) will emerge as "new" (so-called unconventional) hydrocarbon resources because of their tremendous potential. Currently, more than 50% of Canada's oil production is from heavy oils (Alboudwarej et al., 2006; Hinkle and Batzle, 2006).

Such heavy oils are highly viscous, difficult to move in reservoirs, and much more expensive to produce. In addition to mining and other cold production methods, many different techniques (e.g., thermal, chemical, or in situ combustion, etc.) have been applied to mainly reduce viscosity and assist the heavy-oil production. None of these techniques have matured completely yet, and engineering developments are occurring rapidly. These techniques remain expensive in terms of energy and resources used (lots of water) and in terms of efficiency and overall environmental impact. The steam-assisted gravity drainage (SAGD) technique is a current popular technique. In a steam chamber, more than 60% of oil in place can be produced (Caruso, 2005; Gupta, 2005). However, on a reservoir scale, efficiency can be low (approximately15% with different resources). Clearly, seismic techniques hold great potential for assisting reservoir characterization and recovery monitoring. Monitoring has been demonstrated successfully in several fields [Cold Lake (Eastwood et al., 1994) and Duri Field, Indonesia (Jenkins et al., 1997)]. However, to be effective, we must understand the seismic properties of the heavy oils and the heavy-oil sands. This understanding of in situ properties is the key to bridging the seismic response to reservoir properties and changes. Schmitt (2004) provided a general review of rock physics as related to heavy-oil reservoirs. Here, we examine the seismic properties of heavy oils in detail.

Seismic techniques hold great potential for characterization and recovery monitoring of heavy-oil reservoirs. Knowledge of in situ fluid and rock behavior is a key to linking the seismic response to reservoir properties and changes. Definitions of heavy oils differ widely. The U.S. Geological Survey (USGS) defines heavy oil as a dense and viscous oil that is chemically characterized by its content of asphaltenes. According to the USGS (http://pubs.usgs.gov), American Petroleum Institute (API) gravity of heavy oil has been defined from 22° to 10° (for heavy oil) and less than 10° for ultraheavy oil (bitumen). In terms of viscosity definitions, heavy oil is defined as having viscosities of 100–10,000 centipoise (cp) and bitumen is defined as having viscosities greater than 10,000 cp.

High-density heavy oils can be formed by several mechanisms. Heavy "tar mats" can occur in deep reservoirs by chemical precipitation and tend to accumulate at the base of the reservoir. More commonly, heavy oils are formed in shallow deposits by biodegradation of lighter oils; alkane chains and lighter hydrocarbons are consumed by bacteria, leaving a mixture of complex organic compounds (Hunt, 1996). This requires contact with circulating freshwater. Although this mechanism can be different than that forming tar mats, the heavy components may be similar.

Typically, heavy crude oils are classified into four types — saturates, aromatics, resins, and asphaltenes (SARA analysis) — on the basis of solubility classes. Heavy oils contain more resins and asphaltenes with high molecular weight. As described by Batzle et al. (2006), because of complex heavy compounds in heavy oil, the simple empirical trends developed to estimate fluid properties of light oil (Han and Batzle, 2000) may not be appropriate for heavy oils.

In terms of molecular dynamics, the phase transition between liquids and solids can be very complicated. On

---

[1]University of Houston, Department of Earth and Atmospheric Sciences, Houston, Texas, U.S.A.
[2]Colorado School of Mines, Geophysics Department, Golden, Colorado, U.S.A.
This paper appeared in the September 2008 issue of THE LEADING EDGE and has been edited for inclusion in this volume.

the molecular level, fluid, solid, and glass have different structures. Molecules are disordered and not rigidly bound in liquids, ordered in a regular lattice in crystalline solids, and disordered but rigidly bound (amorphous solid) because of high viscosity in glassy solids.

The glass point is often defined as when liquid viscosity equals 10^{15} cp. A fluid with viscosity higher than the glass point is a glassy solid. A drop below the glass point temperature occurs when viscosity exceeds 10^{15} cp. A quasi-solid is a transition phase between a glassy solid and liquid phase for viscous materials.

A crystalline solid has a melting point (temperature) at which the solid absorbs (fusion) heat and transforms into a fluid phase. A glassy solid has no distinct melting point; instead, there is a temperature transition zone called the quasi-solid phase. With increasing temperature, glass gradually softens and eventually liquidizes and transforms into a liquid phase.

Viscosity, a measure of the fluid resistance to flow, is the key property controlling heavy-oil production and, as we shall see, it also has a strong influence on seismic properties. Temperature, composition, and density (API gravity) are dominant influences on heavy-oil properties. Although viscosity has been carefully studied because it controls the economics of production and transportation, there is currently no definitive model for viscosity of heavy oils.

Measured data demonstrate that composition is also a dominant factor controlling viscosity. Although viscosity data show large variations in magnitude, they have similar temperature dependence (Figure 1; Dusseault, 2006). In general, viscosity shows increased temperature dependence at low API gravity (high density). Generally, we can calibrate viscosity with empirical relations over a local area where variations in composition are restricted.

We have conducted numerous density measurements using the constant mass method. Density data are fitted quite well using linear temperature dependence. This is similar to the behavior of light oils. Gas in solution has a small effect for most shallow heavy oils and may be negligible because of the small amount of gas that can go into solution (low gas-oil ratio).

Velocity Models for Light and Heavy Oils

P-wave velocities of light oils (22° API) are independent of frequencies and have no measurable shear velocities. The velocity of light oil is basically controlled by pressure-temperature conditions and velocity-pseudo density, which is derived from API gravity, gas-oil ratio (GOR), and gas gravity (Han and Batzle, 2000). For light oil, velocity and viscosity increase with decreasing temperature. We can linearly correlate velocity to temperature (Figure 2a). No viscosity effect on the velocity of light oil has been observed.

The suite of velocity data used to develop this model includes heavy-oil samples at high temperatures. This suggests that heavy oils at high temperatures are similar to light oils. However, at low temperatures, viscosity of heavy oils drastically increases and heavy oils transform into a viscoelastic state.

We have measured P-wave velocities of heavy-oil samples as functions of temperature, pressure, and GOR. Using the pulse transmission method, the velocity can be calculated as

$$V = L/t, \qquad (1)$$

where L is the length of the sample and t is the traveltime of the P-wave. We can control sample pressure and

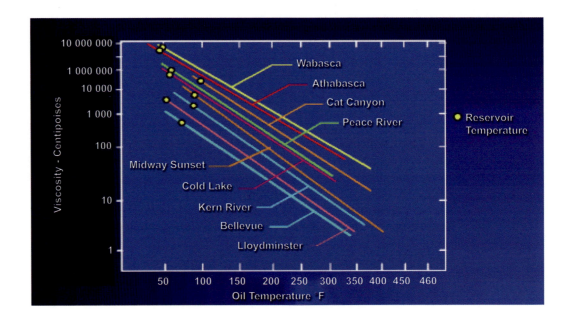

Figure 1. Viscosity temperature trends for various heavy oils. From Dusseault, 2006.

temperature generally within sensor accuracy (<0.5%). With calibration using distilled water, the accuracy of the P-wave velocity is better than 0.5%.

We found that amplitude and frequency of the P-wave signals of heavy oil are very sensitive to temperature. The signal-to-noise ratio decreases significantly at low temperature. For 3-MHz transducers, the P-wave signal of heavy oil is attenuated with decreasing temperature. This amplitude behavior indicates we are approaching a viscous relaxation effect within the heavy oils at lower temperatures.

The velocity behavior of heavy oils is distinctly different from light oils at low temperatures. As can be seen in Figure 2b, below approximately 50°C, the trend becomes nonlinear. This stronger temperature dependence in P-wave velocity indicates that we are beginning to be influenced by the viscous shear properties of the fluid (i.e., it begins to act like a solid).

As mentioned, heavy oils in the quasi-solid phase possess shear rigidity. Several methods are available to obtain shear rigidities (Behura et al., 2007). We measured shear velocity in heavy oil in the glass state with conventional ultrasonic wave-transmission methods. But because with increasing temperature heavy oil transitions into a more conventional fluid, and the transmitted shear signal becomes very noisy and attenuated, we developed an alternative method: We measure the reflected shear wave off a fluid-solid interface, derive the reflection coefficient, then derive the shear impedance and shear velocity for the viscous liquid (Han et al., 2005). This allows measurement of shear velocity of lighter fluids as a function of API gravity, temperature, GOR, and pressure. Basically, we use the reflected amplitude from a buffer-water interface to calibrate the amplitude measured from a buffer-oil interface under the same conditions (Figure 3). If the buffer impedance and oil density are known, we can derive shear velocity. The typical error in shear velocity is approximately 5%. The main sources of errors are from stability of coupling among transducer, buffer, and electronics. In general, the relative error will increase to 20% at a shear velocity less than 100 m/s.

We can now examine the influence of API gravity and temperature on shear velocity. Figure 4 shows the shear velocities measured on several heavy-oil samples ranging from 14.36° to 8.05° API. Most shear velocities were measured at room pressure. There is a general increase in

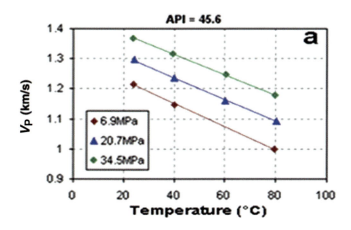

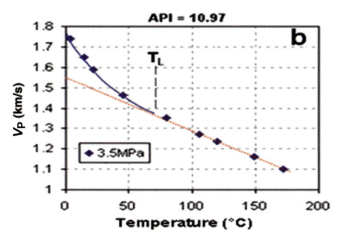

Figure 2. Velocity-temperature measurements for (a) a light oil at different pressures and (b) a heavy oil at different pressures. Note that V_P temperature trends can be fit very well by straight lines. At low temperatures, V_P temperature data show a strong nonlinear trend. The temperature at which this departure begins is the liquid point.

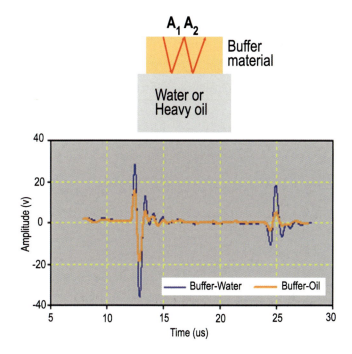

Figure 3. Shear-wave measurements using the reflection of a shear wave off of a solid-liquid interface. Calibration is performed using water (shear modulus = 0).

shear velocity with decreasing API gravity (increasing density). However, factors other than simple density (e.g., specifics of the composition) influence velocity.

Using the measured densities and P- and S-wave velocities, we can extract the bulk (K) and shear (μ) moduli of heavy oils. Figure 5 shows the moduli of a sample with 8° API at a pressure of 0.69 mPa (100 psi). These data show that the bulk modulus decreases rapidly from 3.7 GPa at -8°C to 2.2 GPa at 50°C, then continues decreasing to 1.9 GPa at 76°C with a much lower gradient. The data show clearly that whenever the shear rigidity of heavy oil is negligible, the bulk modulus shows a linear trend with increasing temperature. Similar to the case of velocities, we can define the liquid point as the temperature at which the shear rigidity vanishes and the slope of the bulk modulus-temperature trend changes.

Factors Influencing Velocity

The velocity of shallow heavy oils is a function of temperature, API gravity (density), viscosity, and wave frequency. From the point of view of petroleum engineers and geochemists, the density of oil may have no unique relation to velocity or viscosity because oils with the same density may have very different chemical composition. However, from the data we examined from light- and heavy-oil samples, API gravity is still a dominant influence on velocities and velocity dispersion at temperatures lower than the liquid point. We observed that heavy oils with similar API gravities show systematic differences in velocity, but usually these differences are less than 10%. The compositional dependence is outside of the scope of this paper (see Hinkle et al., 2008).

Pressure effect

Similar to light oil, the P-wave velocity of heavy oil increases with increasing pressure and decreases with increasing temperature. Figure 6a shows velocity data for an 8.6° API dead oil sample. The P-wave velocity was measured at increasing pressure from 0 to 20.7 MPa (3000 psi) and temperature from 3.5°C to 80°C. For each temperature, velocity tends to increase linearly with pressure. With increasing temperature, the velocity gradient with respect to pressure seems to decrease slightly. It is of interest that heavy oil with high velocity shows a slightly higher velocity gradient with respect to pressure. Figure 6b shows relative velocities; that is, all temperature curves are normalized to the velocity at 10 MPa. The increase of the relative velocity with pressure seems to be constant but slightly less than 0.4% per MPa. Overall, the pressure effect on single-phase heavy oil is small for a low-pressure heavy-oil reservoir. The model developed to predict the pressure effect on light oil can be used for heavy oil without correction.

Gas effect

Two conditions determine the influence of gas on heavy oil: in solution and as free gas. For completely dissolved gas, there is only one "live" liquid phase. We measured the dissolved gas effect on velocity for several heavy oils. Figure 7 shows measured velocity on an 11° API heavy-oil sample with GOR near 0 (dead oil), 2, and 37. The data show velocity as a function of temperature

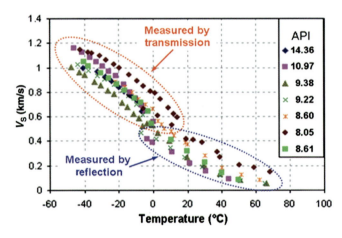

Figure 4. Shear velocity versus temperature for several heavy oils (dead). Measurement techniques were pulse transmission and reflection amplitude.

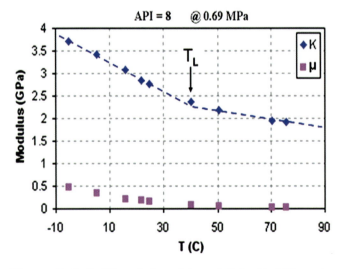

Figure 5. Bulk (K) and shear (μ) moduli for 8° API gravity oil. At a liquid point temperature of 40°C, a shear modulus appears and the bulk modulus-temperature trend changes slope.

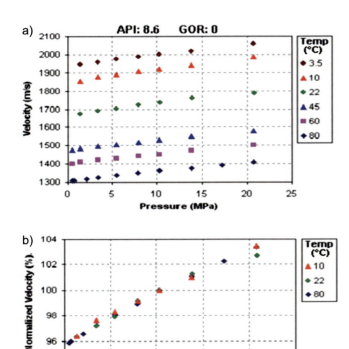

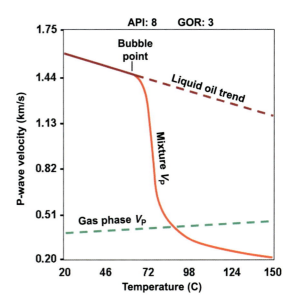

Figure 6. The effect of pressure on the P-wave velocity of (a) an 8.6° API heavy oil at several temperatures. The pressure trend is very consistent so that relations for lighter oils may be used. (b) If velocities for each temperature are normalized to the velocity at 10 MPa, the trends collapse to a single curve.

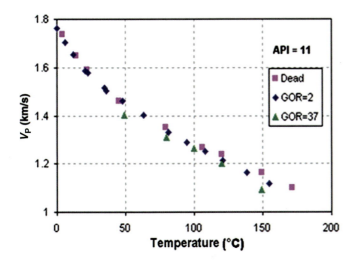

Figure 7. Gas in solution (no free-gas phase) lowers the velocity of heavy oils. The effect is small and usually can be ignored.

at 3.45 MPa (500 psi) for dead and a GOR of 2 and at 24.2 MPa (3500 psi) for a GOR of 37. The velocity of the dead oil is almost the same as that of the live oil with a GOR of 2 and a few percent higher than the live oil

Figure 8. Gas coming out of solution has a dramatic influence on the velocity of the heavy oil-gas mixture. Even for a low GOR of 3, as the temperature raises past the bubble point, velocity drops to a fraction of the value for the single-phase liquid.

with a GOR of 37. Gas dissolved in heavy oil does reduce oil velocity — the higher the GOR, the lower the velocity. However, the low capacity to dissolve gas in heavy oils and low-pressure environment of most heavy-oil reservoirs generally results in a very low GOR. Thus, although gas in solution has an effect on velocity, we usually need not worry about it.

In contrast, small amounts of free gas in heavy-oil reservoirs are very important. During production, pressures often drop below the bubble point. Under these conditions, gas exsolution can generate foamy oil (gas bubbles in heavy oil). It is also possible to cross the bubble point by raising temperature. Figure 8 shows the effect of crossing the bubble point with increasing temperature. Small amounts of gas bubbles can drop the velocity below even that of the free-gas phase itself (this is a density effect). Note that this assumes pressure equilibrium. If heavy oils have viscosity more than 10^6 cp, the gas effect on seismic velocity of heavy oils may be reduced because local pressure in the oil may not reach equilibrium with that of gas.

Temperature effect

Probably the most important parameter for heavy oil is viscosity, which depends largely on API gravity and temperature. API gravity roughly represents the amount of heavy compounds (e.g., resins and asphaltenes) in heavy oils. API gravity is the basis of most published viscosity

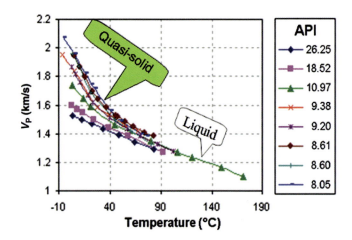

Figure 9. P-wave velocities of several oils as a function of temperature. At low temperatures, lighter oils continue to act like liquids whereas heavy oils enter the quasi-solid phase.

models. In contrast, temperature is an environmental condition. In a shallow, low-temperature environment, heavy oil is in the quasi-solid phase. Heavy molecules tend to interact to resist any relative movement. Increasing temperature lowers the coherent force between heavy molecules and reduces viscosity and velocity. Figure 9 shows measured P-wave velocities of heavy oil samples with different densities in a range of 0.897–1.014 gm/c³ (8–26° API) at low pressures [<6.9 MPa (1000 psi)]. Velocity trends fall into several domains.

When temperature is higher than the liquid point, heavy-oil properties are similar to that of light oil; the velocity gradient with temperature is nearly a constant (approximately 3.0 m/s/°C) and slightly decreases for heavier oils. Velocity of heavy oil with temperature greater than the liquid point (T_L) can be expressed as

$$V = V\{API, T, [\eta(API, T)^* f]\ldots\} \\ \approx V_0(API) + B(API)^* \Delta T \ (T > T_L). \quad (2)$$

At high temperature, viscosity is low and its influence on velocities is negligible. The term $B(API)$ is a function of API gravity for the oil. The pressure and GOR effect on velocity are less important. Velocity linearly relates to temperature.

When temperature is lower than the T_L, heavy oil is in the quasi-solid phase. Viscosity increases rapidly and the viscosity-frequency effect can no longer be ignored. Velocity deviates from the simple trend in the liquid phase (Figure 9). The velocity gradient with respect to temperature is no longer constant. These data were measured at a frequency of approximately 1 MHz. With decreasing temperature, the velocity gradient of heavy oil in the quasi-solid phase increases from that of liquid phase and decreases toward that of the glass-solid phase. This velocity gradient reaches its highest value within the transition zone.

We have observed cases in which the viscosity does not conform to a simple API relationship, but velocity does. As an example, a waxy oil, with 26.25° API (density 0.897 g/c³) is categorized as light oil. Physically, the oil appears solidified and cannot flow at room temperature. With such a high apparent viscosity, we expected a high velocity gradient with respect to temperature near room temperature. However, the measured data show similar behavior to that of a typical light oil. One possible interpretation is that the viscosity of the waxy oil is apparently not really high and room temperature is still not lower than the liquid point. This suggests that, in general, heavy-oil velocity is largely controlled by API gravity. However, for a particular reservoir or region, the composition of the heavy oil may affect velocity.

Liquid point

The nonlinear P-wave velocity of heavy oil can be explained by this viscosity threshold characterized by T_L. We can generalize by re-examining the data in terms of a normalized temperature (T_{nor}) for the various oils, defined as

$$T_{nor} = \frac{T_w - T}{T_w - T_g}, \quad (3)$$

where T_w and T_g represent the temperature when oil viscosity is 1 cp (water) and 10^{15} cp (glass point), respectively. Currently, we apply a viscosity model (Beggs and Robinson, 1975) to calculate normalized temperature. We assume that the normalized temperature is representative for different heavy oils, which will simplify our analysis of heavy-oil velocity data.

Figure 10 shows the P-wave velocity of the eight heavy-oil samples in Figure 9 but now as a function of T_{nor}. With this normalized temperature (based on viscosity), the data for different oils all cross near a single point (P-wave velocity approximately 1.5 km/s and T_{nor} approximately 0.89). If T_{nor} is less than 0.89, velocities show a linear relationship to the normalized temperature. If T_{nor} is greater than 0.89, velocities deviate up from the linear trend. We define this point ($T_L = 0.89$) as the liquid point; it corresponds to a velocity of approximately 1.5 km/s and a viscosity of approximately 900 cp on the basis of the viscosity model from Beggs and Robinson (1975). This is the threshold between the linear and the deviated velocity-temperature trends. Liquid points vary for different API gravity oils. However, our velocity data reveal that the various oils all have a liquid point corresponding to this one viscosity value. Thus, the liquid point as defined by our velocity data represents a phase transition threshold.

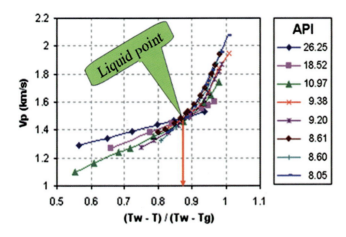

Figure 10. P-wave velocities plotted in terms of normalized temperature. The measured oils have velocities that all cross at the liquid point, which is a function of viscosity.

Frequency effect

The frequency dependence of velocity occurs when oils are in a quasi-solid phase between the glass and liquid points. In practice, heavy oils in the glass phase are considered elastic because viscosity is too high ($>10^{15}$ cp), and molecules in such glassy solids are fixed in location and the material is considered rigid, similar to a crystalline solid. Heavy oils in the liquid phase are also elastic because the viscosity effect on velocity is negligible. Thus, the frequency effect on velocity is coupled to viscosity (Batzle et al., 2006). Therefore, the liquid point temperature and viscosity depends on wave frequencies. This is a relaxation phenomenon in which the effective stiffness will depend on the rate or frequency of deformation. From our data at 1 MHz, viscosity of the liquid point is slightly less than 1000 cp. For a seismic wave with a frequency of 30 Hz, viscosity of the liquid point should be much higher (=lower temperature) because molecules have a much longer time to move relative to one another. The expected behavior for liquid point as a function of oil API gravity and frequency is shown in Figure 11.

Most viscoelastic materials have a correspondence between viscosity and frequency. For moduli, raising the frequency has the same effect as increasing viscosity. This principle allows building a relationship for the velocity of heavy oil expressed now as a function of API gravity, temperature, and a coupled function of viscosity-frequency as shown in

$$V = V\{API,\ T,\ [\eta(API,\ T)^* f]\ldots\}$$
$$\approx V_0(\rho) + B\{API,\ T,\ [\eta(API,\ T)^* f]\}^* \Delta T (T < T_L). \quad (4)$$

This relationship shows why we should expect discrepancies among velocity measurements made at widely

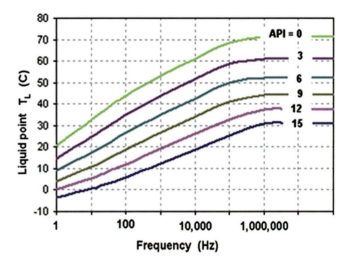

Figure 11. Expected behavior of the liquid point as a function of oil density (API gravity) and the measurement frequency.

different frequencies; for example, laboratory ultrasonics versus field seismic.

Summary

Most heavy oils are biodegraded and found in shallow, relatively low-temperature environments (<1000 m depth). Pressure on heavy oil is normally low (less than 10 MPa). Pressure and GOR dependence for velocity are similar to those of light oils.

On the basis of the data presented in this article, we conclude that, in general, heavy-oil velocity is largely controlled by API gravity similar to the case with light oils. However, after comparing a wide variety of distinct reservoirs, we find that the composition of heavy oil can affect velocity, but this needs to be locally evaluated and calibrated.

The temperature effect on velocity is of critical importance for heavy oils. This temperature dependence can be divided into three parts (Figure 12). When temperature is higher than the liquid point, velocity decreases linearly with increasing temperature, as is the case with light oils. When temperature is between liquid and glass points, heavy oil is in a quasi-solid phase. The velocity gradient with temperature of quasi-solid oil increases from that of the fluid phase, reaches a maximum, then decreases to approach that of the glass phase. And when temperature drops below the glass point, heavy oil is more like a solid. With decreasing temperature, P- and S-wave velocity will continue to increase but with a low temperature gradient.

Finally, the velocity data reported here were measured at ultrasonic frequencies (MHz range). The velocities of

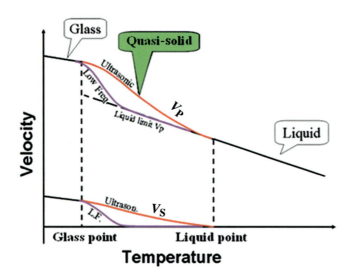

Figure 12. P-wave velocities plotted in terms of normalized temperature. This wide variety of oils has velocities that all cross at the liquid point, which is a function of viscosity.

heavy oils in the quasi-solid phase are strongly frequency dependent, as is the liquid point. Therefore, measurement and modeling are required to explore velocities of heavy oil in seismic and sonic logging frequencies.

References

Alboudwarej, H., J. Felix, S. Taylor, R. Badry, C. Bremner, B. Brough, C. Skeates, A. Baker, D. Palmer, K. Pattison, M. Beshry, P. Krawchuk, G. Brown, R. Calvo, J. Triana, R. Hathcock, K. Koerner, T. Hughes, D. Kundu, J. Cardenas, and C. West, 2006, Highlighting heavy oil: Oilfield Review, Summer, 31–53.

Batzle M., R. Hofmann, and D. Han, 2006, Heavy oils, seismic properties: The Leading Edge, **25**, 750–755.

Beggs, H. D., and J. R. Robinson, 1975, Estimating the viscosity of crude oil systems: Journal of Petroleum Technology, **27**, 1140–1141.

Behura, J., M. Batzle, R. Hofmann, and J. Dorgan, 2007, Heavy oils: their shear story: Geophysics, **72**, no. 5, E175–E183.

Caruso G., 2005, When will oil production peak?: Annual Asia Oil and Gas Conference.

Dusseault, M., 2006, Mechanics of heavy oil: U.S. Society of Rock Mechanics short course.

Eastwood, J., P. Lebel, A. Dilay, and S. Blakeslee, 1994, Seismic monitoring of steam-based recovery of bitumen: The Leading Edge, **13**, 242–251.

Gupta. S., 2005, Unlocking planet's heavy oil and bitumen resource — a look at SAGD: SPE Distinguished Lecturer Series.

Han, D., and M. Batzle, 2000, Velocity, density and modulus of hydrocarbon fluids: Empirical modeling: 70th Annual International Meeting, SEG, Expanded Abstracts, 2091–2094.

Han, D., J. Liu, and M. Batzle, 2005, Measurement of shear wave velocity of heavy oil: 75th Annual International Meeting, SEG, Expanded Abstracts, 1514–1517.

Hinkle, A., and M. Batzle, 2006, Heavy Oils: a worldwide overview: The Leading Edge, **25**, 742–749.

Hinkle, A., E.-J. Shin, M. W. Liberatore, A. M. Herring, and M. L. Batzle, 2008, Correlating the chemical and physical properties of a set of heavy oils from around the world: Fuel, **87**, 3065–3070.

Hunt, J. M., 1996, Petroleum geochemistry and geology: W. H. Freeman and Co.

Jenkins, S., M. Waite, and M. Bee, 1997, Time-lapse monitoring of Duri steamflood: a pilot and case study: The Leading Edge, **16**, 1267–1275.

Schmitt, D. R., 2004, Rock physics of heavy oil deposits: Annual CSEG Convention, Extended Abstract.

Chapter 3

Modeling Studies of Heavy Oil — In Between Solid and Fluid Properties

Agnibha Das[1] and Mike Batzle[1]

Introduction

Rocks filled with heavy oil do not comply with established theories for porous media. Heavy oils demonstrate a blend of purely viscous and purely elastic properties, also referred to as viscoelasticity. They have a nonnegligible shear modulus that allows them to support shear-wave propagation depending on frequency and temperature. These oils behave as solids at high frequencies and low temperatures and as fluids at low frequencies and high temperatures. The solid-like properties of heavy oils violate Gassmann's equation, the most common and widely used fluid-substitution technique in the industry.

Few instances of elastic property modeling for heavy-oil-saturated rocks have been reported. Most previously reported work has involved modeling without comparison with measured data, or modeled results on simple grain-fluid aggregates with comparison to measured ultrasonic data. We have modeled the viscoelastic properties of heavy-oil-saturated rock samples using the Hashin–Shtrikman (HS) bounds and the frequency-dependent complex shear modulus of the heavy oil. The two studied rock samples are very different in terms of lithology and consolidation state. In our exercise, we have extended the HS bounds to incorporate complexities such as intragranular porosity and the contribution of heavy oil to rock matrix properties. By considering the complex shear modulus of the heavy oil in our HS calculations, we have been able to estimate attenuation. We also tested the applicability of Ciz and Shapiro's (2007) form of the generalized Gassmann's equations in predicting the saturated bulk and shear moduli of the heavy-oil-saturated rock samples.

Comparison of modeled results with measured data shows a good agreement between the two over a wide range of frequencies spanning the seismic and ultrasonic band.

Uvalde Heavy-oil Rock and the Canadian Tar Sands

The studied rock samples are Uvalde rock from Texas and tar sands from Canada. The two samples exhibit different lithologies and consolidation states. The Uvalde is a carbonate consisting of consolidated calcite grains with heavy oil in the pore spaces. It has a porosity of approximately 25% and permeability of 550 mD. The heavy oil has an American Petroleum Institute (API) density of -5 (approximately 1.12 g/cm^3) and an extremely high viscosity that is highly dependent on temperature. A scanning electron microscope (SEM) image of the Uvalde heavy-oil rock (Figure 1) shows that the carbonate grains themselves are porous; hence, the total rock porosity can be divided into intragranular porosity (the porosity within the grains) and intergranular porosity (the porosity between adjacent grains). The frequency dependence of the shear modulus of the Uvalde heavy oil can be approximated using a Cole-Cole type distribution, essentially an empirical relationship between the complex shear modulus, the "static" and "infinite frequency" shear moduli, angular frequency, and relaxation time. Figure 2a shows the Cole-Cole approximations to the shear modulus of the Uvalde heavy oil at different temperatures as a function of frequency, along with measured shear modulus data at 20°C. The black triangles represent the measured data, and the solid lines represent the Cole-Cole approximations. Figure 2b shows the temperature and frequency dependence of the velocity measured on the Uvalde heavy-oil-saturated rock. We used measured velocities at 20°C for comparison with modeled results. The Canadian tar sands are unconsolidated quartz grains held together by viscous heavy oil (Figure 3). They typically have very high porosities in the range of 36–40%. Figure 4 shows measured shear modulus of the heavy oil extracted

---

[1]Colorado School of Mines, Department of Geophysics, Golden, Colorado, U.S.A.
This paper appeared in the September 2008 issue of THE LEADING EDGE and has been edited for inclusion in this volume.

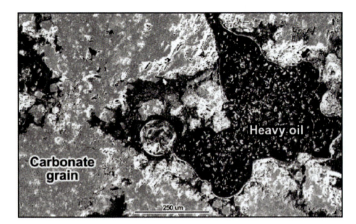

Figure 1. SEM photograph of the Uvalde heavy-oil rock. From Batzle et al., 2006.

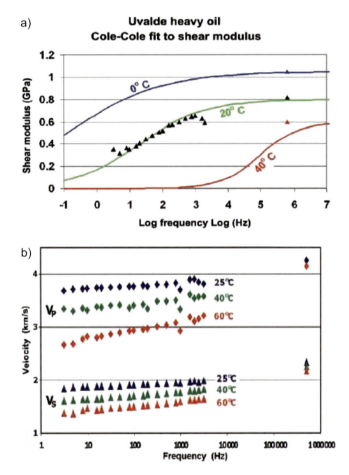

Figure 2. Measured elastic and acoustic properties of the Uvalde heavy-oil rock and the extracted heavy oil.
(a) Measured shear modulus with Cole-Cole fit.
(b) Frequency and temperature dependence of velocities.
From Batzle et al., 2006.

from a Canadian tar sand sample as a function of frequency and Cole-Cole fit to the data. The blue circles represent the measured data, and the green line represents the Cole-Cole fit to the data. We used the Cole-Cole fit for our modeling.

Figure 3. SEM photograph of the Canadian tar sands. The quartz grains are light and the intermediate black regions are the heavy oil.

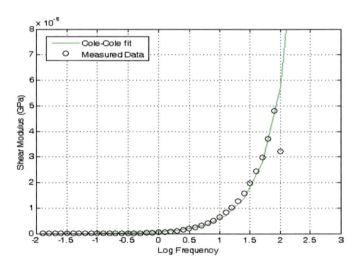

Figure 4. Measured elastic and acoustic properties of the Canadian tar sand sample and the extracted heavy oil. Real part of measured shear modulus at 20°C (from Hinkle, 2007) with Cole-Cole fit.

Elastic Property Estimation Using HS Bounds

The HS bounds give ranges for the effective bulk and shear moduli of isotropic two-phase composites as a function of the volume fraction of the constituents. They can be extended to include multiple phases in which case the bounds have to be computed more than once, two phases at a time. Using the complex modulus in the HS expressions in place of the real modulus, the bounds can be used for viscoelastic material. Bounds of bulk and shear moduli of viscoelastic materials using HS expressions have been calculated by others (Hashin, 1970; Gibiansky and Lakes, 1997; Torquato et al., 1998). However, such calculations and estimates have been done for materials such as cellular solids (honeycomb structures) and sand-epoxy mixtures, and the method needs to be tested for actual rock-oil systems by comparing with measured data.

We used the HS bounds and complex shear modulus of heavy oil to calculate bounds for the complex effective bulk ($K^* = K' + iK''$) and shear modulus ($G^* = G' + iG''$) for the Uvalde rock and the Canadian tar sands. This has also enabled us to calculate the P- and S-wave attenuation, Q_P^{-1} and Q_S^{-1} by taking a ratio between the real and imaginary parts, as follows

$$Q_P^{-1} = \frac{K''}{K'}; Q_S^{-1} = \frac{G''}{G'}. \quad (1)$$

HS modeling of the Uvalde heavy-oil rock

The SEM picture of the rock (Figure 1) shows the presence of intragranular porosity in the calcite grains. This would cause the grains to have bulk and shear moduli less than that of pure calcite. We separately calculated the bulk and shear moduli of the grains using the HS bounds, assuming an intragranular porosity of 5% and considering the intragranular spaces to be dry. Because the porosity is built inside of the grains, the stiffer calcite mineral encloses the softer pore spaces, and so the effective modulus of the grains would be closer to the upper bound. Hence, we took values that were 70% of the upper bounds and not average properties. To compute the bulk modulus of the fluid saturating the intergranular pore space, we considered a two-component fluid phase, heavy oil and air (with heavy-oil saturation S_{HO} of 0.9), and used the upper HS bounds. The Ruess bound, commonly used to calculate effective bulk modulus of fluid mixtures, was not used in this case because the Uvalde heavy oil at 20°C is a solid, and fluid-like behavior is observed only at high temperatures. We used the fluid shear modulus as predicted by the Cole-Cole fit to measured data from Batzle et al. (2006). The obtained solid and fluid moduli were used to compute the effective modulus of the aggregate. Figure 5 compares the modeled bulk and shear modulus with actual data measured at 20°C and 1000 psi (~7 MPa) differential pressure. Differential pressure is the difference between confining pressure and pore pressure. The measured data closely

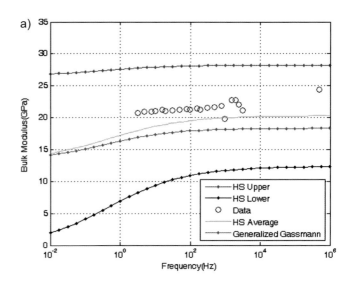

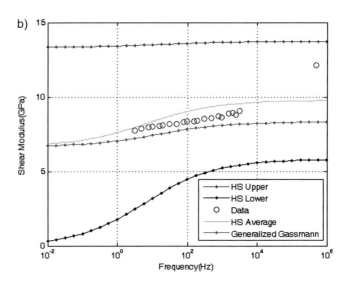

Figure 5. HS bounds for (a) bulk and (b) shear moduli as a function of frequency for a system of porous calcite grains and heavy oil. Heavy oil is not considered part of the matrix. The blue circles represent measured data at 20°C from Batzle et al., 2006. The green line represents average HS properties.

follow the average HS line. Figure 6 compares the modeled and measured velocities.

So far, we have considered the heavy oil only as a pore-filling fluid. However, we could also consider it part of the matrix; that is, as a coating on the grains and acting as a viscous cementing material and therefore contributing to matrix properties. To simulate this situation, we separately calculated the bulk and shear moduli of the matrix (grains and heavy oil) and the fluid phase. We computed the matrix properties using the average HS bounds and 5% cement saturation. Cement saturation denotes the fraction of the intergranular pore space occupied by heavy oil acting as cement and contributing to the matrix properties. Figure 7 compares the modeled and

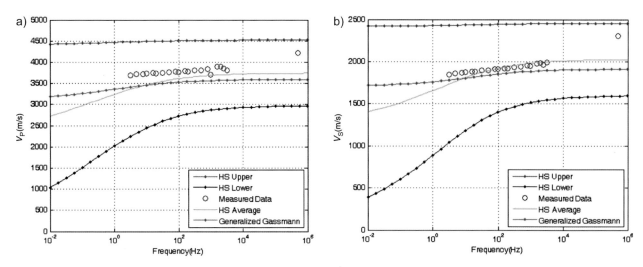

Figure 6. HS bounds for (a) V_P and (b) V_S as a function of frequency for a system of porous calcite grains and heavy oil. Heavy oil is not considered part of the matrix. The blue circles represent measured data at 20°C from Batzle et al., 2006. The green line represents average HS properties.

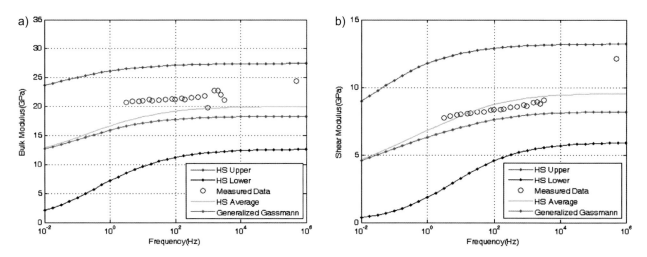

Figure 7. HS bounds for (a) bulk and (b) shear moduli as a function of frequency for a system of porous calcite grains and heavy oil, and considering part of the heavy oil as contributing matrix properties. The blue circles represent measured data at 20°C from Batzle et al., 2006. The green line represents average HS properties.

measured data. The predicted values in this case are slightly lower than when heavy oil is not considered part of the matrix. This is because the introduction of heavy oil as a matrix component lowers the bulk and shear moduli of the matrix. Another observation is that the modeled bulk and shear moduli are significantly lower at low frequencies, compared with the previous case. This is because at low frequencies the heavy oil behaves more like a fluid, which means it has a much lower shear modulus. The low shear modulus then contributes toward significant lowering of the effective bulk and shear moduli of the matrix and, hence, that of the rock. Figure 8 compares the modeled and measured velocities. It can be observed that, in this case and the previous one, the differences between predicted and measured values are smaller for velocities than moduli.

We calculated shear attenuation from the computed complex effective shear modulus of the rock using equation 1. Figure 9 shows the estimated and measured shear-wave attenuation for the two scenarios discussed above. The estimated attenuation values are higher than actual measurements.

Generalized Gassmann's equations for a solid infill of the pore space

We also calculated the saturated bulk and shear moduli of the Uvalde heavy-oil sample using the generalized

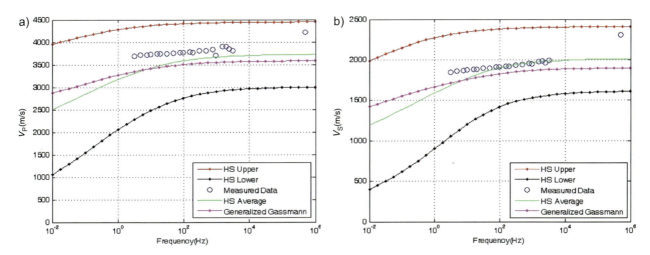

Figure 8. HS bounds for (a) V_P and (b) V_S as a function of frequency for a system of porous calcite grains and heavy oil and considering part of the heavy oil as contributing to matrix properties. The blue circles represent measured data at 20°C from Batzle et al., 2006. The green line represents average HS properties.

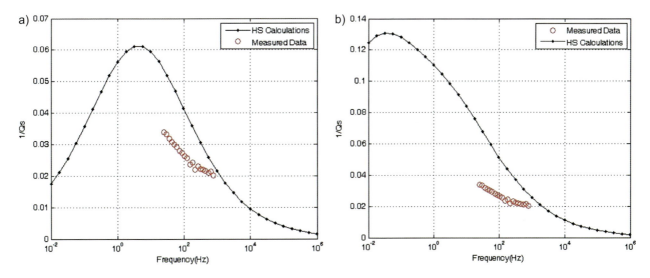

Figure 9. Attenuation estimation as a function of frequency from HS calculations for a system of porous calcite grains and heavy oil. The red circles represent measured data at 20°C from Gautam et al., 2003. (a) Shear-wave attenuation from HS calculations. Heavy oil is not considered a part of the matrix. (b) Shear-wave attenuation from HS calculations. Heavy oil is considered a part of the matrix. Cement saturation is 5%.

Gassmann's equations given by Ciz and Shapiro (2007). The advantage of using this set of equations is that they can account for the nonnegligible shear modulus of the heavy oil filling the pore spaces. The saturated shear modulus computed with these equations is, unlike Gassmann's equation, significantly different from the dry case.

The generalized Gassmann's equations for the saturated bulk and shear modulus as given by Ciz and Shapiro (2007) are

$$K_{sat}^{*-1} = K_{dry}^{-1} - \frac{\left(K_{dry}^{-1} - K_{gr}^{-1}\right)^2}{f\left(K_{if}^{-1} - K_{\phi}^{-1}\right) + \left(K_{dry}^{-1} - K_{gr}^{-1}\right)}, \quad (2)$$

and

$$\mu_{sat}^{*-1} = \mu_{dry}^{-1} - \frac{\left(\mu_{dry}^{-1} - \mu_{gr}^{-1}\right)^2}{f\left(\mu_{if}^{-1} - \mu_{\phi}^{-1}\right) + \left(\mu_{dry}^{-1} - \mu_{gr}^{-1}\right)} \quad (3)$$

where K_{sat}^* and μ_{sat}^* are the saturated bulk and shear moduli, K_{dry} and μ_{dry} are the dry frame bulk and shear moduli, K_{gr} and μ_{gr} are the bulk and shear moduli of the grain material of the frame, K_{ϕ} and μ_{ϕ} are the bulk and shear moduli associated with the pore space, and K_{if} and μ_{if} are the bulk and shear moduli of the solid infill of the pore space. For a monominerallic homogeneous porous frame, $K_\phi = K_{gr}$ and $\mu_\phi = \mu_{gr}$. K_{dry} and μ_{dry} were obtained from average HS properties for a solid-air composite. We considered two cases: one with no heavy oil in the matrix and the other with some heavy oil in the matrix and contributing to the solid properties. The saturated bulk and shear moduli calculated from equations 2 and 3 compare very well with average HS properties (Figures 5–8).

HS modeling of Canadian tar sand

It is relatively simple to compute the HS bounds for the Canadian tar sands by modeling it as a homogeneous isotropic mixture of quartz grains and viscous heavy oil. We considered the tar-sand aggregate to be quartz grains and heavy oil with an overall porosity of 36%. To compute the bulk modulus of the heavy oil, we considered a two-component fluid phase, heavy oil and air (with S_{HO} of 0.9), and used the upper HS bounds. We used the heavy oil shear modulus as predicted by the Cole-Cole fit to measured data from Hinkle (2007). We also calculated moduli and velocities of the tar sands using Leurer and Dvorkin's (2006) viscoelastic model for unconsolidated sediments with viscous cement. Figure 10 shows a comparison between the two models. The moduli values calculated by Leurer and Dvorkin's viscoelastic model are lower than the HS average and lie closer to the lower HS

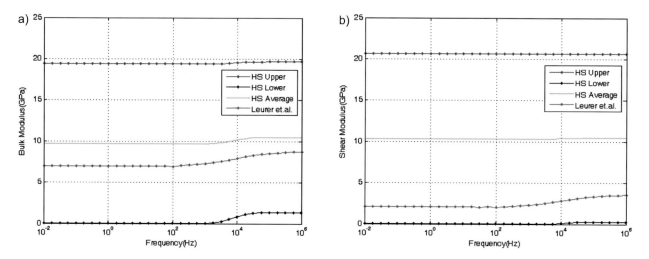

Figure 10. HS bounds for (a) bulk and (b) shear moduli as a function of frequency for a system of quartz grains and heavy oil. Heavy oil is not considered part of the matrix. The blue circles represent measured data at 20°C. The green line represents average Hashin-Shtrikman properties.

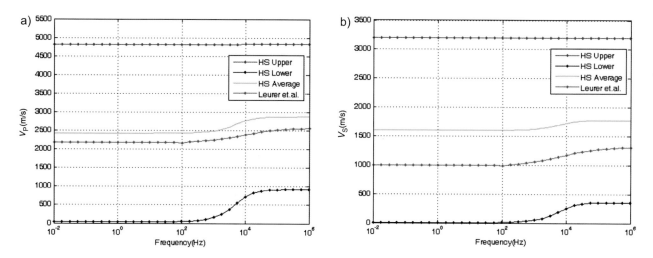

Figure 11. HS bounds for (a) V_P and (b) V_S as a function of frequency for a system of quartz grains and heavy oil. Heavy oil is not considered part of the matrix. The blue circles represent measured data at 20°C. The green line represents average HS properties.

bounds, implying that the aggregate behaves more like a system of sand grains covered with heavy oil and hence is very soft. Figure 11 shows a comparison of the velocities.

Conclusions

We have been able to use the HS bounds to estimate the elastic and acoustic properties of two very different rock types in terms of lithology and physical properties. In this exercise, we have been able to account for complexities such as intra- and intergranular porosities of the grains and complex frequency-dependent shear modulus of the saturating heavy oil. The comparison between measured and predicted data is good. We also tested the applicability of Ciz and Shapiro's form of the generalized Gassmann's equations for predicting saturated moduli and velocities and obtained a reasonably good match with measured data.

Acknowledgments

This work was supported by the Fluids and DHI Consortium of the Colorado School of Mines and the University of Houston. We thank Merrick Johnston for the SEM picture of the Canadian tar sand and Klaus Leurer for providing numerical code for computing moduli using the viscoelastic model.

References

Batzle, M., R. Hofmann, and D. H. Han, 2006, Heavy oils-seismic properties: The Leading Edge, **25**, 750–756.

Ciz, R., and S. A. Shapiro, 2007, Generalization of Gassmann equations for porous media saturated with a solid material: Geophysics, **72**, no. 6, A75–A79.

Gibiansky, L. V., and R. Lakes, 1997, Bounds on the complex bulk and shear moduli of a two-dimensional two-phase viscolastic composite: Mechanics of Materials, **25**, 79–95.

Hashin, Z., 1970, Complex moduli of viscoelastic composites — 1: General theory and application to particulate composites: International Journal of Solids and Structures, **6**, 539–552.

Hinkle, A., 2007, Relating chemical and physical properties of heavy oils: M.S. thesis, Colorado School of Mines.

Kumar, G., Batzle, and R. Hofmann, 2003, Effect of fluids on attenuation of elastic waves: 73rd Annual International Meeting, SEG, Expanded Abstracts, 1592–1595.

Leurer, K. C., and J. Dvorkin, 2006, Viscoelasticity of precompacted unconsolidated sand with viscous cement: Geophysics, **71**, no. 2, T31–T40.

Torquato, S., L. V. Gibiansky, M. J. Silva, and L. J. Gibson, 1998, Effective mechanical and transport properties of cellular solids: International Journal of Mechanical Sciences, **40**, 71–82.

Chapter 4

Correlating the Chemical and Physical Properties of a Set of Heavy Oils from around the World

Amy Hinkle,[1] Eun-Jae Shin,[2] Matthew W. Liberatore,[2] Andrew M. Herring,[2] and Mike Batzle[1]

Introduction

Heavy oil has recently become an important resource as conventional oil reservoirs have limited production and oil prices rise. More than 6 trillion barrels of oil in place have been attributed to the world's heaviest hydrocarbons (Curtis et al., 2002). Therefore, heavy-oil reserves account for more than 3 times the amount of combined world reserves of conventional oil and gas. Of particular interest are the large heavy-oil deposits of Canada and Venezuela, which together may account for approximately 55%–65% of the known less than 20° American Petroleum Institute (API) gravity oil deposits in the world (Curtis et al., 2002).

Heavy oils cover a large range of API gravities, from 22° for the lightest heavy oils to less than 10° for extra-heavy oils. This wide range of values means that heavy oils vary greatly in their physical properties. Thus, extensive research is required before the properties of heavy oil can be properly understood. Several prevailing issues are seen repeatedly in various fields around the world, including how to make measurements on unconsolidated sandstone cores, production of sand with oil and its effect on formation, exsolution gas drive of heavy oil, understanding the control of viscosity and other physical properties of heavy oils, and monitoring of steam recovery processes. Simply, the high viscosity of heavy oils limits its extraction by traditional methods.

Two important distinctions must be made between API gravity and viscosity. First, viscosity determines how well oil will flow, whereas density more closely relates to the yield from the distillation. Additionally, temperature and paraffin content significantly affect viscosity values, whereas API density is relatively unaffected by these parameters (Curtis et al., 2002). Heavy oils usually begin as lighter oils (30°–40° API) that are then altered, often by biodegradation.

Argillier et al. (2001) conducted a rheological study of several heavy oils and concluded that the asphaltene content was a controlling factor for viscosity. Their data indicate that when the asphaltenes passed a critical weight fraction (approximately 10%), viscosity increased dramatically. They speculate that the long asphaltene chains begin to conglomerate and tangle. In contrast, increased resin content actually decreased viscosity. However, a recent analysis by Hossain et al. (2005) found no strong viscosity correlation with asphaltene content. Because viscosity is correlated with shear modulus for heavy oils, it partly controls our seismic velocities. The influence of asphaltenes and resins will be examined more thoroughly in this work.

Previous reports identify variations in asphaltene content as the primary determinant for the large spread of viscosities observed in heavy oils (Argillier et al., 2001; Henaut et al., 2001; Al-Mamaari et al., 2006). If asphaltene concentration determines heavy-oil viscosity, it should also directly relate to the heavy-oil shear modulus. Here, a comprehensive suite of measurements on seven heavy-oil samples from around the world addresses the chemical and physical properties of heavy oils in the context of the solubility classifications, such as asphaltenes and resins. A rapid screening technique for alternative hydrocarbons is needed to be more rapid, reliable, and meaningful than the traditional saturate-aromatic-resin-asphaltene (SARA) analysis (Kharrat et al., 2007). Pyrolysis-molecular-beam mass spectrometry (pyrolysis-MBMS) is a method that can rapidly generate large data sets of chemical information on complex substances (Windig et al., 1982; Windig and Meuzelaar, 1984; Evans and Milne, 1987a, 1987b). Pyrolysis-MBMS can be optimized to crack and volatilize the

---

[1]Colorado School of Mines, Department of Geophysics, Golden, Colorado, U.S.A.
[2]Colorado School of Mines, Department of Chemical Engineering, Golden, Colorado, U.S.A.

entire sample in a few minutes, leading to chemical information much more rapidly than nuclear magnetic resonance (NMR) or SARA analysis. Furthermore, because the entire sample is introduced into the molecular-beam mass spectrometer, the chemical information can be correlated with bulk properties of the original oil. When pyrolysis-MBMS is correlated with the signature of a particular species or class of molecules (in this case asphaltenes), it becomes a very powerful predictive tool. Rheometric, ultrasonic, and pyrolysis-MBMS measurements will lead to a predictive tool correlating chemical signatures to the viscosity and shear modulus of heavy oils.

Experimental

Heavy-oil samples

Heavy oil is defined by the U. S. Department of Energy as having API gravities that fall between 10.0 and 22.3° (Nehring et al., 1983). Extra-heavy oils are defined as having API gravities <10.0°. Heavy oils are classified as such using API gravity rather than viscosity values. API gravity can be expressed as

$$API = \frac{141.5}{\rho_f} - 131.5, \quad (1)$$

where ρ_f is the specific gravity of the fluid at 0.1 MPa and 15.6°C.

Heavy crude oils are often characterized geochemically using a process called SARA fractionation. The crude oil can be separated into four components on the basis of solubility classes. These four components are saturates, aromatics, resins, and asphaltenes. Heavy oils tend to be rich in the high-molecular-weight components, which are resins and asphaltenes. Unfortunately, there are numerous issues that render SARA fractionation problematic (Kharrat et al., 2007). Procedures used within testing laboratories vary widely. Normally, resins are the fraction soluble in pentane but are insoluble in propane. Asphaltenes dissolve in solvents such as carbon disulfide, but various laboratories precipitate the material in different light alkanes. Some use pentane, others use heptane, and still others use iso-octane. The molecular weight of the fluids used has a major influence on the results (Kharrat et al., 2007). In addition, the techniques used to wash and filter the precipitants vary significantly. To reduce some of these variabilities, all of our analyses were performed at a single commercial laboratory. Still, variations in reported values of resins or asphaltenes for any single oil can easily vary by 10%.

Seven heavy-oil samples are investigated here: three samples from Canada, one sample from Venezuela, one Alaskan sample (Ugnu), one Utah sample (Asphalt Ridge), and one western Texas sample (Uvalde). Humble Geochemical performed SARA analysis for all of the samples (Table 1). The fluid densities were determined by dividing mass by volume of the heavy oils. The Canadian, Venezuelan, and Alaskan heavy oil was donated from various companies. The Utah and Texas samples came from rocks collected at the outcrop, and the oil was extracted. All of the samples were dead oils, or gas free.

Rheology

Low-frequency viscosity and shear modulus measurements were collected in a range from 0.01 to 100 Hz on a TA Instruments AR-G2 rheometer (New Castle, DE). A small amount of heavy oil (approximately 1.5 mL) was loaded between a Peltier plate and a 40-mm aluminum plate. A sample gap of 1 mm was used for all measurements. Isothermal experiments were completed at temperatures from 0 to 80°C (±0.1°C). A sinusoidal torsional stress over the frequency range of 0.1–100 Hz was applied to the sample and the resulting sinusoidal strain was measured. The shear modulus (G') is the ratio of the stress in phase with the strain to the strain magnitude, and the loss modulus (G'') is the ratio of the stress 90° out of phase with strain to the strain magnitude (Tschoegl, 1989; Macosko, 1994). The complex viscosity is also recorded during a stress sweep measurement. Amplitude sweeps (i.e., a measurement of the moduli as a function of stress

Table 1. Properties of seven heavy oils.

Sample Name	Saturate (wt%)	Aromatic (wt%)	Resin (wt%)	Asphaltene (wt%)	Density (g/cc)	API Gravity
Alaska	23	22	35	18	0.997	10.4
Canada 1	18	33	30	20	1.014	8.09
Canada 2	18	27	27	15	1.991	11.3
Canada 3	15	23	19	10	1.003	9.56
Texas	4	17	37	43	1.119	−5.00
Utah	19	14	46	20	1.000	10.0
Venezuela	19	32	29	18	1.013	8.05

at a constant frequency) were recorded for each sample to verify that the sample was in the linear viscoelastic region for the stress sweep measurements.

Ultrasonic measurements

High-frequency shear modulus measurements were collected in a range of 0.5–1 MHz using a standard ultrasonic pulse technique. Compressional and shear-wave velocities of the material are determined from the travel-time of an ultrasonic pulse through the heavy oil. Once the velocities are determined, they can be related to shear and bulk moduli using the following equations:

$$V_S = \sqrt{\frac{G'}{\rho}}, \quad (2)$$

and

$$V_P = \sqrt{\frac{K + 4/3G'}{\rho}}, \quad (3)$$

where V_S is the shear-wave velocity, V_P is compressional velocity, ρ is the fluid density, K is the bulk modulus, and G' is the shear (or storage) modulus. It is difficult to differentiate between noise and the shear-wave signal above approximately 20°C. For this reason, the shear data collected reside in a temperature range between $-25°C$ and $20°C$.

Pyrolysis-MBMS

Pyrolysis of the heavy-oil samples was carried out to study their chemical composition and correlate chemical information with SARA analysis. A molecular-beam mass spectrometer was used to detect chemical species because it allowed direct and real-time sampling from the pyrolysis system. All reactions were carried out under atmospheric pressure in a quartz tube reactor. All samples (approximately 30 mg) were contained in a quartz holder, or "boat," and inserted into flowing, preheated helium carrier gas. The carrier gas (10 L/min) was introduced through the end of a reactor consisting of a quartz tube with an inner diameter of 2.5 cm. The reactor was heated using an electric furnace set at 550°C and coupled to a molecular-beam mass spectrometer for product detection (Shin et al., 2001, 2003). The residence time of the pyrolysis vapors in the reactor pyrolysis zone was estimated to be approximately 100 ms, which is short enough to minimize secondary cracking reactions at this temperature. Total pyrolysis time was 3 min. The short time frame indicates the rapid nature of this method and the potential for screening many samples. Vapors exiting the reactor flow through the sampling orifice of the molecular-beam mass spectrometer with subsequent formation of the molecular beam, which provides rapid sample quenching and inhibits condensation and aerosol formation. The molecular-beam mass spectrometer provides universal detection of all sampled products and the molecular-beam sampling ensures that representative products from the original molecules are detected (Evans and Milne, 1998; Shin et al., 2003).

In this work, a mass range of 50–350 amu and ionization energy of 25 eV was used. The pyrolysis-MBMS method of sample analysis is rapid (2–10 min) and can generate data from 100 samples per day depending on analytical conditions. A detailed description of techniques and methodologies is given in previous work (Shin et al., 2001, 2003, 2004). The pyrolysis-MBMS results were used as a basis of multiple predictor variables for least-squares regression analysis to build a predictive model for geophysical properties of heavy oils such as SARA. The software package used was Unscrambler (DeVries and TerBraak, 1995; Faber and Kowalsky, 1996; Faber et al., 2003). The partial least-squares regression analysis can be viewed as a two-stage procedure. First, the model is constructed using training samples for which the predictor and predicted variables are known or measured. Next, the model is validated by comparing the predictions against reference values for samples that were not used for model building. After the model is validated, it is used for the prediction of the response variable for unknown samples (DeVries and TerBraak, 1995; Faber and Kowalsky, 1996; Faber et al., 2003). Detailed description of the analysis is out of the scope for this work and only pertinent results are reported.

Results

Rheology

The measured values of complex viscosity, shear modulus, and loss modulus as collected on the G2 rheometer provide insight into the changing viscoelastic nature of heavy oils as a function of temperature and chemical makeup. As expected, complex viscosity shows a strong dependence on temperature. Viscosity increases by orders of magnitude as the temperature is decreased linearly for the Canada 1 heavy oil (Figure 1). The Canada 1 heavy oil is representative of the basic rheological response of the seven oils examined. Also, the viscosity decreases with increasing frequency (i.e., shear thinning) for temperatures below 40°C. At higher temperatures, the Canada 1 heavy oil behaves like a Newtonian fluid (i.e., no frequency dependence). Overall, the shear thinning response of the heavy oils is more pronounced at high frequencies and low temperatures.

Because the storage and loss moduli (G' and G'', respectively) are directly related to the complex viscosity,

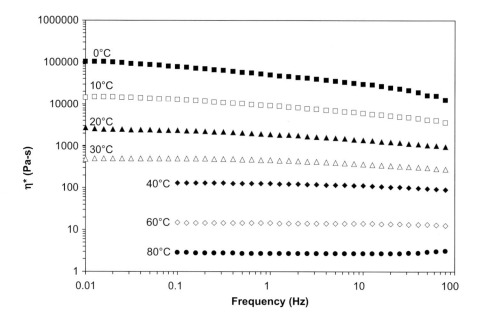

Figure 1. Complex viscosity as a function of frequency at various temperatures for Alaska heavy oil.

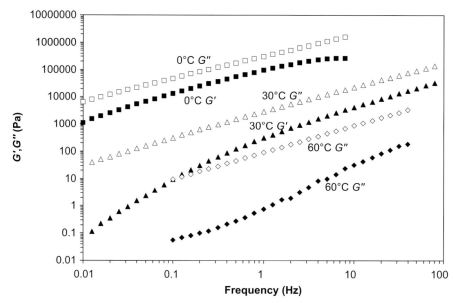

Figure 2. Storage modulus (G') and loss modulus (G'') measured as a function of temperature and frequency for Alaska heavy oil.

the heavy-oil samples exhibit strong dependence on temperature and frequency (Figure 2). The Canada 1 heavy oil again serves as a typical example of the seven oils studied. The moduli increase with decreasing temperature and increasing frequency. At all temperatures and frequencies examined, the loss modulus is larger than the storage modulus. The larger loss modulus indicates that the viscoelastic response of the fluid is dominated by the liquid-like (or out-of-phase) contribution to the stress. The coupling of the viscosity and moduli is further probed using ultrasonic measurements (discussed in the next section).

The temperature dependence of all seven heavy oils is clearly demonstrated in an Arrhenius-type plot (Figure 3). A linear correlation between the logarithm of the viscosity and inverse temperature is quantified by an Arrhenius-type relationship (Glasstone et al., 1941), $\eta^* = e^{-E_{vis}/RT}$.

The viscosity activation energies (E_{vis}) range from 73 kJ/mol (Canada 3 heavy oil) to 120 kJ/mol for (Texas heavy oil). The viscosity activation energy of a fluid can be related to the fluid's heat of vaporization (Glasstone et al., 1941) and will be investigated in future work.

The dependence of the viscosity of lighter, conventional crude oils strongly correlates with the API gravity. However, the viscosity of these heavy oils is not dependent on fluid density (Figure 4). Therefore, the next logical step is to correlate the viscosity of the heavy oils to their chemical makeup. One chemical control suggested in the literature is asphaltene content (Argillier et al., 2001; Henaut et al., 2001; Al-Mamaari et al., 2006). To test this possibility, viscosities for each sample (at 20°C and 1 Hz) were plotted as a function of asphaltene content (Figure 5). The choice of 20°C and 1 Hz is arbitrary, but

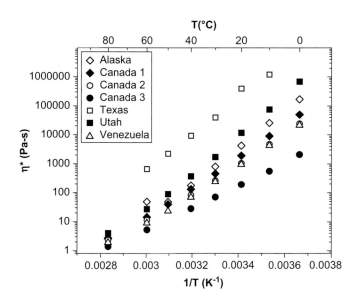

Figure 3. Arrhenius plot of complex viscosity as a function of inverse temperature (all data at 1 Hz).

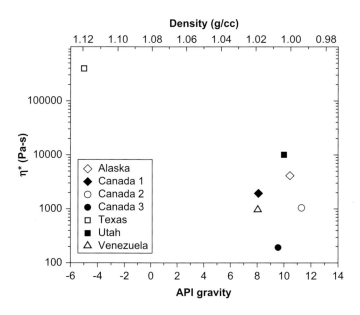

Figure 4. Complex viscosity as a function of API gravity for seven heavy oils (all data at $T = 20°C$, 1 Hz).

the trends seen under this condition hold at other temperatures and frequencies. No apparent correlation between viscosity and asphaltenes is evident. Because asphaltenes do not correlate strongly with viscosity, viscosity was then plotted with respect to resin content (Figure 6). A linear relationship between the logarithm of the viscosity and resin content is possible, but there is one far outlying point. The extremely high viscosity of the Texas heavy-oil sample does not scale with resin content in the same way as the other six oils.

Because the viscosity seems dependent on the resin and asphaltene content, the correlation between the sum of

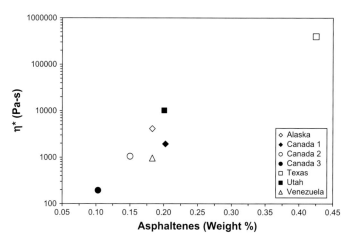

Figure 5. Complex viscosity as a function of asphaltene content for seven heavy-oil samples (all data at $T = 20°C$, 1 Hz)

these two heavier chemical components and viscosity may lead to a more definitive relationship. A linear relationship between the log of the complex viscosity and the total resin + asphaltene content encompasses all seven oils (Figure 7). The empirical relationship represented by the solid line in Figure 7 is $\log(\eta^*) = 1.63 e^{14.6(wt\%A+R)}$, where wt %A + R is the combined weight percent of asphaltenes and resins. The correlation coefficient (R^2) value for the relationship between viscosity and combined asphaltene and resin content for the oil samples was 0.95 (Figure 7). The correlation for asphaltenes + resins is much better than the relationships for asphaltenes or resins alone. Because there is a definitive correlation between viscosity and combined asphaltene and resin content, the shear modulus from ultrasonic measurements should also depend on combined resin and asphaltene content.

Ultrasonic measurements

The correlation of the ultrasonic data with the various chemical components of the SARA analysis is analogous to the rheological studies in the previous section. In general, the shear modulus measured for these oils ranges from 0.2 to 0.9 GPa at $-7°C$. Additionally, the measured shear modulus of the Canada 1 heavy oil increases by six times as the temperature is decreased 30°C, which is analogous to Figure 1 from the rheology section. Reproducible measurements were not obtained for two samples (Canada 2 and Canada 3) and thus are missing from this discussion. Again, the property of interest, shear modulus in this case, was plotted as a function of API gravity, resin content, and asphaltene content with little or no correlation between the measured heavy oils. For brevity, these plots are omitted but are available (Hinkle, 2007). However, a power-law correlation is observed for these heavy oils when correlated with the total resin + asphaltene content (Figure 8). The empirical power-law

Figure 6. Complex viscosity as a function of resin content for seven different heavy-oil samples (all data at $T = 20°C$, 1 Hz).

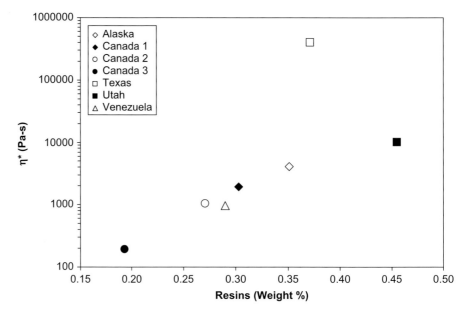

Figure 7. Complex viscosity as a function of combined resin and asphaltene content for seven different heavy-oil samples (all data at $T = 20°C$, 1 Hz). The error in the total resin + asphaltenes is somewhat lower because underestimates in resin content are usually compensated for in asphaltene values and vice versa.

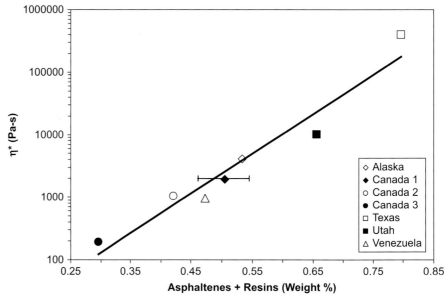

Figure 8. Ultrasonic shear modulus as a function of combined resin and asphaltene content. ($T = -10°C$)

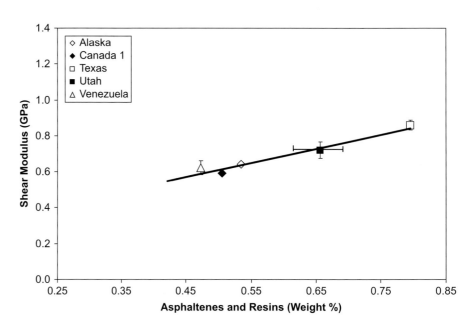

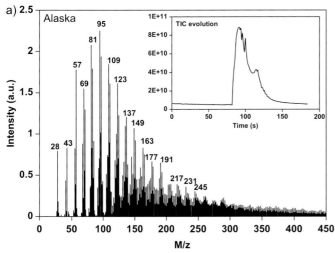

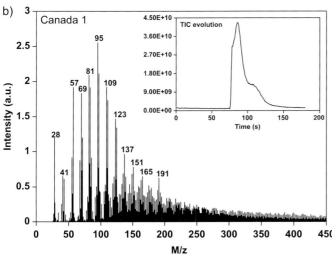

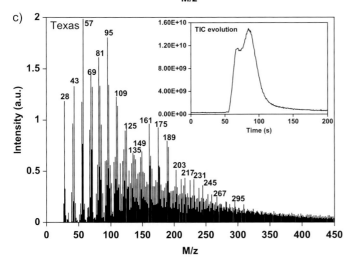

Figure 9. Average spectra of products (with background correction) detected by pyrolysis-MBMS resulting from the pyrolysis of (a) Alaska, (b) Canada 1, and (c) Texas heavy oil at 550°C with residence time of approximately 0.1 s.

relationship for these heavy oils is $G' = 0.98x^{0.67}$, where x is the combined weight percent of asphaltenes and resins. The R^2 for this relationship is 0.94.

Pyrolysis-MBMS

The chemical analysis of the heavy oils using pyrolysis-MBMS provides new insight and convenience beyond the traditional SARA fractionation analysis. An average spectrum from the pyrolysis of selected heavy-oil samples (Alaska, Canadian 1, and Texas heavy oils) provides the raw chemical information from this experiment (Figure 9a, b, and c, respectively). The short time frame of typical product evolution as shown in the insets of Figure 9a, b, and c indicates the rapid nature of this method and the potential for screening many heavy-oil samples for characterization. Simply, the average spectra are very complex, and it is very difficult to identify chemical similarities among the different heavy oils. However, common products for the three heavy oils are found at m/z 57, 71, and 85 (saturated alkene fragment ions); m/z 95, 109, 123, and 137 (two double-bond species); m/z 68, 82, 96, 110, 124, 138, 152, 166, and 180 (ions with one triple-bonded fragment); and m/z 56, 70, 84, 98, and 112 (alkene species), with a systematic 14-amu growth for each group of species. Although these similarities among the spectra are observable, there are still differences in other portions of the spectra when comparing different oils. We used these pyrolysis-MBMS multivariate data as predictor variables to build a predictive model for SARA properties. If geochemical properties can be predicted by running many samples in a short period of time using pyrolysis-MBMS, it would be a very powerful and convenient characterization tool for heavy oils.

In Figure 10, the result of partial least-squares regression using pyrolysis-MBMS to predict one of the SARA

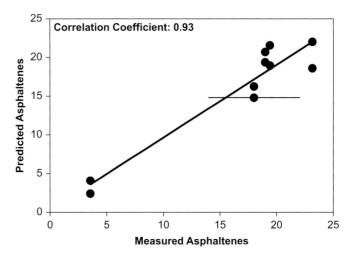

Figure 10. Measured value versus prediction for asphaltenes using pyrolysis-MBMS data as predictor variables.

Figure 11. Correlation coefficient (R^2) of pyrolysis-MBMS variables with asphaltenes resulting from partial least-squares regression analysis.

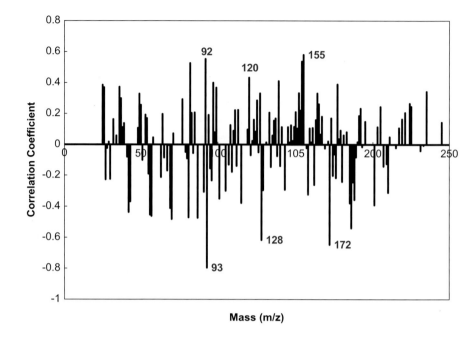

properties, asphaltene, is shown. Only five of seven samples were used for this analysis because those five samples had duplicates; R^2 is 0.93. It is notable that with a limited number of samples, it was possible to have a decent predictive model for one of the SARA components from the rapid pyrolysis-MBMS data. In Figure 11, mass variables that were positively and negatively correlated with asphaltenes are shown. Detailed studies of the masses are needed for a more complete understanding of the chemical characteristics of asphaltenes. This analysis proves the rapid screening capability of pyrolysis-MBMS, which could be used for building predictive models for heavy oils in the near future.

Conclusions

Heavy oils are viscoelastic materials. The parameters that control heavy-oil viscosity will also control heavy oil shear modulus. Chemistry was shown to be essential to understanding variations in viscosity and shear modulus. Previously, the viscosity of heavy oils was often correlated with asphaltenes weight content alone. However, almost all of the published papers studied a single heavy-oil sample where saturates, aromatics, and resins contents were constant while asphaltenes content varied (Argillier et al., 2001, 2002; Henaut et al., 2001; Luo and Gu, 2005). In this work, various naturally occurring heavy-oil samples were studied, and they indicate that viscosity depends on the combined asphaltenes and resin concentration.

In the petroleum industry, viscosity mapping of heavy-oil fields is currently based on API gravity measurements. As was demonstrated in this work, and has been reported by other researchers (Al-Mamaari et al., 2006), there is no good correlation between API gravity and viscosity for heavy oil. Combined resin and asphaltene content could be used to better map viscosity variations across heavy-oil fields. Combined asphaltenes and resins could also be helpful for predicting shear modulus variations in heavy-oil fields.

In addition, pyrolysis-MBMS techniques were demonstrated as a tool for rapid characterization of heavy oil. An empirical relationship was established between pyrolysis-MBMS measurements and one of the available heavy-oil geochemical measurements, asphaltenes. Results suggest that pyrolysis-MBMS may provide a more consistent compositional analysis than SARA fractionation. In addition, with more complete calibration to establish more robust relationships, a model using pyrolysis-MBMS data could be developed to predict viscosity and shear modulus for heavy oils for which no data are available. In closing, the merging of field measurements with extracted fluid properties is the subject of current research with the objective of identifying more efficient and effective methods for economic extraction of the abundant heavy-oil resource.

Acknowledgments

Financial support by the Geophysical Properties of Fluids/Direct Hydrocarbon Indicators consortium made this work possible.

References

Al-Mamaari, R. S., O. Houache, and S. A. Abdul-Wahab, 2006, New correlating parameter for the viscosity of heavy crude oil: Energy and Fuels, **20**, 2586–2592.

Argillier, J. F., L. Barre, F. Brucy, J. L. Douranaux, I. Henaut, and R. Bouchard, 2001, Influence of asphaltene content and dilution on heavy oil rheology: SPE Paper 69711.

Argillier, J. F., C. Coustet, and I. Henaut, 2002, Heavy oil rheology as a function of asphaltene and resin content and temperature: SPE Paper 79496-MS.

Curtis, C., R. Kopper, E. Decoster, A. Guzman-Garcia, C. Huggins, L. Knauer, M. Minner, N. Kupsch, L. Linares, H. Rough, and M. Waite, 2002, Heavy-oil reservoirs: Oilfield Review, **14**, 30–51.

DeVries, S., and C. J. F. TerBraak, 1995, Prediction error in partial least squares regression: A critique on the deviation used in The Unscrambler: Chemometrics and Intelligent Laboratory Systems, **30**, 239–245.

Evans, R. J., and T. A. Milne, 1987a, Molecular characterization of the pyrolysis of biomass: 1. Fundamentals: Energy and Fuels, **1**, 123–137.

Evans, R. J., and T. A. Milne, 1987b, Molecular characterization of the pyrolysis of biomass. 2. Applications: Energy and Fuels, **1**, 311–319.

Evans, R. J., and T. A. Milne, 1998, Chemistry of tar formation and maturation in the conversion of biomass: Fuel and Energy Abstracts, **39**, 197–198.

Faber, K., and B. R. Kowalsky, 1996, Prediction error in least squares regression: Further critique on the deviation used in The Unscrambler: Chemometrics and Intelligent Laboratory Systems, **34**, 283–292.

Faber, N. M., X. H. Song, and P. K. Hopke, 2003, Sample-specific standard error of prediction for partial least squares regression: Trends in Analytical Chemistry, **22**, 330–334.

Glasstone, S., K. J. Laidler, and H. Eyring, 1941, The theory of rate processes: The kinetics of chemical reactions, viscosity, diffusion and electrochemical phenomena, *in* L. P. Hammett, ed., International chemical series: McGraw-Hill.

Henaut, I., L. Barre, J. F. Argillier, F. Brucy, and R. Bouchard, 2001, Rheological and structural properties of heavy crude oils in relation with their asphaltenes content: SPE Paper 65020.

Hinkle, A., 2007, Relating chemical and physical properties of heavy oils: M.S. thesis, Colorado School of Mines.

Hossain, M. S., C. Sarica, H. Q. Zhang, L. Rhyne, and K. L. Greenhill, 2005, Assessment and development of heavy oil viscosity correlations: SPE Paper 97907.

Kharrat, A. M., J. Zacharia, V. J. Cherian, and A. Anyatonwu, 2007, Issues with comparing SARA methodologies: Energy and Fuels, **21**, 3618–3621.

Luo, P., and Y. Gu, 2005, Effects of asphaltene content and solvent concentration on heavy-oil viscosity: SPE Paper 97778.

Macosko, C. W., 1994, Rheology principles, measurements, and applications: Wiley-VCH.

Nehring, R., R. Hess, and M. Kamionski, 1983, The heavy oil resources of the United States: U. S. Department of Energy, R-2946-DOE.

Shin, E. J., M. Nimlos, and R. J. Evans, 2001, A study of the mechanisms of vanillin pyrolysis by mass spectrometry and multivariate analysis: Fuel, **80**, 1689–1696.

Shin, E. J., M. R. Hajaligol, and F. Rasouli, 2003, Characterizing biomatrix materials using pyrolysis molecular-beam mass spectrometer and pattern recognition: Journal of Analytical and Applied Pyrolysis, **68–69**, 213–229.

Shin, E. J., M. R. Hajaligol, and F. Rasouli, 2004, Heterogeneous cracking of catechol under partially oxidative conditions: Fuel, **83**, 1445–1453.

Tschoegl, N. W., 1989, The phenomenological theory of linear viscoelastic behavior: An introduction: Springer-Verlag.

Windig, W., and H. L. C. Meuzelaar, 1984, Nonsupervised numerical component extraction from pyrolysis mass spectra of complex mixtures: Analytical Chemistry, **56**, 2297–2303.

Windig, W., P. G. Kistemaker, and J. Haverkamp, 1982, Chemical interpretation of differences in pyrolysis mass spectra of simulated mixtures: Journal of Analytical and Applied Pyrolysis **3**, 199–212.

Chapter 5

Measuring and Monitoring Heavy-oil Reservoir Properties

Kevin Wolf,[1] Tiziana Vanorio,[1] and Gary Mavko[1]

Introduction

The level of interest in heavy-oil and bitumen reservoirs has dramatically increased in recent times. Increased production of these reservoir types has stimulated research on the properties of these reservoirs under various conditions to aid in the initial characterization of the reservoir and monitor production strategies in situ utilizing seismic data. Rock physics provides the crucial link between the physical properties of the reservoir and seismic properties that can be remotely measured; however, to this point there is not a robust model that can be used to predict or infer the properties of heavy-oil or bitumen sands from seismic data, nor is there sufficient experimental data to calibrate such models. We present a methodology to characterize and monitor heavy-oil reservoirs by inverting converted-wave seismic data to obtain P-to-S converted-wave elastic impedance (PSEI) estimates as a function of angle. By examining these data in "PSEI space" (crossplots of PSEI values obtained at different angles), we can infer the conditions in the reservoir and possibly relate them to physical properties of the reservoir through a reliable rock physics model. This methodology points out the need for better defined rock physics models that need to be calibrated to a large, robust data set. Experimental measurement of heavy-oil sands is challenging, and to meet these challenges, we have designed an ultrasonic pulse transmission system that has been optimized for use with heavy-oil sand samples. These samples provide several unique challenges in the laboratory that are not typically encountered when measuring traditional hard rocks such as carbonates or sandstones. Although the system has several specialized components, we will focus on the design of the transducers used in the system. The transducers are uniquely designed so that they closely match the impedance of the bitumen sand over a wide temperature range, resulting in sharp first arrivals while maximizing the received amplitude.

Converted PSEI

PSEI was defined by Gonzalez et al. (2003) as follows.

$$PSEI(\theta_P) = \rho^c V_s^d, \quad (1)$$

where

$$c = \frac{K \sin \theta_P}{\sqrt{\frac{1}{K^2} - \sin^2 \theta_P}} \left(2\sin^2 \theta_P - \frac{1}{K^2} - 2\cos \theta_P \sqrt{\frac{1}{K^2} - \sin^2 \theta_P} \right), \quad (2)$$

and

$$d = \frac{4K \sin \theta_P}{\sqrt{\frac{1}{K^2} - \sin^2 \theta_P}} \left(\sin^2 \theta_P - \cos \theta_P \sqrt{\frac{1}{K^2} - \sin^2 \theta_P} \right). \quad (3)$$

K is the average V_P/V_S, θ_P is the incident P-wave angle at the conversion point, ρ is the density, and V_S is the shear-wave velocity. Analyzing the behavior of exponents c and d reveals that at small incidence angles, V_S and ρ contribute to the obtained PSEI value. However, at larger angles, the density term begins to dominate in its contribution to the PSEI value. This behavior allows for the separation of density and shear-wave velocity trends in PSEI space. Figure 1 shows an example of this separation for a hypothetical situation in a well. The original well log data are plotted in black, and the data were then systematically altered to reveal the shear-wave velocity and density trends. The warm colors (red to yellow) reveal the V_S trend as V_S is decreased to 50% its original value in 10% decrements. Similarly, the cool colors (green to blue) reveal the density trend as density is

---
[1]Stanford University, Department of Geophysics, Stanford, California, U.S.A.
This paper appeared in the September 2008 issue of THE LEADING EDGE and has been edited for inclusion in this volume.

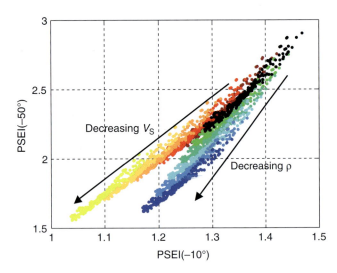

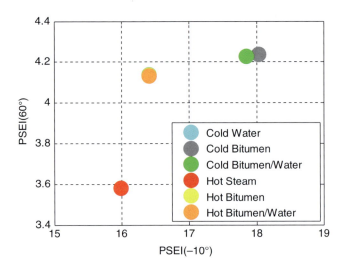

Figure 1. Crossplot of near-angle PSEI ($-10°$) versus far-angle PSEI ($-50°$) for a well log from a bitumen reservoir. The original well log data are shown in black. The warm colors represent changing shear-velocity measurements by decrements of 10% to a minimum of 50% of their original value. The cool colors represent changing density measurements by decrements of 5% to a minimum of 75% of their original value.

Figure 2. PSEI analysis as applied to reservoir states for monitoring of heated zones and steam chambers. Note that cold water and hot bitumen data points are mostly obscured by the hot bitumen/water data point.

decreased to 75% of its original value by 5% decrements. The ability to discriminate between these two trends is important for heavy-oil reservoirs because they are commonly produced thermally via steam injection. As a reservoir is heated, the shear velocity of the heavy-oil sand decreases dramatically, and at the point of injection, steam will begin to displace the heavy oil in the pore space, which will decrease the reservoir density in that area. Knowing that we can identify these trends with PSEI, we can therefore use it to help characterize and monitor the production of thermally produced heavy-oil sands. This is illustrated in Figure 2, which shows where various synthetic reservoir states plot in PSEI space. These reservoir states show that discriminating between cold bitumen bearing zones, heated bitumen bearing zones, and steam saturated zones is possible in PSEI space provided near- and far-angle PSEI attributes can be determined.

PSEI inversion for reservoir characterization and monitoring

To illustrate the utility of PSEI for characterization, we have created a simple three-layer synthetic earth model. It consists of a uniform shale unit overlying a 30-m thick reservoir interval that extends from a depth of 470–500 m and is underlain by a uniform limestone basement. The inversion has been performed on the reservoir when it is in three distinct states: (1) the original cold reservoir, (2) the reservoir containing a heated zone that is 15 m wide and 15 m thick at its widest and thickest portions, and (3) the reservoir containing the same heated zone and a steam-saturated zone 6 m wide and 6 m thick across its widest and thickest dimensions. Figure 3 shows the distribution of reservoir properties for each reservoir state. Converted-wave synthetic seismic data were then computed through the earth models using a 125-Hz Ricker wavelet at offsets corresponding to P-wave incidence angles of 10° and 50° at the top of the reservoir. The inversion procedure used was a model-based inversion with the initial model for each inversion created using three "wells" — one on each end of the earth model and one located at its center. To build a model that is consistent with the converted-wave seismic, the wells must first be converted to pseudowells as proposed by Gonzalez et al. (2004). These pseudowells can then be used to build the initial model, and the model-based inversion can be carried out. The results of the inversion at 10° (left column) and 50° (right column) are shown for the three earth models in Figure 4. The results from the cold reservoir inversion (top row) clearly show the top and bottom of the reservoir in the 10° inversion. However, the base of the reservoir is poorly resolved in the 50° angle data. This is due to the character of the seismic data at the reservoir-limestone interface. The angle of incidence for this case is postcritical and the resulting low-amplitude seismic reflections are not sufficient for the inversion to clearly resolve the boundary between the reservoir and the limestone basement. Similar patterns with regard to incident angle can be seen in the inversion results from the heated reservoir (middle row). Again, the 50° inversion does not adequately define the

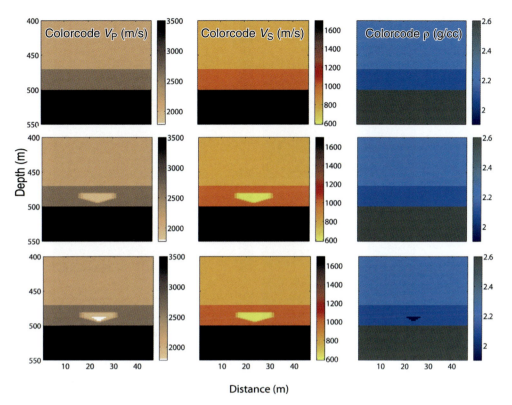

Figure 3. Synthetic earth models for cold reservoir (top), heated reservoir (middle), and heated reservoir with steam chamber (bottom). The shale overburden and limestone basement rock are the same in all cases, and the reservoir spans the depth from 470 to 500 m. The P-wave velocity is shown in the first column, S-wave velocity in the middle column, and density in the right column.

base of the reservoir. However, examining the 10° inversion result, not only are the top and base of the reservoir delineated, but the heated zone is also clearly visible and located in the correct position. At 50°, the inversion no longer sees the heated zone in the reservoir; however, this result is not unexpected. Because the angle is increased, the sensitivity of PSEI to shear-wave velocity is decreased, and the sensitivity to density is increased. Because the heated reservoir only contains contrasts in the shear-wave velocity, it should appear homogeneous at far offsets. For the heated reservoir with steam chamber (bottom row), the 10° inversion again clearly defines the heated zone. Also, the 50° inversion result shows a low-impedance zone in the heated zone that corresponds to the location of the steam chamber. This is the result of the density contrast created by steam replacing the bitumen in the pore space. These results are very encouraging and demonstrate that the proposed methodology can provide a robust method for tracking heated zones within a bitumen reservoir as well as the formation of steam chambers provided that good-quality converted-wave seismic data are available. However, there are also some artifacts in the inversion that must be taken into account when interpreting the results. One such example is the high-impedance zone directly overlying the heated reservoir in the 10° inversion results. This high-impedance zone is likely a result of wavelet effects or tuning or is possibly related to limited frequency content of the inversion. Also, with real seismic data there will be a lower signal-to-noise ratio than in our idealized synthetic case, which will have an effect on the quality of the inversion. All of these effects should be taken into consideration when interpreting inversion results.

Data in PSEI space

The results of the near- and far-angle PSEI inversion for reservoir characterization appear promising. However, to fully realize the utility of the inversion it is necessary to plot the data in PSEI space (Figure 5). A realistic way of looking at the data is to imagine them as being snapshots of a reservoir at different times during production. The cold reservoir (Figure 5a) is representative of the untouched reservoir as it is in the subsurface before production has commenced. There is a clear separation in data points that correspond to the overlying shale (magenta), underlying limestone (green), and the reservoir interval we are interested in (blue). The next snapshot (Figure 5b) corresponds to the reservoir after steam has been injected into the reservoir to heat the reservoir and stimulate production. By heating the reservoir, the shear modulus of the bitumen is reduced, which in turn lowers the shear-wave velocity in the heated zone. As expected, the location of the heated reservoir points (red) shift to the left in PSEI space, indicating that the reservoir zones corresponding to those data points have been heated. Note that some of the blue data points also have migrated in the same direction, but this is because the seismic inversion tends to smear boundaries between different zones instead of providing an extremely sharp boundary.

Figure 4. Inversion results for cold reservoir (top), heated reservoir (middle), and heated reservoir with steam chamber (bottom). The left column shows the 10° inversion, and the right column shows the 50° inversion. Dashed black lines show the top and bottom of the reservoir as picked from the synthetic seismic sections.

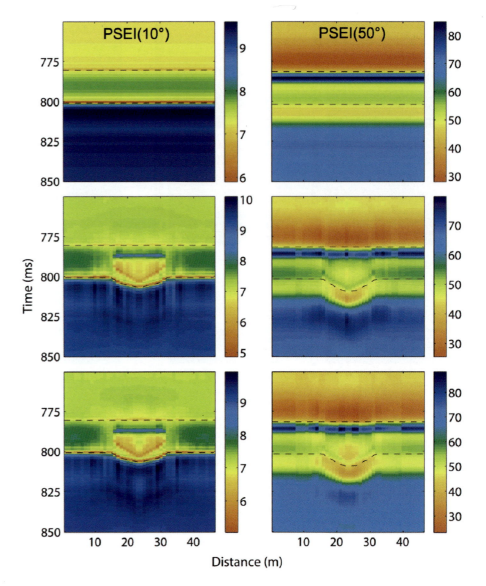

Also, the models themselves have a transition between reservoir properties instead of a sharp boundary. The next snapshot (Figure 5c) occurs later in time, when there has been a sufficient amount of steam injected into the reservoir for a long enough time to permit the formation of a steam chamber around the injection point. The effect of this is to lower the density of the reservoir in the steam chamber zone and decrease the P-wave velocity. As density decreases, we expect data points to migrate downward in PSEI space. This can be seen in Figure 5c, where the distribution of data points corresponding to the steam chamber (yellow) generally plot below the heated zone data points (red). This is encouraging; however, comparing the location of the yellow points in Figure 5c with the location of heated data points in Figure 5b, there is some overlap between the points. This suggests that the formation of a steam chamber may not be sufficient to delineate a clear separation of data points corresponding to heated reservoir zones versus steam-filled reservoir zones. However, the relative distribution of the points corresponding to the heated and steam-filled zones does allow one to discriminate which data points, and hence which reservoir zones, are more likely to be steam saturated. Also, if the size of the steam chamber grows larger in dimension than is presented in this example, the effect of smearing the boundaries in the inversion will be less pronounced, which may allow for a clearer distinction between heated and steam-saturated sections of the reservoir.

Laboratory Measurements of Heavy-oil Reservoirs

The preceding inversion example shows the possibilities that exist for using PSEI to monitor thermal production methods. However, without robust rock physics models to link the observed data to the physical properties of the reservoir these results can only be used in a qualitative description of the reservoir. To facilitate a quantitative description, there is the need for a robust rock physics

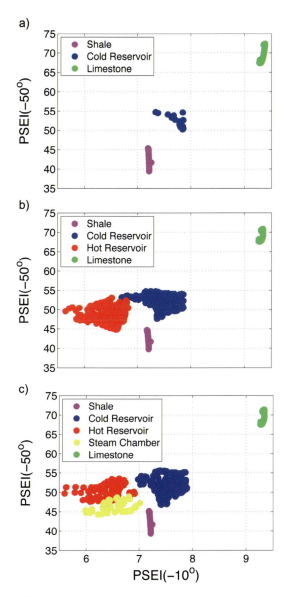

Figure 5. (a) Cold reservoir, (b) heated reservoir, and (c) heated reservoir with steam chamber inversion results plotted in PSEI space.

model or transform. As such, there is a strong need for a massive set of robust laboratory measurements of the general properties, including acoustic and elastic properties, of bitumen and heavy-oil sands to establish well-defined rock physics relationships for these materials.

Experimental rock physics has conducted a great deal of research on room temperature measurements of the properties of hard rocks such as sandstones, carbonates, etc., to establish empirical and theoretical relationships between several rock properties. Despite their inherent complexities, these types of rocks are typically much simpler to make measurements on than bitumen sands for several reasons. Hard rocks have a high impedance and higher quality factor than the soft sediments that typically make up bitumen reservoirs. The relatively low impedance and quality factor of bitumen sand samples make it more difficult to obtain large-amplitude P- or S-wave signals through them. In addition, bitumen sands are generally poorly consolidated sands that are bound together by the bitumen filling the pore space; this leads to properties that can vary quite drastically once the samples are heated. These characteristics make the measurement of bitumen sand velocities via ultrasonic pulse transmission experiments particularly challenging. To attempt to overcome these challenges, we have designed a unique ultrasonic pulse transmission system specifically for the measurement of bitumen sands. The system requires many unique features, and in this chapter we will focus on what we consider to be the most crucial element — the transducers.

Transducer design

Traditionally, ultrasonic velocity measurements have been made with high-impedance end caps with minimal effort put into the design of the transducers themselves. This has not been a major impediment for most measurements because most rock physics measurements have been made on hard rocks (e.g., sandstones, carbonates, and some shales). The properties of these hard rocks are not as variable as unconsolidated sediments, especially unconsolidated sediments that are fixed by bitumen. Our work has paid particular attention to the design of the transducers to ensure that the best signal possible is recorded under all operational conditions.

Basic transducer design must take into account all vital pieces of a transducer. This includes the piezoelectric crystal (or crystals for generating multiple wave types or polarizations), the backing behind the crystals, the end cap or matching layer between the crystals, and the load, as well as the load that the generated waves will travel through — bitumen sand in this case. Figure 6 illustrates a general schematic for an ultrasonic transducer with two piezoelectric crystals — one to generate P-waves and one to generate S-waves. To realize the optimum design, a transducer modeling code has been written that utilizes the KLM equivalent circuit (Krimholtz et al., 1970) to measure the response of the transducer given different transducer and sample properties. The advantage of using the KLM model is that it easily allows for the addition of any number of piezoelectric crystals, matching layers, etc., to be added into the transducer model. This allows for the modeling of many types of transducers, including those with multiple crystals for making anisotropic measurements, etc. The ABCD matrix representation of the individual components and loads in the KLM model is also used. Each component of the KLM model can be represented as a two-port network that can then be characterized using matrix methods (Ramo et al., 1984). This formulation allows the input and output voltage and current in each component to be related to each other through a 2 × 2 matrix. Then, to calculate the response of the entire transducer, the entire chain of matrices is multiplied together, which then relates the input voltage and current to the output force and velocity from the transducer. This matrix representation makes the calculation of the transducer response fast and easier to code. The responses of various transducer designs under various conditions are shown in Figure 7. The effect of varying the end cap or matching layer material is shown in Figure 7a. Traditional hard-rock transducers utilize stainless steel end caps. For hard-rock samples, this does not create a large mismatch in impedance between the transducer and the sample because the sample itself has a high impedance. However, when dealing with unconsolidated or other soft samples, having a high-impedance end cap creates a large impedance contrast between the transducer and the samples, resulting in less energy being transferred to the sample itself. Zimmer (2003) attempted to overcome these problems for measurements on unconsolidated sands by using a glass-filled polycarbonate end cap. This type of end cap has an impedance more similar to unconsolidated sand than traditional steel end caps. However, this material cannot withstand the elevated temperatures (200°C) needed for our experiment. The choice was made to use Torlon end caps because Torlon has a low impedance similar to that of unconsolidated sand and its properties are reasonably stable with temperature. Figure 7a shows the impulse response of the transducer at room temperature with two different end-cap materials. The traditional steel end cap is shown in blue, and the Torlon end cap is shown in red. Figure 7b shows the exact same transducer; however, the modeled response at elevated temperatures is shown. Clearly the Torlon provides a cleaner, higher amplitude signal than stainless steel in this case, especially for the shear wave. Another important consideration is the thickness of the end cap or matching layer. Ideally, the thickness of the end cap or matching layer should be precisely one-quarter of a wavelength thick. This ensures that any reflected energy will constructively interfere and lead to the best possible impulse response of the system at the frequency at which it is

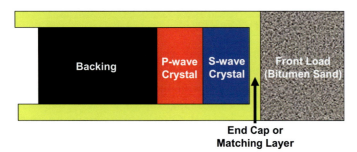

Figure 6. Schematic for the generalized transducer modeled. It consists of a crystal backing, two piezoelectric crystals (one to generate P-waves and one to generate S-waves), an end cap or matching layer, and the load at the front of the transducer, which is bitumen sand in this case.

Figure 7. The effect of varying transducer properties on the generated signal: (a) varying end-cap material at room temperature, stainless steel in blue and Torlon in red; (b) varying end-cap material at elevated temperature, stainless steel in blue and Torlon in red; (c) varying Torlon end-cap thickness; and (d) varying backing material used, stainless steel in blue, Dow Epoxy Resin 332 in green, and lead metaniobate in red.

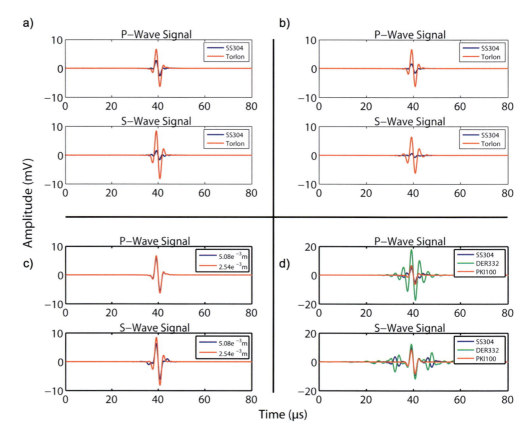

designed to operate. However, in this case compressional and shear waves need to be generated, and each wave type has its own associated wavelength. As such, a compromise must be reached between the two ideal thicknesses for each type of wave. As can be seen in Figure 7c, a thickness of 2.54×10^{-3} m provides a large amplitude, broadband impulse response. If the thickness is increased or decreased, the impulse response is adversely affected. For increased thickness, shown in blue, the first arrival of the shear wave is not as prominent. In addition, the maximum amplitude is slightly decreased. The backing material used in the transducer also has a large effect on the impulse response of the transducer as a whole. The backing should ideally be a very lossy and high-impedance material compared to the crystals used to generate the signal. The high impedance will ensure that energy radiated out of the back of the crystal is mostly reflected back into the sample. The backing also needs to be very lossy so that any energy that is transmitted into the backing does not reflect off the rear of the backing and adversely affect the signal that is generated. One fairly common procedure for producing high-impedance lossy backings is to combine tungsten and epoxy in the correct proportion to obtain the desired characteristics. However, this practice poses several difficulties for our system. The first is that the epoxy typically used is not able to withstand elevated temperatures. The second is that the P- and S-wave impedances need to be matched to specific values, which would involve much time-consuming research to get the correct proportions and impedances. To overcome these problems, we used the same piezoelectric material that is used for the crystal as the backing. This ensures that the backing has the correct matching impedance for the P- and S-waves. Figure 7d compares the impulse response for three transducers that are identical except for the backing material used. It is immediately obvious that the most broadband transducer is the one with the perfectly matched backing of lead metaniobate. It will be easiest to clearly identify the first arrival of this signal given the clear first break on the P- and S-wave impulse response. It is also apparent that the perfectly matched backing has a lower overall amplitude. This is because the energy leaving the back of the piezoelectric crystal is not reflected back by the backing. Although this is not necessarily desirable, the clean nature of the signal makes up for the loss in amplitude that is suffered. Also, it is important to keep in mind that the backing is not lossy in any of the three cases presented in Figure 7d. This means that any energy passed into the backing will reflect off the rear of the backing and interfere with the desired signal. An easy way to overcome this problem is to ensure that the backing is long enough that the reflected wave will not arrive at the far end of the sample before the first break of the desired signal. Our final design includes this consideration. The frequency at which the measurements are made has also been modified from traditional ultrasonic transducers used for consolidated rocks. Two-hundred-fifty-kilohertz broadband piezoelectric crystals are used instead of standard 1-MHz crystals; this was found to reduce energy absorption and scattering as the signals pass through the samples (Zimmer, 2003). Also, using a frequency of 250 kHz ensures that we will have at least one wavelength within our samples to obtain robust measurements. As is evident from the preceding discussion, much thought and planning has been placed into building the most effective piezoelectric transducers possible for the purpose of carrying out ultrasonic pulse transmission experiments on bitumen sands at high temperatures. This planning, although time-consuming, should ensure that we obtain the cleanest and largest possible signal through our samples, which should ensure the most accurate results possible.

Conclusions

A methodology utilizing PSEI attributes has been developed for characterizing and monitoring bitumen reservoirs undergoing thermal production. The example shown demonstrates the ability of the method to discriminate between nonreservoir and reservoir rocks, as well as zones within the reservoir that have been heated or have had reservoir fluids replaced by steam. The lack of a well-defined rock physics model or transform for heavy-oil or bitumen sands precludes the development of a quantitative interpretation at this point; however, with the design heavy-oil-specific laboratory equipment, we can now begin to accrue needed data for the development for such a model. This will enable a clearer link between measured seismic properties and the physical properties and conditions present within the reservoir.

References

Gonzalez, E. F., G. Mukerji, and G. Mavko, 2003, Near and far offset P-to-S elastic impedance for discriminating fizz water from commercial gas: The Leading Edge, **22**, 1012–1015.

Gonzalez, E. F., G. Mavko, and T. Mukerji, 2004, A practical procedure for P-to-S 'elastic' impedance (PSEI) inversion: Well log and synthetic seismic examples for identifying partial gas saturations: 74th Annual International Meeting, SEG, Expanded Abstracts, 1782–1785.

Krimholtz, R., D. A. Leedom, and G. L. Matthaei, 1970, New equivalent circuits for elementary piezoelectric transducers: Electronics Letters, **6**, 398–399.

Ramo, S., J. R. Whinnery, and T. Van Duzer, 1994, Fields and waves in communications electronics: J. Wiley and Sons.

Zimmer, M. A., 2003, Seismic velocities in unconsolidated sands: measurements of pressure, sorting and compaction effects: Ph.D. thesis, Stanford University.

Chapter 6

Seismic Rock Physics of Steam Injection in Bituminous-oil Reservoirs

Evan Bianco,[1] Sam Kaplan,[1] and Douglas Schmitt[1]

Introduction

This case study explores rock physical properties of heavy-oil reservoirs subject to the steam-assisted gravity drainage (SAGD) thermal-enhanced recovery process (Butler and Stephens, 1981; Butler, 1998). Previously published measurements (e.g., Wang et al., 1990; Eastwood, 1993) of the temperature-dependent properties of heavy-oil saturated sands are extended by fluid substitutional modeling and wireline data to assess the effects of pore fluid composition, pressure, and temperature changes on the seismic velocities of unconsolidated sands. Rock physics modeling is applied to a typical shallow McMurray formation reservoir (135–160 m depth) encountered within the bituminous Athabasca Oil Sands deposit in Western Canada to construct a rock-physics-based velocity model of the SAGD process. Although the injected steam pressure and temperature control the fluid bulk moduli within the pore space, the effective stress-dependent elastic frame moduli are the most poorly known yet most important factors governing the changes of seismic properties during this recovery operation. The results of the fluid substitution are used to construct a 2D synthetic seismic section to establish seismic attributes for analysis and interpretation of the physical SAGD process. The findings of this modeling promote a more complete description of 11 high-resolution, time-lapse, 2D seismic profiles collected over some of the earliest steam zones.

The SAGD process has been adopted as the recovery method of choice for producing bitumen from the Athabasca Oil Sands in Western Canada and it has changed relatively little since the first test installation and experiment at the underground test facility in the early 1990s. The invention of Geophysics Research, Department of Physics, this thermal-enhanced oil recovery and horizontal drilling technology was born out of the challenges associated with producing extremely dense and viscous oil [American Petroleum Institute (API) density $<10°$] from shallow siliciclastic reservoirs. Steam carries a significant portion of its energy as latent heat, and it is much more efficient at transferring heat to the reservoir than merely circulating hot water. Engineering models of this thermal process assume that steam chamber growth is symmetric about the well pairs, but because of lithologic heterogeneities and steam baffles, this is most certainly not the case (Figure 1). Before seismic profiling can be optimized as a tool for tracking the movement of steam in the reservoir, it is important to understand the behavior of oil-sands material when subjected to elevated temperatures, pore pressure, and fluid saturation conditions inflicted by SAGD.

Figure 2 shows a scanning electron micrograph (SEM) image of typical McMurray oil sand. Oil sand, by definition, lacks or has very little cementation. As such, its moduli (bulk and shear) are largely dependent on grain-to-grain contacts. These contacts are held in place by confining pressure, and any reduction in effective pressure will result in a reduction in effective moduli. The micrographs display a subtle interlocked texture characterized by relatively high incidences of long and interpenetrative grain contacts. This suggests that the shallow oil sands were once much deeper than they are today, although lithification did not occur. Furthermore, the bitumen has undergone biodegradation and is highly viscous; it may actually support the sand grains in much of this material and act as partial cement when strained at seismic frequencies.

Dipole sonic logs and density logs can provide local measurements of effective elastic properties. Effective rock properties (i.e., effective P-velocity, V_P, effective S-velocity V_S, and effective density, ρ) are extracted from wireline data to establish reasonable elastic parameters (κ_{eff}, μ_{eff}, ρ) for input to Gassmann's equation for fluid

---

[1]University of Alberta, Institute for Geophysical Research, Department of Physics, Edmonton, Alberta, Canada
This paper appeared in the September 2008 issue of THE LEADING EDGE and has been edited for inclusion in this volume.

Figure 1. Schematic representation of the plumbing system for oil drainage into horizontal well injector/producer pairs in a typical SAGD process. Steam is pumped into the top well (red arrows), and heated oil drains along the edges of a steam chamber under gravity toward a producing well all along the length of the wellbore (green arrows).

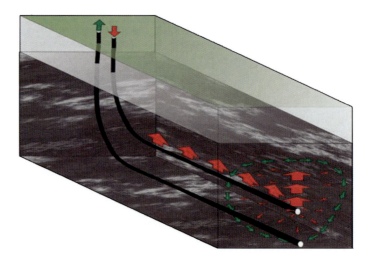

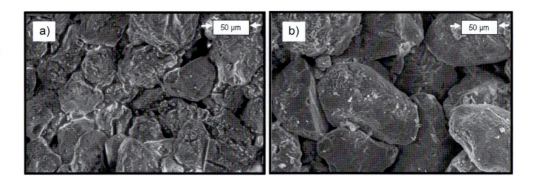

Figure 2. SEM image of (a) uncleaned oil-sands material, and (b) oil sands material with organic components removed (cleaned). The reflective and resinous material in cracks and pores of (a) is bitumen, and trace amounts of clay can be seen in (a) and (b).

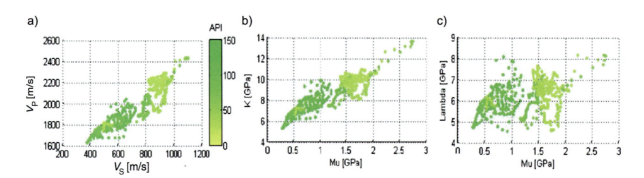

Figure 3. Scatterplot of (a) V_P versus V_S, (b) bulk modulus (κ) versus shear modulus (μ), and (c) incompressibility versus shear modulus (μ) computed from dipole-sonic log measurements through the McMurray formation and the overlying mudstones of the Clearwater formation. The color scale for each panel is cast to the gamma-ray reading (API) for each depth.

substitution (Figure 3). Upon injecting steam into the reservoir, a depleted oil zone expands gradually and replaces the initial bitumen saturation with a combination of water, steam, and residual oil at elevated temperatures and pressures. Also, oil production rates are increased if the steam is injected at high pressures because the unconsolidated oil sand is deformed although there is currently little understanding of what actually takes place within a steam zone. Permeability has been reported to increase as a result of geomechanical changes (e.g., Wong, 2003; Li and Chalaturnyk, 2006), and the portion of the reservoir that is touched by steam is completely physically altered from its original state. In that sense, unconsolidated reservoirs cannot be treated with simple fluid substitution calculation and the pressure-dependent effects of the frame bulk modulus must be taken into account.

The fluid is also complicated. Ternary diagrams aid in studying the effective P-velocity response of this three-component fluid ensemble (oil, water, steam) for various temperature and pore-pressure scenarios (Figure 4). The result is a more accurate description of the in situ elastic parameters within an idealized steam chamber, because the frame has likely been completely altered by increased pore pressure and temperature. Ternary diagrams also present a visual representation of the stability of a given effective property of interest. In this reservoir, we assume that a considerable amount of oil remains irreducibly trapped within the steam zone; the initial oil saturation of 88% falls to 62%, corresponding to a modest recovery factor of 30% within the depleted zone. If we assume that recovery is 60%, then the resultant change in effective P-velocity of this new material is only a few percent higher. Compared to an initial velocity of 2150 m/s and a final velocity of 1506 m/s (a 30% decrease in velocities), the error associated with such a saturation range is negligible. The ternary diagram can be used to evaluate a property of interest for all possible saturation combinations of a three-component mixture.

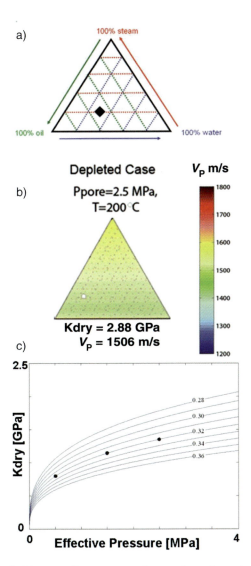

Figure 4. Ternary diagrams can be used to plot mixture properties of three-component mixtures. The black diamond in (a) indicates a pore fluid containing 20% steam, 30% water, and 50% oil. The effective P-wave velocity has been calculated for all permutations of oil + water + steam saturation scenarios. Typical reservoir temperatures and pressures were used to estimate the change in P-wave velocity of oil sands from 2150 m/s before steam injection to 1506 m/s after steam injection. The pressure-dependent decrease in frame bulk modulus (Kdry) is computed using a modified Hertz–Mindlin formulation extrapolated from unconsolidated spherical grains (c). Curves in (c) represent typical porosities for unconsolidated materials.

Integrating Rock Physics and Reservoir Parameters for Improved Synthetic Modeling

A background or baseline acoustic velocity model was created from a composite of logs from the general area. This "high-resolution" velocity model (Figure 5) was impregnated with the idealized velocity anomaly shown in the middle panel of Figure 6.

To explain the propagation of seismic energy through an oil-depleted steam chamber surrounded by cold, untouched, virgin reservoir, a finite-difference algorithm was used to calculate the wavefield generated through the acoustic velocity model. Shots were collected every 2 m along the profile, and receivers were placed every 2 m on either side of the shot. The poststack image was created using conventional normal moveout (NMO) correction and stacking with a resulting common midpoint (CMP) spacing of 1 m and is shown in the bottommost panel of Figure 6.

The top of the steam zone is marked by a trough event (Figure 6), as expected, because of a strong negative reflection coefficient at the top. The base of the anomaly is marked by a strong peak event produced by an increased reflection coefficient at the bottom of the reservoir. The steam zone yields a large increase in amplitude and a significant time delay that may serve as a proxy for steam chamber thickness. Furthermore, this example also shows that scattering plays a significant role in the seismic response with diffraction hyperbolas, complicated reverberations, and multiple reflections from within the steam zone (Figure 6) apparent. The perturbation in the wavefield is localized about the steam zone, and internal multiples persist beneath its thickest part (approximately 108–118 m along the profile in Figure 6) for several cycles beneath the reservoir. Furthermore, the top and base of the steam zone are seen to produce independent sets of remnant diffractions that remain after NMO correction and stacking have been applied. The two sets of diffractions actually have

Figure 5. High-resolution velocity model used as input for finite-difference forward modeling. The background horizontal layers are derived from closely spaced wells at the underground test facility where observation wells are spaced 40 m apart. This extremely high-density wireline sampling allows for an intricate and high-resolution cross section through a steam zone.

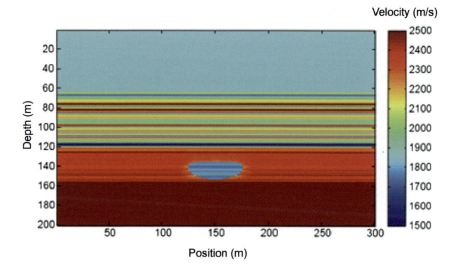

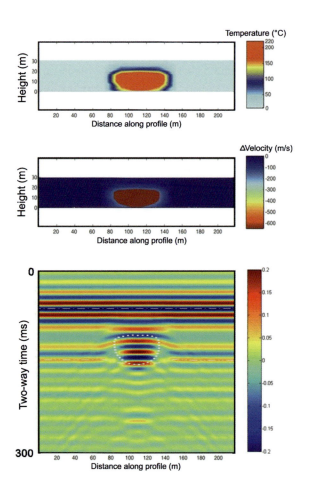

Figure 6. The top panel shows a temperature cross section, and the middle panel shows the computed change in P-wave velocity from a developed steam chamber. This velocity anomaly has been inserted into a background cross-sectional velocity model for the input to the finite-difference algorithm used for generating synthetic seismic data (shown in the bottom panel). These are poststack seismic data that have undergone conventional processing, velocity analysis, NMO correction, and stacking.

different curve shapes; the top set is faster (shallow hyperbolas), and the bottom set is slower (steeper hyperbolas). It is possible that the bottom set of diffractions are time-retarded propagation within the low-velocity steam zone, whereas the top set is not retarded and is merely scattering from the outside surface of the steam zone. These observations promote the need for 2D migration. Fomel et al. (2007) have demonstrated the ability to perform poststack migration on data using diffraction information alone and there may be some promise to using their method or other standard migration procedures to actually extract quantitative time-lapse information from such features. The absence of diffractions and internal multiple events leads to a simpler and ideal image of the steam chamber. For comparison, an ideal image of the reservoir and steam zone is shown in Figure 7, where the noisy events of 2D wave propagation are not incorporated. Here, under ideal conditions, the exact shape of the steam zone can be traced out and the geometry and volume of the depleted reservoir can be accurately determined. The same is not true for the seismic section generated by finite difference. There is of course more detail, and the physics is 2D (instead of the less complete 1D convolutional model), but the final image is plagued with higher ambiguity. Here, the brightest amplitudes do not extend across the whole width of the steam zone. Direct mapping of the traveltime delay or amplitude anomaly in this case would lead the interpreter to conclude that the steam zone is actually smaller than it really is. The edge effects and resolution limits of small-scale velocity anomalies such as the one presented here must be modeled on a case-by-case basis, and simple Fresnel zone arguments will not be sufficient in determining the maximum horizontal resolution (Schmitt, 1999). Recently, Zhang et al. (2007) have presented an extensive study that shows how important adequate resolution is to characterizing such reservoirs.

Chapter 6: Seismic Rock Physics of Steam Injection in Bituminous-oil Reservoirs

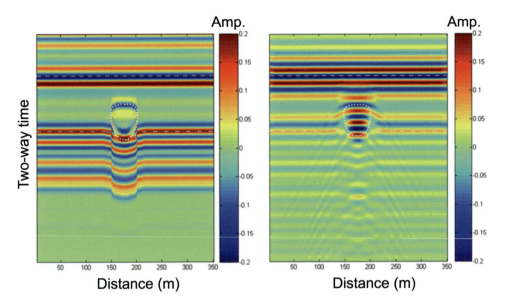

Figure 7. Comparison between an "ideal" seismic profile (left) using simple 1D convolution, and a more realistic seismic profile over a steam zone (right) shown in Figure 7 that was acquired using a finite-difference algorithm. The image on the right was highly sampled, conventionally processed, and stacked.

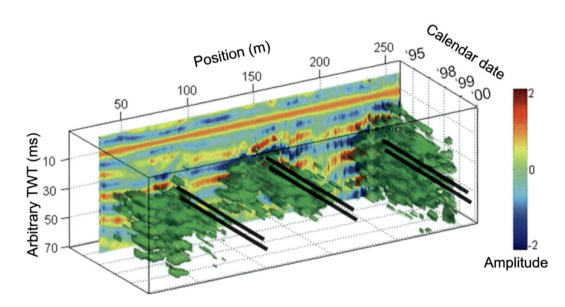

Figure 8. Three-dimensional representation of repeated 2D seismic (time-lapse) data collected over three steaming horizontal well pairs at the underground testing facility. Position and two-way traveltime are plotted on the *x*- and *y*-axes respectively, and the volume of data is given a "depth" perspective by stacking the repeated sections along the *z*-axis (in ascending calendar date). The "brightest" amplitudes (positive and negative) have been rendered as semitransparent "isosurfaces." These isosurfaces are thought to be indicators of the lateral extent of the steam chambers. The reverberations are proportional to the magnitude of steam in the reservoir and coincide with the modeled reverberations in Figure 4. The approximate location of the well pairs are indicated by the black lines; however, their size and vertical separation are not to scale.

Comparison with Real Seismic Data and 2D Time-lapse Imaging

Over a 5-year period, a series of 11 high-resolution 2D lines were collected by the University of Alberta at a SAGD site in northern Alberta. Because the steam zones were relatively shallow, an exceptionally small 1-m CMP spacing with offsets ranging from 48 to 142 m (48 channels were available and spaced 2 m apart) was used. Care was taken in the repositioning of the sources and the receivers during these tests, and the final sections exhibit good repeatability. Figure 8 shows a volume CMP position by traveltime by calendar date visualization of these time-lapse data; the "brightest" amplitudes have been highlighted as green "isosurfaces." This image shows two interesting results that are consistent with the numerical seismic experiment: (1)

The amplitude increase caused by the steam is very large and detectable, and (2) the steam anomalies are similar to the reverberation and scattering symptoms as modeled. However, the steam anomalies are not symmetric about the well pairs (in particular, the leftmost steam chamber is entirely asymmetric). Furthermore, there appears to be a sharp lateral dropoff in the amplitudes between the steam zone, and this could be attributed to the interference of diffraction energy off the top of the steam zones.

Conclusions

The oil sands of Western Canada are truly unique materials and represent an immense resource that is only in the early stages of exploitation. To progress the understanding of rock physics in these settings, theoretical rock physics relationships such as Biot–Gassmann theory must be modified to account for the unconsolidated form and high viscosity of the oils. Additionally, there is much that remains unknown about the viscoelastic behavior of oil sand and the effect that it has on seismic wave propagation. Ideally, our synthetic full-field finite-difference experiment should be adapted to treat shear-wave and attenuation information. Such modeling would provide a more realistic depiction of wave propagation; however, there are still too many uncertainties about attenuation and viscosity characteristics at seismic frequencies to promote full viscoelastic modeling at the same scale of spatial resolution.

The results from this analysis suggest that the feasibility of seismic monitoring does not only depend on the thermal and mechanical related changes associated with fluid substitution, but also on the scale of the steam anomaly itself. Thorough rock physics modeling can aid in the long-term survey design for monitoring reservoir depletion. Parameters such as spatial sampling, optimal fold, repeat time intervals, and source type, to name only a few, can all be evaluated before first steam. Future work is required to correlate these seismic measurements to the thermocouple profiles from observation wells (e.g., Birrell, 2003), after which time the full extent and utility of such highly sampled data can be ascertained.

Acknowledgments

This research was funded by National Sciences and Engineering Research Council and earlier contracts from Alberta Oil Sands Technology and Research Authority.

References

Birrell, G., 2003, Heat transfer ahead of a SAGD steam chamber: a study of thermocouple data from phase B of the underground test facility (Dover project): Journal of Canadian Petroleum Technology, **42**, 40–47.

Butler, R., 1998, SAGD comes of age!: Journal of Canadian Petroleum Technology, **37**, 9–12.

Butler, R. M., and D. J. Stephens, 1981, The gravity drainage of steam-heated heavy oil to parallel horizontal wells: Journal of Canadian Petroleum Technology, **20**, 90–96.

Eastwood, J., 1993, Temperature-dependent propagation of P- and S-waves in Cold Lake oil sands: comparison of theory and experiment: Geophysics, **58**, 863–872.

Fomel, S., E. Landa, and M. T. Taner, 2007, Poststack velocity analysis by separation and imaging of seismic diffractions: Geophysics, **72**, no. 6, U89–U94.

Li, P., and R. J. Chalaturnyk, 2006, Permeability variations associated with shearing and isotropic unloading during the SAGD process: Journal of Canadian Petroleum Technology, **45**, 54–61.

Schmitt, D. R., 1999, Seismic attributes for monitoring of a shallow heated heavy oil reservoir: a case study: Geophysics, **64**, 368–377.

Wang, Z., and A. Nur, 1990, Effect of temperature on wave velocities in sands and sandstones with heavy hydrocarbons: SPE Reservoir Engineering, **3**, 158–164.

Wong, R. C. K., 2003, A model for strain-induced permeability anisotropy in deformable granular media: Canadian Geotechnical Journal, **40**, 95–106.

Zhang, W. M., S. Youn, and Q. Doan, 2007, Understanding reservoir architectures and steam-chamber growth at Christina Lake, Alberta, by using 4D seismic and crosswell seismic imaging: SPE Reservoir Evaluation & Engineering, **10**, 446–452.

Chapter 7

Prediction of Pore Fluid Viscosity Effects on P-wave Attenuation in Reservoir Sandstones

Angus Best,[1] Clive McCann,[1] and Jeremy Sothcott[1]

Introduction

Seismic wave attenuation (absorption or intrinsic attenuation) is underutilized by the petroleum industry mainly because of the difficulty of measuring it with sufficient accuracy using seismic reflection methods. However, the recent trend toward time-lapse 3D seismic monitoring of reservoirs means that absolute attenuation measurements are no longer so important. Instead, temporal changes in seismic amplitude and wavelet frequency content could hold the key to interpreting changes in reservoir fluid properties in response to certain production strategies. Seismic monitoring of heavy-oil reservoirs would be particularly appropriate because of the known link between seismic attenuation and pore fluid viscosity (Batzle et al., 2006a).

Seismic wave attenuation in porous rocks is thought to be dominated by viscous fluid flow mechanisms such as those described by the unified Biot and Squirt (BISQ) model (Dvorkin and Nur, 1993; Dvorkin et al., 1994). By definition, the magnitude of attenuation is related to the viscosity of the pore fluid and the frequency of the elastic wave, as well as to other pore fluid parameters (density, bulk modulus) and pore geometry parameters (porosity, permeability, tortuosity, squirt flow length). This theoretical link between attenuation and pore fluid viscosity could be exploited if changes in heavy-oil viscosity caused by thermal stimulation, typically in the range 1–1000 cP (1 cP = 10^{-3} Pa·s), give rise to sufficiently large changes in elastic wave attenuation for detection and monitoring by surface seismics. Also, an accurate model of frequency/viscosity-dependent attenuation as a function of lithology could help tie in sonic well logs at frequencies of 10–20 kHz to surface seismics at less than 200 Hz for the benefit of reservoir characterization.

In this chapter, a laboratory ultrasonic data set of P-wave velocity and attenuation measured as a function of pore fluid viscosity in three reservoir sandstones is compared to the BISQ model. The three sandstones (Elgin, Berea, Northern North Sea) represent a typical spread of porosity, permeability, and intrapore clay contents for reservoir sandstones (see Table 1). The importance of clay content for increasing ultrasonic attenuation in sandstones is well established for P-waves (Klimentos and McCann, 1990) and S-waves (Best et al., 1994), but in the absence of a validated elastic wave model, it is difficult to predict how this behavior will manifest itself on surface seismic data sets. Hence, there is a need for careful crosschecking of theoretical models against well-constrained experimental data sets. The BISQ model is a strong candidate for further investigation because it uses input parameters that are readily available to the reservoir engineer or petrophysicist. The only ambiguity in the BISQ model concerns the value of the characteristic squirt flow length. However, in practice, this is found by matching model outputs to velocity or attenuation measurements at a given frequency (Dvorkin and Nur, 1993). The resultant calibrated model can then be used to explore frequency- and viscosity-dependent velocity and attenuation effects for a given lithology.

After determining the BISQ squirt flow lengths for the three sandstones reported here, there was reasonable agreement between predicted and observed attenuation magnitudes. The BISQ model was then used to predict the likely sonic and seismic attenuation magnitudes for each sandstone as a function of pore fluid viscosity. The results indicate that significant attenuation changes due to thermal stimulation may only be detectable in clay-rich sandstones with low permeabilities at seismic and sonic logging frequencies.

---

[1]University of Southampton Waterfront Campus, The National Oceanography Centre, Southampton, United Kingdom

Table 1. Average lithological properties of the sandstone samples used in the ultrasonic experiments.

Parameter	Elgin	Berea	Northern North Sea
Porosity (%)	10.3 ± 1.9	20.5 ± 1.9	14.8 ± 1.1
Permeability (mD)	152 ± 53	519 ± 93	5.0 ± 1.3
Clay content (%)	2.7	7.4	14.1
Frame bulk modulus (GPa)	25.92 ± 1.12	15.35 ± 1.85	20.87 ± 2.75
Frame shear modulus (GPa)	25.80 ± 0.94	13.67 ± 0.70	17.93 ± 1.29
Mineral bulk modulus (GPa)	36.5	36.5	36.5
Mineral density (kg·m^{-3})	2632	2634	2645

Laboratory Measurements

Laboratory experiments were conducted on seven different reservoir sandstones using the ultrasonic pulse-echo method (Winkler and Plona, 1982; Best, 1992; McCann and Sothcott, 1992). Recordings of arrival times and amplitudes of tone burst pulses (approximately 800 kHz) reflected from the Perspex/sample interfaces were used to calculate velocity (±0.3%) and attenuation coefficient (±20 dB/m) from which quality factor Q and its inverse Q^{-1} were derived. For each lithology, six rock samples (5 cm diameter, 2–3 cm long) were saturated with a different viscosity pore fluid, giving P-wave velocity (V_P) and attenuation (Q_P^{-1}) results over a viscosity range of 0.3–1000 cP. The results, reported in Best and McCann (1995) for a differential pressure of 50 MPa, showed complex interactions with pore fluid viscosity and lithology (porosity, permeability, clay content). There appeared to be evidence for two competing attenuation mechanisms; namely, global and squirt viscous fluid flow. Global flow, as described by the Biot model (Biot, 1956a; Biot, 1956b), seemed to predict the attenuations of high permeability sandstones over only part of the viscosity range, and not at all in low-permeability sandstones. The relatively high attenuations seen in the clay-rich sandstones not predicted by the Biot theory were thought to be indicative of a clay-related squirt flow loss mechanism. Squirt flow losses are caused by local fluid flow in and out of compliant pores (Mavko and Nur, 1975; Mavko and Nur, 1979; Palmer and Traviolia, 1980; Murphy III et al., 1986). The observed increases in velocity with viscosity in clay-rich sandstones were consistent with a squirt flow mechanism. Similar P-wave velocity V_P and attenuation Q_P^{-1} results for a differential pressure of 60 MPa are presented in Figure 1 with average rock and pore fluid properties given in Table 1 and 2, respectively.

Figure 1a shows a net increase in V_P with viscosity from 0.3 to 1000 cP for all three sandstones, most notable for the Northern North Sea sandstone. A small drop in velocity between 0.3 and 1 cP is seen for the Elgin sandstone but not for the other two sandstones; this is probably attributable to a reasonably large jump in rock sample porosity from 7.2% to 11.9% (see Best and McCann, 1995). The Elgin sandstone also shows a small decrease in velocity above approximately 60 cP, whereas this occurs above approximately 300 cP for the Berea and North Sea sandstones. Taking net velocity dispersions between 1 and 1000 cP gives values of 3.0% for Elgin, 2.9% for Berea, and 5.8% for the Northern North Sea sandstone. The direction of the velocity dispersion is consistent with a squirt flow type mechanism (see Mavko et al., 1998).

Figure 1b shows complex variations in Q_P^{-1} with similar shaped curves for the Berea and North Sea sandstones indicating possible attenuation peaks at approximately 1, 60, and 800 cP. By contrast, only one attenuation peak is seen for the Elgin sandstone at approximately 330 cP. The attenuation peaks have quite high magnitudes of Q_P^{-1} values greater than 0.02, equivalent to a quality factor Q of less than 50.

Comparison with BISQ Theory

A unified theory of the Biot and squirt flow models was introduced by Dvorkin and Nur (1993) for P-waves in rocks at high pressures (no microcrack dependence). This so-called BISQ model accounts for the interaction of the two mechanisms as a function of frequency and pore fluid viscosity among other rock input parameters (porosity, permeability, etc.). If this model can explain the experimental observations discussed above, then it could also be used to predict the behavior of these and similar reservoir rocks at seismic and sonic frequencies during thermal recovery of heavy oil.

Because different samples were used for each pore fluid, the corresponding porosity, permeability, and dry frame moduli were used for each rock sample in the BISQ model. The dry frame moduli were calculated using Gassmann's model (Gassmann, 1951) by matching the Gassmann saturated

Chapter 7: Prediction of Pore Fluid Viscosity Effects on P-Wave Attenuation

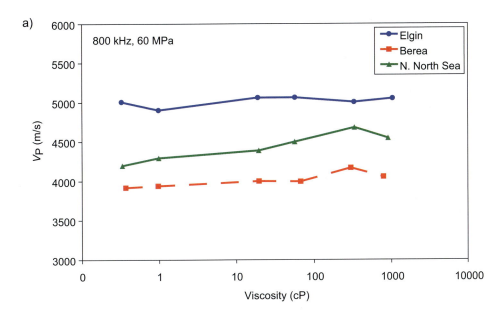

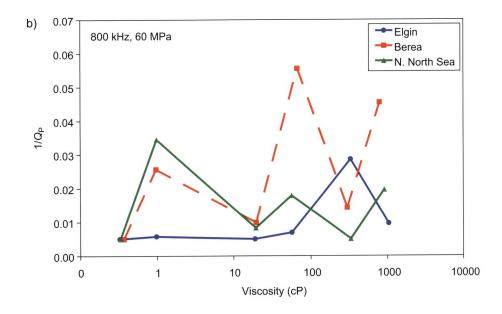

Figure 1. Ultrasonic (a) P-wave velocity V_P and (b) attenuation Q_P^{-1} as a function of pore fluid viscosity at a differential pressure of 60 MPa for three reservoir sandstones: Elgin, Berea, and Northern North Sea.

velocities (V_P, V_S) to the observed saturated velocities. The temperature varied between experiments, so the pore fluid properties were calculated for the correct experimental temperature of each rock. Because the pore fluid viscosity was changed by using different pore fluids (hexane, distilled water, and four different strength solutions of glycerol/water; see Table 2), it is possible that each fluid will have different chemical softening effects on the rock, but this was not quantified.

The BISQ model requires knowledge of the squirt flow length R (mm). In practice, this was achieved by matching the BISQ velocity to the observed velocity at 800 kHz by varying the squirt flow length for each sample/pore fluid combination. The resulting squirt flow lengths are shown in Figure 2, plotted against fluid mobility $M = permeability/viscosity$ (mD/cP) (see Batzle et al., 2006a). In general, squirt flow length decreases with increasing fluid mobility for each sandstone, changing by three orders of magnitude for these rocks. All squirt flow lengths fall on the same trend apart from the two highest fluid mobilities for Berea; this suggests a certain degree of consistency between different lithologies with a common squirt flow loss mechanism, lending weight to the theory.

Results

The BISQ predicted attenuations (Q_P^{-1}) are shown for each of the three rock types in Figures 3–5 as a function of pore fluid viscosity at 800 kHz together with the experimental values. The experimental results indicate relaxation peaks within the measured range of viscosities (0.3–1000 cP). This is clearest for Elgin sandstone, in which a single smooth peak is seen at approximately 300 cP. The experimental results for Berea and North Sea sandstones are more complex because, in addition to dominant peaks in Q_P^{-1} at approximately 70 and 1 cP, respectively, the Q_P^{-1} values oscillate between successive high and low values at other viscosities in Figures 4 and 5. The question is: does the BISQ model capture the essential features of the experimental attenuation observations in these typical reservoir rocks?

In fact, the range of magnitudes of the Q_P^{-1} values at 800 kHz predicted by the BISQ model is very close to the observed range over the viscosity interval for all three rocks. This is a significant point because observed attenuation magnitudes are notoriously difficult to predict. Furthermore, differences between the model and observations seem to be restricted to the position of the main attenuation peak, the height of the side lobes, and the lack of any secondary attenuation peaks. This could be a result of incorrect selection of the dominant squirt flow length (unlikely given that the squirt flow length was matched to the experimental velocities) or failure of the model to account for the interaction between multiple squirt flow lengths acting at the same time.

Another observation that may not be coincidence is that the viscosity offset between the model and experimental attenuation peaks varies systematically with clay content. There is good agreement for the North Sea sandstone (clay content 14.1%), underprediction by approximately 30 cP for the Berea sandstone (7.4%), and by approximately 300 cP for the Elgin sandstone (2.7%). This assumes that the largest attenuation peak seen in the experimental data corresponds to the peak predicted by BISQ in this viscosity range (BISQ predicts two peaks for a single squirt flow length: one due to squirt flow and the other to Biot global flow). It is also possible that the BISQ model predictions correspond to one of the other attenuation peaks seen in the Berea data (e.g., with similar magnitudes). These results lend weight to the previous assertion that clay-related squirt flow is an important loss mechanism at these frequencies. For example, the interaction between fluid flowing between compliant porosity associated with grain contacts and fluid flowing from compliant clay pores could explain the possibility of multiple squirt flow lengths. However, more experimental data points at intermediate viscosities are required to verify this.

Implications for Seismic Monitoring of Heavy-oil Reservoirs

Whatever the details of attenuation behavior at ultrasonic frequencies, the value of models such as the BISQ

Table 2. Pore fluid properties at 20°C.

Fluid Parameter	Viscosity (cP)	Density (kg/m^3)	Bulk Modulus (GPa)
Hexane	0.33	658	0.75
Water	1.0	998	2.19
Glycerol/water 3	23	1183	3.95
Glycerol/water 4	74	1212	4.30
Glycerol/water 2	456	1246	4.72
Glycerol/water 1	943	1256	4.85

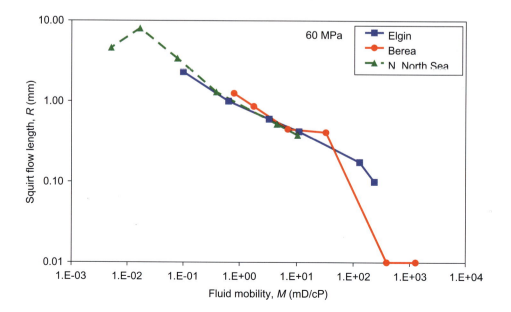

Figure 2. Squirt flow length R against fluid mobility M (note log-log scale).

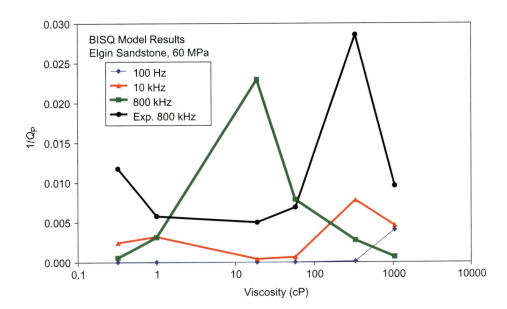

Figure 3. Elgin sandstone BISQ model predictions for P-wave attenuation Q_P^{-1} against pore fluid viscosity at frequencies of 100 Hz, 10 kHz, and 800 kHz. Experimental data at 800 kHz are indicated.

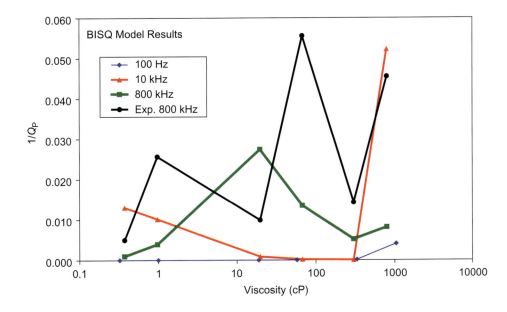

Figure 4. Berea sandstone BISQ model predictions for P-wave attenuation Q_P^{-1} against pore fluid viscosity at frequencies of 100 Hz, 10 kHz, and 800 kHz. Experimental data at 800 kHz are indicated.

model is their ability to accurately predict changes in attenuation at seismic (100 Hz) and sonic logging (10 kHz) frequencies (Dvorkin and Nur, 1993). Although further experimental studies are required to validate the BISQ model over the full frequency range of interest, the generally favorable comparison with the ultrasonic data above suggests it can be used with some confidence at lower frequencies.

Heavy oils typically have American Petroleum Institute (API) gravities less than 20°, with viscosities that range from greater than 10,000 cP at 0°C to less than 2 cP at 200°C (using equations given in Batzle et al., 2006b). Although heavy oils themselves have their own significant attenuation-viscosity behavior (Batzle et al., 2006a), it is worth speculating on the effect on reservoir monitoring of the squirt flow mechanism in the absence of intrinsic fluid attenuation. Hence, the following results may be more appropriate for intermediate API gravity oils in which intrinsic fluid attenuation is negligible. Again, further experimental studies are needed to quantify the actual effect of thermal stimulation on real heavy-oil reservoirs.

Figures 3–5 show the BISQ model results for 100 Hz and 10 kHz for fully saturated rocks. For these rocks at 100 Hz, no significant changes in Q_P^{-1} over the viscosity range of 1–10,000 cP would be seen; the maximum value of Q_P^{-1} is approximately 0.01 ($Q_P = 100$) for the North Sea sandstone, which would be barely detectable on reflection seismic data. However, larger changes in Q_P^{-1} could be expected at higher viscosities for all rocks. By

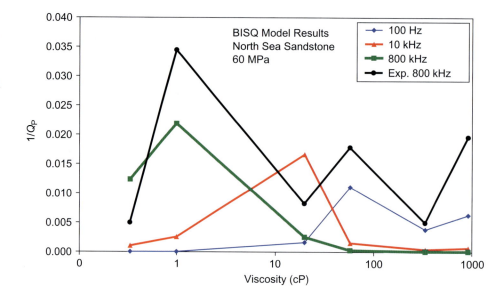

Figure 5. North Sea sandstone BISQ model predictions for P-wave attenuation Q_P^{-1} against pore fluid viscosity at frequencies of 100 Hz, 10 kHz, and 800 kHz. Experimental data at 800 kHz are indicated.

contrast, significant and detectable variations in P-wave attenuation ($Q_P^{-1} > 0.01$) would be observed at 10 kHz on wireline logs in North Sea- and Berea-type rocks over this viscosity range. This could have important implications for tying in 4D seismic survey data to wells if this frequency dependence were not taken into account.

There is one unexpected outcome from this exercise: the results indicate possibly larger attenuation variations with viscosity in the seismic and sonic ranges in reservoir rocks with significant clay contents such as the North Sea sandstone than in clean reservoir rocks such as the Elgin sandstone.

Conclusions

Laboratory ultrasonic measurements of P-wave velocity and attenuation were obtained on three reservoir sandstones at 60 MPa (Elgin, Berea, and North Sea) at six different pore fluid viscosities in the range of 0.3–1000 cP (only homogeneous saturated rocks considered). The results showed complex variations of Q_P^{-1} with viscosity. The unified Biot and squirt flow theory of Dvorkin and Nur (1993), known as the BISQ model, was able to predict the range of Q_P^{-1} magnitudes seen in each sandstone over the experimental viscosity range. There were some discrepancies, in particular, differences in the viscosity of the main attenuation peaks for the Elgin and Berea sandstones, although the BISQ model predicted the right viscosity for the North Sea sandstone. There was some evidence for multiple squirt flow lengths in the Berea and North Sea sandstones, perhaps related to the presence of significant levels of intrapore clay minerals (7.4% and 14.1%, respectively).

Extension of these ultrasonic-frequency (800 kHz) results to possible reservoir monitoring scenarios at 100 Hz (4D seismics) and 10 kHz (sonic well logs) showed a negligible effect of viscosity (in the range of 0.3–1000 cP, corresponding to the lighter end of the heavy-oil spectrum) on Q_P^{-1} for Elgin and Berea sandstones. However, the BISQ-predicted changes in Q_P^{-1} with viscosity for the North Sea sandstone were significant and most probably detectable on field seismic/sonic data. The BISQ results for all three sandstones suggest significant changes in Q_P^{-1} at viscosities greater than 1000 cP (corresponding to heavy oils and tars).

More experimental data are needed to validate the BISQ (and other possible) models over the range of viscosities of heavy oils seen during thermal recovery and for different reservoir lithologies (including carbonates and sands). Nevertheless, these preliminary results indicate that seismic attenuation could become an important reservoir monitoring parameter for detecting changes in pore fluid viscosity in the future. The most important result of this study is that Q_P^{-1} in clay-rich reservoir rocks could be more sensitive to viscosity changes than Q_P^{-1} in clean reservoir rocks with implications for detecting low-permeability (clay-rich) zones.

Acknowledgments

Funding was provided by the U. K.'s Natural Environment Research Council. Angus Best collected the experimental data during his Ph.D. studies at the University of Reading, United Kingdom, between 1989 and 1992.

References

Batzle, M. L., D. H. Han, and R. Hofmann, 2006a, Fluid mobility and frequency-dependent seismic velocity — direct measurements: Geophysics, **71**, no. 1, N1–N9.

Batzle, M. L., R. Hofmann, and D. H. Han, 2006b, Heavy oils — seismic properties: The Leading Edge, **25**, 750–756.

Best, A. I., 1992, The prediction of the reservoir properties of sedimentary rocks from seismic measurements. Ph.D. thesis, University of Reading.

Best, A. I., and C. McCann, 1995, Seismic attenuation and pore-fluid viscosity in clay-rich reservoir sandstones: Geophysics, **60**, 1386–1397.

Best, A. I., C. McCann, and J. Sothcott, 1994, The relationships between the velocities, attenuations and petrophysical properties of reservoir sedimentary rocks: Geophysical Prospecting, **42**, 151–178.

Biot, M. A., 1956a, Theory of propagation of elastic waves in a fluid-saturated porous solid: I. Low-frequency range: Journal of the Acoustical Society of America, **28**, 168–178.

Biot, M. A., 1956b, Theory of propagation of elastic waves in a fluid-saturated porous solid. II. Higher frequency range: Journal of the Acoustical Society of America, **28**, 179–191.

Dvorkin, J., R. Nolen-Hoeksema, and A. Nur, 1994, The squirt-flow mechanism: macroscopic description: Geophysics, **59**, 428–438.

Dvorkin, J., and A. Nur, 1993, Dynamic poro-elasticity: a unified model with the squirt and the Biot mechanisms: Geophysics, **58**, 523–533.

Gassmann, F., 1951, Elastic waves through a packing of spheres: Geophysics, **16**: 673–685.

Klimentos, T., and C. McCann, 1990, Relationships between compressional wave attenuation, porosity, clay content, and permeability of sandstones: Geophysics, **55**, 998–1014.

Mavko, G., T. Mukerji, and J. Dvorkin, 1998, The rock physics handbook: tools for seismic analysis in porous media: Cambridge University Press.

Mavko, G., and A. Nur, 1975, Melt squirt in the asthenosphere: Journal of Geophysical Research, **80**, 1444–1448.

Mavko, G., and A. Nur, 1979, Wave attenuation in partially saturated rocks: Geophysics, **44**, 161–178.

McCann, C., and J. Sothcott, 1992, Laboratory measurements of the seismic properties of sedimentary rocks, in A. Hurst, P. F. Worthington, and C. M. Griffiths, eds., Geological applications of wireline logs 2. Special Publication of the Geological Society of London 65, The Geological Society, 285–297.

Murphy III, W. F., K. W. Winkler, and R. L. Kleinberg, 1986, Acoustic relaxations in sedimentary rocks: dependence on grain contacts and fluid saturation: Geophysics, **51**, 757–766.

Palmer, I. D., and M. L. Traviolia, 1980, Attenuation by squirt flow in undersaturated gas sands: Geophysics, **45**, 1780–1792.

Winkler, K. W., and T. J. Plona, 1982, Technique for measuring ultrasonic velocity and attenuation spectra in rocks under pressure: Journal of Geophysical Research, **87**, 10776–10780.

Chapter 8

Elastic Property Changes in a Bitumen Reservoir during Steam Injection

Ayato Kato,[1] Shigenobu Onozuka,[2] and Toru Nakayama[3]

Introduction

The Hangingstone steam-assisted gravity drainage (SAGD) operation of Japan Canada Oil Sands Limited (JACOS) is approximately 50 km south-southwest of Fort McMurray in northern Alberta, Canada. JACOS started the operation in 1997 and has produced bitumen since 1999. The oil-sands reservoirs in Hangingstone occur in the Lower Cretaceous McMurray formation and are approximately 300 m in depth. The sedimentary environments are fluvial to upper estuarine channel fill, and the oil-sands reservoirs correspond to vertically stacked incised valley fill with very complex vertical and horizontal distribution.

A time-lapse seismic survey was conducted to monitor the steam movement. Identical processing was used for the 2002 baseline survey and the 2006 monitoring survey. A related chapter by Nakayama et al. in this book investigates differences between the data sets in the reservoir zone and shows significant difference in the seismic response around the production areas where steam was injected. Figure 1 shows seismic sections from both surveys. The large difference in the seismic response and time delay are interpreted as effects of steam injection.

We measured ultrasonic P- and S-wave velocities on core samples from the oil-sands reservoir. The oil sands are easily collapsed at room temperature because of poorly consolidated packing. The cores were frozen at the well site so that the higher viscous bitumen strengthened the packing. X-ray tomography (CT) images were acquired to investigate structures within the cores and detect weak portions for plugging. The oil sands are high-porous clean sands with small amounts of clay. We cut the cores into blocks and carefully whittled four plug samples from the blocks by a lathe with liquid nitrogen as the cutting fluid. The plug samples were trimmed to be held firmly with transducers. The four plug samples were 1.5 inches (3.81 cm) in diameter and approximately 1 inch (2.54 cm) in length (Table 1).

During the ultrasonic measurement, the plug sample was pushed into a rubber jacket and mounted by two pedestals on which P- and S-wave piezoelectric transducers were oppositely attached. A unit of the two pedestals sandwiching the plug sample was loaded into a pressure vessel that can separately control temperature and pressure. P- and S-waves with 0.5-MHz frequency were propagated through the sample and recorded. Finally, the velocities were calculated from the sample length and the propagation time with correction for the system delay.

Pressure and Temperature Dependence

The P- and S-wave velocities in the oil sands were measured under several pressure conditions at a constant temperature of 10°C, which nearly corresponds to the reservoir temperature before the steam injection. Figure 2 shows the velocities as a function of differential pressure; that is, confining pressure (900 psi or 6205 kPa)) minus pore pressure. The P- and S-wave velocities gradually decrease with decreasing differential pressure. Using the natural logarithm as a fitting curve, we obtained a relationship between the velocities and pressure:

$$V_P = 0.0593^3 \cdot Log(900 - P_{pore}) - 0.375 + V_{P0}$$
$$V_S = 0.0780^3 \cdot Log(900 - P_{pore}) - 0.495 + V_{S0} \quad (1)$$

where P_{pore} is pore pressure (psi); V_P and V_S are P- and S-wave velocities (km/s), respectively; and V_{P0} and V_{S0}

---

[1]University of Houston, Department of Earth and Atmospheric Sciences, Houston, Texas, U.S.A.
[2]JOGMEC, Chiba, Japan
[3]JAPEX, Tokyo, Japan
This paper appeared in the September 2008 issue of THE LEADING EDGE and has been edited for inclusion in this volume.

Figure 1. Example of seismic sections with interpreted horizons. Six horizons including Top Devonian, Top Wabiskaw, and other seismic events are shown. Top Devonian is regarded as the reservoir bottom (Base McMurray), and Top Wabiskaw is approximately 5 m shallower than the reservoir top (Top McMurray).

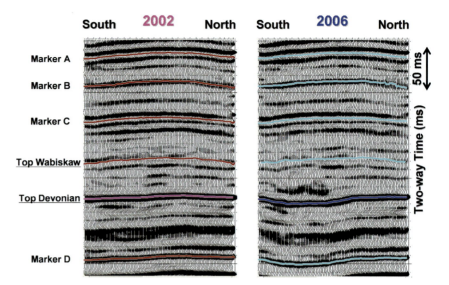

Table 1. List of plug samples of the oil sands.

	Length (inch)	Diameter (inch)	Weight (g)	Bulk Density (g/cc)
#2	0.88	1.51	48.27	1.86
#3	1.04	1.49	56.43	1.90
#7	1.21	1.50	63.81	1.82
#10	0.89	1.49	41.99	1.65

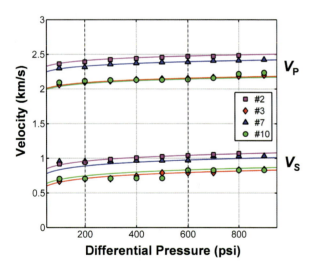

Figure 2. P- and S-wave velocities of the oil sands as a function of differential pressure at a constant temperature of 10°C. The solid lines represent velocities calculated by equation 1. The numbers (2, 3, 7, and 10) represent samples of the oil-sands core.

are P- and S-wave velocities (km/s) at the initial conditions (pore pressure of 300 psi (2068 kPa) and temperature of 10°C). The solid lines in Figure 2 represent the velocities calculated by equation 1. The correlation coefficients between measured and calculated V_P and V_S are 0.99 and 0.95, respectively. The increase of pore pressure (300 to 700 psi [2068 kPa to 4826 kPa]) induced by the steam injection causes a decrease in velocity of 65 m/s for V_P and 86 m/s for V_S.

Figure 3 shows the P- and S-wave velocities as a function of temperature at a constant pore pressure of 700 psi and confining pressure of 900 psi. A slope of the velocities to temperature significantly changes at approximately 30°C. Whereas the P- and S-wave velocities steeply decrease at the lower temperatures, P-wave velocity more gently decreases and S-wave velocity nearly remains constant at the higher temperatures. In the lower temperature range, Han et al. (2008) pointed out that bitumen filling in the pores works as a quasi-solid to stiffen the rock grain frame moduli. We divided velocities at a given temperature by the initial velocity at 10°C to obtain normalized velocities (Figure 4). The normalized velocities were fitted with two lines that connect at 30°C.

T < 30 °C

$$V_P/V_{P1} = -0.00550 \cdot T + 1.06$$
$$V_S/V_{S1} = -0.0190 \cdot T + 1.19$$

(2)

T ≥ 30 °C

$$V_P/V_{P1} = -0.00168 \cdot T + 0.940$$
$$V_S/V_{S1} = -0.000640 \cdot T + 0.639$$

where T is temperature (°C) and V_{P1} and V_{S1} are P- and S-wave velocities (km/s), respectively, at a temperature of 10°C and pore pressure of 700 psi. The black solid lines in Figure 4 represent equation 2. The correlation coefficients between measured and calculated V_P and V_S are

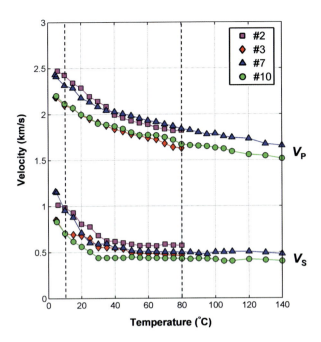

Figure 3. P- and S-wave velocities of the oil sands as a function of temperature at a pore pressure of 700 psi and a confining pressure of 900 psi.

0.97 and 0.94, respectively. In accordance with equation 2, the normalized P- and S-wave velocities at 80°C are 0.89 (V_P decrease of 11% compared with V_{P1}) and 0.62 (V_S decrease of 38% compared to V_{S1}), respectively.

Application of the Gassmann Equation

Can the Gassmann equation properly predict the changes in oil-sands velocity induced by the steam injection? Or at what temperature range can it work? One implicit assumption in the Gassmann equation is that pore fluids have no rigidity. Han et al. (2008) measured ultrasonic velocities on several bitumen samples in a wide temperature range and observed that the bitumen properties are similar to conventional oils at temperatures higher than its liquid point. Dry frame moduli were calculated from the ultrasonic laboratory measurement data on the basis of the Gassmann equation to investigate its validity for predicting velocity changes caused by the steam injection. In the calculation, we assumed that the pore fluids have no rigidity. Porosity and water saturation were determined to be 37% and 19.6%, respectively, by petrophysical analysis of well logs. Mineral bulk modulus was assumed to be 35.5 GPa, taking into account the small amounts of clay. Bulk modulus and density of the pore fluids were calculated by the FLAG program, which had been developed by the Fluids/DHI consortium for precisely estimating fluid properties at in situ conditions. Figure 5 shows the calculated dry frame moduli for

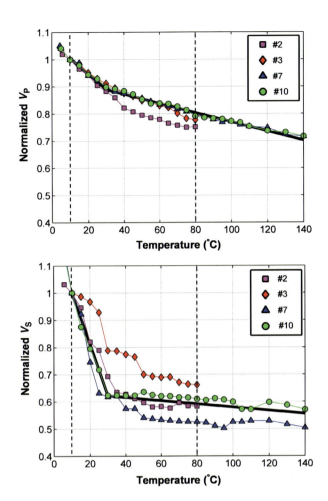

Figure 4. Normalized P- and S-wave velocities of the oil sands as a function of temperature at a pore pressure of 700 psi and a confining pressure of 900 psi. Black solid lines represent equation 2.

oil-sands sample 7 at a pore pressure of 700 psi. The calculated shear modulus rapidly decreases with increasing temperature from 10°C to 40°C, and then gradually decreases from 40°C to 80°C. The bulk modulus obviously decreases at lower temperatures than 80°C. The significant depression of the dry frame moduli at the lower temperatures can be interpreted as (1) the bitumen with higher viscosity stiffens the grain contacts; (2) the Gassmann equation fails because the bitumen has substantial rigidity; or (3) there is an effect of velocity dispersion between the low frequency of the Gassmann equation and the high frequency of the ultrasonic measurement. On the other hand, the dry frame moduli do not deviate largely from one value (bulk and shear moduli are approximately 0.80 and 0.46 GPa, respectively) at temperatures greater than 80°C. The unvaried dry frame moduli implies that the Gassmann equation would be applicable for predicting velocity changes at the higher temperatures at which the bitumen neither stiffens the grain contact nor has rigidity and the velocity dispersion is small enough to be neglected.

To justify the above hypothesis, we calculated velocity changes caused by fluid property changes on the basis of the Gassmann equation. In the calculation, dry frame moduli are assumed constant (although temperature varies) and are calculated from ultrasonic laboratory measurements at 80°C on the basis of the Gassmann equation. For a given new temperature, the original pore fluids at 80°C are assumed to be replaced by new fluids that equilibrate with the new temperature. Figure 6 compares measured and calculated velocities. The calculations are fairly consistent with the measurements at temperatures higher than 80°C. This result encourages the use of the Gassmann equation for predicting velocity changes caused by any property changes in pore fluid (due to temperature changes as well as water saturation and phase changes) at the higher temperatures until the rock frame undergoes direct damage from the steam injection.

The dry rock frame is considered to have poorly consolidated grain packing. Thus, we calculated the theoretical dry frame moduli for an idealized spherical random grain packing and compared them with the calculated values by the Gassmann equation. In the theoretical calculation, the ideally smooth sphere case of the Walton model (Walton, 1987) was applied. Bulk and shear moduli of the grain material were assumed to be 35.6 and 39.2 GPa, respectively. Porosity and average number of contacts per grain (coordination number) were assumed to be 37% and 9.03, respectively. Figure 7 shows the dry frame moduli at the hydrostatic confining pressure of 200 psi. The dry frame bulk and shear moduli at 37% porosity are 0.78 and 0.47 GPa, respectively, and nearly consistent with the dry frame moduli at temperatures greater than 80°C calculated by the Gassmann equation.

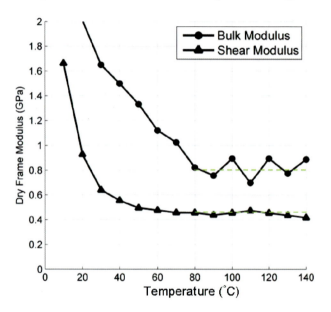

Figure 5. Dry frame moduli of the oil sands (sample 7) at a pore pressure of 700 psi calculated from the ultrasonic laboratory measurements on the basis of the Gassmann equation. Pore fluids are assumed to have no rigidity. Porosity, water saturation, and mineral bulk modulus are 37%, 19.6%, and 35.5 GPa, respectively.

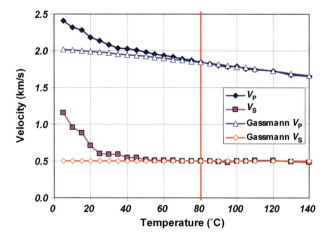

Figure 6. Comparison between the laboratory-measured velocities and calculated velocities by the Gassmann equation for the oil sands (sample 7). Porosity, water saturation, and mineral bulk modulus in the Gassmann equation are the same as Figure 5.

Sequential Rock Physics Model

We combined the laboratory measurement results to obtain a sequential rock physics model that can predict

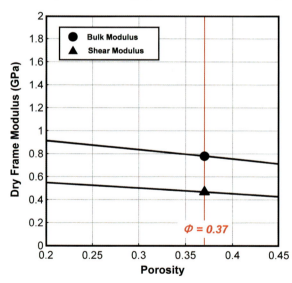

Figure 7. Theoretical dry frame moduli at a hydrostatic confining pressure of 200 psi (1379 kPa) for a poorly consolidated grain packing calculated by the Walton model (ideally smooth sphere case). In the calculation, bulk and shear moduli of grain material and coordination number are assumed to be 35.6 GPa, 39.2 GPa, and 9.03, respectively.

sequential velocity changes induced by the steam injection.

P_{pore} 300 psi → 700 psi (at $T = 10°C$)

$$V_P = 0.0593 \cdot Log(900 - P_{pore}) - 0.375 + V_{P0}$$
$$V_S = 0.0780 \cdot Log(900 - P_{pore}) - 0.495 + V_{S0} \quad (3)$$

T 10°C → 30°C (at $P_{pore} = 700$ psi)

$$V_P = (-0.00550 \cdot T + 1.06)V_{P1}$$
$$V_S = (-0.0190 \cdot T + 1.19)V_{S1} \quad (4)$$

T 30°C → 80°C (at $P_{pore} = 700$ psi)

$$V_P = (-0.00168 \cdot T + 0.940)V_{P1}$$
$$V_S = (-0.000640 \cdot T + 0.639)V_{S1} \quad (5)$$

and

$$T \geq 80°C \text{ (at } P_{pore} = 700 \text{ psi)} \quad (6)$$

where the Gassmann equation can predict velocity changes, and fluid properties are based on the FLAG program. V_{P1} and V_{S1} are P- and S-wave velocities (km/s), respectively, at a pore pressure of 700 psi and a temperature of 10°C and can be calculated from the first relations between the velocities and pressure (equation 3). Once steam is injected into the reservoir, the pressure front rapidly spreads to the periphery, and the temperature front follows it. The P- and S-wave velocities decrease because of the pore-pressure increase in a natural logarithm relationship with differential pressure (equation 3). After the pressure change, the velocities decrease with increasing temperature as a linear relation with a change in the slope at 30°C as shown in Figure 4 (equations 4 and 5). The velocities at temperatures greater than 80°C can be calculated via the Gassmann equation with the FLAG program (equation 6).

Velocity Dispersion

Significant frequency differences exist between ultrasonic laboratory measurements and surface seismic. Schmitt (1999) showed that sonic velocities in the heavy-oil zone are much higher than surface seismic frequency velocities collected by vertical seismic profile. We also observed significant velocity dispersion in this field (Figure 8). The seismic interval velocities were calculated from two-way time differences between two seismic horizons (Top Wabiskaw and Top Devonian in Figure 1) at well locations that nearly correspond to the reservoir top and bottom. The sonic interval velocities were calculated from time-depth relations of the corresponding reservoir interval. The interval velocities were not affected by the steam injection because the surface seismic and well log data used for the calculation were acquired before it. Figure 8 shows that the seismic velocities are lower than the sonic velocities. For the eight wells calculated, the interval velocity difference is 7.6% on average.

We propose a practical method for calibrating the velocity dispersion (Figure 9). As stated before, the Gassmann equation is applicable at temperatures greater than

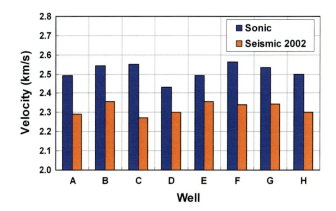

Figure 8. Interval velocity of the oil-sands reservoir at well locations from the surface seismic and the sonic log data. The seismic interval velocities were calculated from two-way time differences between two seismic horizons that nearly correspond to the reservoir top and bottom. The sonic interval velocities were calculated from time-depth relations of the corresponding reservoir interval. Letters A–H represent wells used for the calculation.

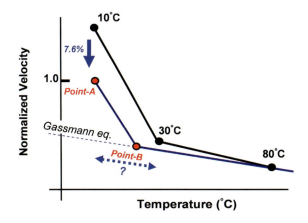

Figure 9. Illustration of a practical method for calibrating the velocity dispersion. Point B represents the temperature limitation of the Gassmann equation for the low-frequency band of surface seismic. Point A represents surface seismic velocity at initial conditions that are calculated from well sonic velocity with the average velocity dispersion. Point A and point B are transformed into normalized velocity and connected with each other by a line.

80°C for high-frequency ultrasonic laboratory measurements. When we apply the Gassmann equation to the surface seismic data, we assume that the temperature limitation of the Gassmann equation can extend to lower temperatures until the dry frame moduli change or the bitumen has substantial rigidity. In addition, we have the average value of the interval velocity difference between the surface seismic and the sonic data at the initial condition. The seismic velocity at the initial condition is estimated from the sonic velocity with the average velocity dispersion. We transform them into normalized velocities and connect the initial point (point A) with the temperature limitation of the Gassmann equation (point B) by a line in a fashion similar to Figure 4. However, we do not know the temperature limitation of the Gassmann equation. Fortunately, point B can only move in a small range from 10°C to 30°C. Assuming the temperature limitation is 25°C, we modified the coefficients associated with the temperature dependence in equation 4 and obtained a new sequential rock physics model that is adapted for the low-frequency band of the surface seismic.

P_{pore} 300 psi → 700 psi (at $T = 10°C$)

$$V_P = 0.0593 \cdot Log(900 - P_{pore}) - 0.375 + V_{P0}$$
$$V_S = 0.0780 \cdot Log(900 - P_{pore}) - 0.495 + V_{S0}$$
(7)

T 10°C → 25°C (at $P_{pore} = 700$ psi)

$$V_P = (-0.00433 \cdot T + 1.04)V_{P1}$$
$$V_S = (-0.0239 \cdot T + 1.24)V_{S1}$$
(8)

and

$$T \geq 25°C \ (at \ P_{pore} = 700 \ psi)$$
(9)

where the Gassmann equation can predict velocity changes and fluid properties are based on the FLAG program.

Sequential Elastic Property Changes

To understand elastic property changes of the oil-sands reservoir during the steam injection, we represented the sequential reservoir conditions using 23 steps (Figure 10). Pore-pressure changes occur in steps 1–5, and temperature changes occur in steps 5–23. In addition, adjacent to the injector well, the movable bitumen is largely replaced by hot water at step 18 and water phase changes from liquid to steam at step 21.

- *Step 1–5:* Pore pressure increases (from 300 to 700 psi).
- *Step 6–23:* Temperature increases (from 10°C to 300°C).
- *Step 18:* Bitumen is replaced by hot water at 200°C (water saturation S_w from 20% to 80%).
- *Step 21:* Phase changes from hot water to steam at 260°C.

Assuming that the sonic P- and S-wave velocities at the initial condition (step 0) are 2.4 and 1.0 km/s, respectively, the seismic velocities (step 1) are calculated from the sonic velocities with the average velocity dispersion. The P- and S-wave velocities slightly decrease during the pressure changes from step 1 to step 5. In the temperature increase from step 5 to step 8, the P- and S-wave velocities

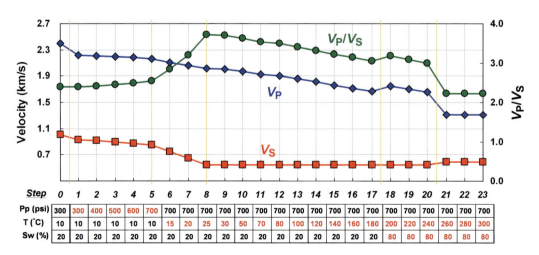

Figure 10. Sequential P- and S- wave velocities and V_P/V_S ratio changes induced by steam injection. Sequential reservoir condition changes are represented by 23 steps. Pore-pressure changes occur from step 1 to step 5, and temperature changes from step 5 to step 23. In addition, adjacent to the injector well, the movable bitumen is largely replaced by hot water at step 18 and water phase changes from liquid to steam at step 21. The sonic P- and S-wave velocities at the initial condition (step 0) are assumed to be 2.4 and 1.0 km/s, respectively.

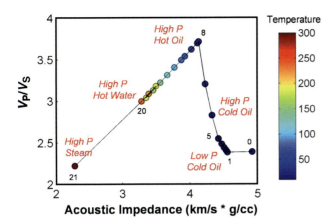

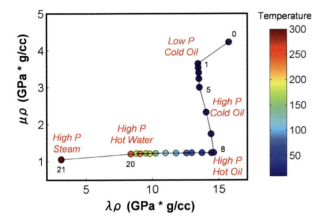

Figure 11. Sequential elastic property changes induced by the steam injection on the basis of the reservoir conditions of Figure 10. Crossplot between acoustic impedance and V_P/V_S ratio (top). Crossplot between $\lambda\rho$ and $\mu\rho$ (bottom). Numbers in the crossplots represent the step in Figure 10.

decrease and the V_P/V_S ratio significantly increases because the amount of decrease of the S-wave velocity is relatively larger than the P-wave velocity. After step 8 (25°C), the P-wave velocity continues to decrease until step 18, whereas the S-wave velocity is virtually constant (because the shear modulus is constant). At step 18 (200°C), at which the movable bitumen is largely replaced by the hot water (S_w from 20% to 80%), the P-wave velocity slightly increases because the hot water has a faster P-wave velocity than the bitumen. Finally, the water phase changes from liquid to steam at step 21 (260°C), leading to a significant P-wave velocity drop (whereas the S-wave velocity slightly increases because of density decrease).

Other elastic property changes of the oil sands can be calculated from the velocity and density changes. The velocity changes can be calculated by the rock physics model (equations 7–9), and density changes can be calculated by the FLAG program. Figure 11 shows examples of the sequential elastic property changes on the basis of the conditions of Figure 10. With increasing pore pressure (steps 1–5), acoustic impedance and $\mu\rho$ slightly decrease, and the V_P/V_S ratio slightly increases whereas $\lambda\rho$ is virtually constant. The temperature increase from 10°C to 25°C (steps 5–8) causes a significant change in the V_P/V_S ratio (increase) and $\mu\rho$ (decrease). At temperatures higher than 25°C, acoustic impedance, V_P/V_S ratio, and $\lambda\rho$ decrease, except in steps 18–20, whereas $\mu\rho$ is virtually constant. The water phase change from hot water to steam causes a significant decrease in acoustic impedance, V_P/V_S ratio, and $\lambda\rho$.

Conclusions

We measured and analyzed the ultrasonic velocities of the oil-sands cores acquired from the SAGD operations and constructed the sequential rock physics model. The practical method for calibrating the velocity dispersion was proposed, and the sequential model was modified to be adapted for the low-frequency band of the surface seismic. We predicted elastic property changes induced by the steam injection. The S-wave velocity can be useful for distinguishing a gently warmed area front because it significantly decreases at approximately 20°C. The V_P/V_S ratio more dramatically changes with the reservoir condition change. The V_P/V_S ratio significantly increases at the gently warmed area front and then significantly drops at the steam front.

Acknowledgments

This study was conducted as a joint project between JACOS and Japan Oil, Gas and Metals National Corporation (JOGMEC). We thank Japan Petroleum Exploration Company Limited (JAPEX), JACOS, and JOGMEC for permission to publish these data.

References

Han, D. H., J. Liu, and M. Batzle, 2008, Seismic properties of heavy oil — measured data: The Leading Edge, **27**, 1108–1115.

Schmitt, D. R., 1999, Seismic attributes for monitoring of a shallow heated heavy oil reservoir: Geophysics, **64**, 368–377.

Walton, K., 1987, The effective elastic moduli of a random packing of spheres: Journal of the Mechanics and Physics of Solids, **35**, 213–226.

Section 2

Geologic/Geophysical Characterization

Chapter 9

The Devonian Petroleum System of the Western Canada Sedimentary Basin — with Implications for Heavy-oil Reservoir Geology

Hans G. Machel[1]

Introduction

The Western Canada Sedimentary Basin is one of the most prolific and best researched petroliferous basins in the world. Hydrocarbons were first discovered in commercial quantities in the Mississippian Turner Valley field southwest of Calgary in 1914, but this field was relatively small and economically insignificant in the long run. The first prolific oil field was discovered in a Devonian reef in 1947 near the town of Leduc close to Edmonton. This discovery started an economic boom in the province of Alberta that is ongoing to this day, on the basis of oil and gas revenue. Although agriculture, forestry, and high-technology industries have considerably reduced the importance of petroleum revenue in the last 20 years, Alberta still derives approximately half (when oil prices are high, even more) of its gross domestic revenue from oil and gas.

Exploration spread to the neighboring provinces of Saskatchewan, Manitoba, and British Columbia in the 1950s and 1960s, when it became clear that oil and gas also occur in the eastern extensions of the Western Canada Sedimentary Basin. Today, after approximately 60 years of intensive exploration, during which very large petroleum resources were found also in various Mesozoic systems, Alberta remains the province with the most petroleum resources, although the emphasis has shifted from conventional to unconventional resources. Most commercial accumulations of hydrocarbons occur in two petroleum systems, the Devonian and the Cretaceous, with relatively minor accumulations in between (Figure 1). The Alberta Basin, which is the western part of the Western Canada Sedimentary Basin, contains approximately 8.2×10^9 m^3 (52×10^9 billion barrels [bbls]) of conventional initial oil in place (IOP), compared with 1.8×10^9 m^3 (12×10^9 bbls) conventional IOP in the Williston Basin, which is the eastern part of the Western Canada Sedimentary Basin (Figures 1–3). These conventional oil and gas resources, which have sustained the local economy and petroleum industry for approximately 50 years, are in rapid decline and now nearly exhausted (AEUB, 2007). Thus the unconventional resources are now of much greater importance, economically and geopolitically. The Alberta Basin hosts supergiant reserves of unconventional oil in Cretaceous sands and Devonian carbonates. The major three oil sand (also called tar sand) deposits alone (Athabasca, Cold Lake, Peace River) contain an estimated 267×10^9 m^3 of oil (Creaney et al., 1994; Hay, 1994; see Figure 1; AEUB Report ST98-2007 cites 270×10^9 m^3, which is within the margin of error). In addition, the underlying carbonates (mainly the Upper Devonian Grosmont Formation) host at least an additional 50×10^9 m^3 of oil (Creaney et al., 1994; Hay, 1994; see Figure 1), whereas AEUB Report ST98-2007 and Alvarez (2008) cite up to 71×10^9 m^3 of oil in the carbonates. There obviously is a much larger margin of error in the estimates of the carbonates, and future resource evaluations may well push the estimates even higher.

The Cretaceous oil sands have been under production and rendered profits for approximately 30–40 years, and they are the main driving force of the Alberta economy today. As of March 2007, 43 companies are known to operate 162 recovery schemes in the Alberta oil sands (Hein and Marsh, 2008). Countless engineering and geologic reports and scientific articles have been published on these deposits, including a series of geologic assessments. Succinct recent summaries are Stanton (2004), Ranger and Gingras (2006), and Hein and Marsh (2008).

[1] University of Alberta, Edmonton, Alberta, Canada

In contrast, the literature concerning the heavy-oil carbonate reservoirs in Alberta is fairly sparse, thematically much less comprehensive, and relatively dated. The most detailed studies published so far are from the early- to mid-1990s (Luo et al., 1994; Luo and Machel, 1995; Dembicki and Machel, 1996). One reason is that these carbonates are not yet under production because so far no commercially viable technology has been available. However, the acquisition of substantial land holdings in the Grosmont area by Shell International in the spring of 2006 has sparked strong interest in these heavy-oil carbonates by numerous companies, several of which are now actively pursuing resource assessments and studying reservoir properties, as reflected in recent conference presentations (Barrett et al., 2007, 2008; Alvarez et al., 2008).

Considering that the published literature on the Cretaceous heavy-oil sand deposits is exhaustive compared with that of the Devonian heavy-oil carbonates, and the strong recent interest in the latter, the objective of this chapter is to provide a relatively short but nevertheless comprehensive overview of the Devonian petroleum system in the Western Canada Sedimentary Basin, with emphasis on Alberta, where most of the oil and gas of this system are found. The Grosmont reservoir, which is by far the largest of the heavy-oil (tar, bitumen) carbonate reservoirs, is thus presented in the context of the light-oil, sweet and sour gas reservoirs of the entire Devonian petroleum system. This chapter necessarily begins with a general discussion of the basin evolution; that is, the three major factors that determined and controlled the reservoir characteristics in the Devonian petroleum system: structure in the Precambrian basement, stratigraphic development, and diagenesis (especially during intermediate to deep burial). Many of the characteristics of this system are typical for Devonian petroleum systems elsewhere in the world, such as in China. However, some features are quite unique to western Canada.

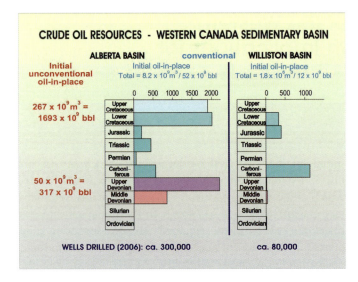

Figure 1. Initial crude oil-in-place in the Western Canada Sedimentary Basin, compiled from Creaney et al. (1994) and Hay (1994). The total numbers of wells drilled are from the Energy Utilities Board of the Alberta government updated from unpublished company reports in 2006. Most oil is located in Alberta. The unconventional resources (heavy oil, tar) exceed the conventional resources by a wide margin. Horizontal scales = billion barrels.

Figure 2. Tectonic (mega-) units of the Canadian Cordillera, also showing the location, outline, and size of the province of Alberta. The limit of the disturbed belt is the outer limit of the Laramide tectonic deformation, with tectonically undisturbed strata to the east. Map is modified from Price (1994).

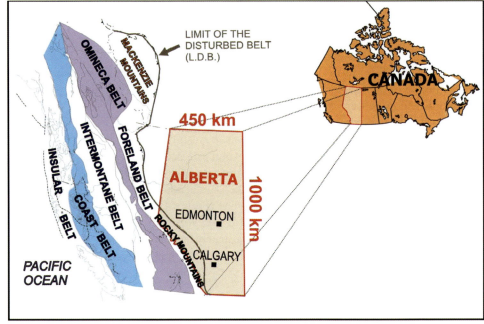

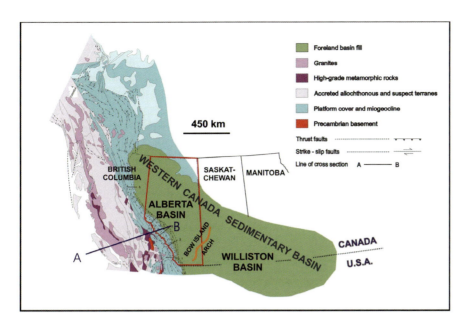

Figure 3. Location of the Western Canada Sedimentary Basin relative to the Canadian Cordillera. The Western Canada Sedimentary Basin stretches across several provinces and into the northern United States. It is divided into the Alberta Basin and the Williston Basin. The Alberta Basin forms the foreland basin of the Rocky Mountains. Cross section A–B: see Figure 4. Map is modified from Price (1994).

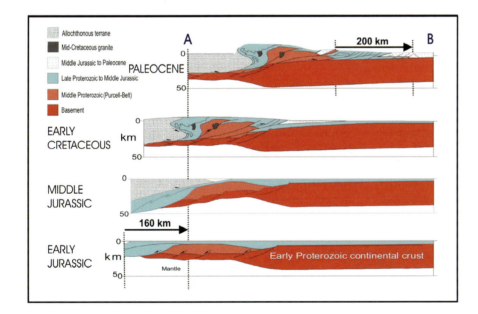

Figure 4. Cross sections A-B across the Canadian Cordillera into the western part of the Alberta Basin (see Figure 3 for location). The four sections show the deformation and approximate extent of tectonic shortening between the early Jurassic and the Paleocene. West-to-east compression shortened the Cordillera by approximately 160 km in the west and by approximately 200 km in the Rocky Mountains in the east. Map is modified from Price (1994).

Basin Evolution and Structure

The Western Canada Sedimentary Basin is a large, elongate feature that is located mainly in Alberta next to the Rocky Mountains and to a minor extent in the adjacent provinces of Saskatchewan and Manitoba to the east (with a minor excursion into the northern United States) and in British Columbia to the west and north (Figures 2 and 3). The Western Canada Sedimentary Basin is divided into the Alberta Basin and the Williston Basin, the border of which roughly coincides with the Bow Island Arch, which is a Precambrian structural high that crosses the Alberta-Saskatchewan border in the south (Figure 3).

The structural evolution of the Western Canada Sedimentary Basin is well known throughout the Phanerozoic and intimately related to the tectonic evolution of the Rocky Mountains (e.g., Price, 1994). The Devonian section was deposited on the passive margin of the ancestral North American continent. Four orogenies affected the region: Antler (Devonian–Carboniferous), Sonoma (Late Permian), Columbian (Jurassic–Early Cretaceous), and Laramide (Mid-Late Cretaceous–Tertiary). The Antler and the Laramide orogenies significantly shaped the size, depth, accommodation space, subsidence, and uplift of the Western Canada Sedimentary Basin (Figure 4). Today, the Alberta Basin constitutes the foreland basin of the Rocky Mountains. Because of the asymmetry of the Alberta

Basin, the Devonian strata form a basin-wide, structural homocline that gently dips from outcrops in northeastern Alberta to nearly 7 km in southwestern Alberta next to the Rocky Mountain fold and thrust belt.

Paleomagnetic, gravimetric, and seismic surveys have identified several structural highs (Figure 5) and two prominent fault systems that strike approximately northwest–southeast and south-southeast–north-northwest (Figure 6) in the Precambrian basement (Switzer et al., 1994; Edwards and Brown, 1995). These structural features controlled deposition of the Devonian sedimentary strata to some, albeit debatable, degree. For example, deposition of the elongate Rimbey-Meadowbrook reef trend was probably controlled by a lineament in the Precambrian basement, the Rimbey Arc (Figure 5). Furthermore, at least some of the many basement faults that are located in this region (Figure 6) appear to have been active during the Paleozoic and Mesozoic, influencing fluid flow and petroleum migration, as discussed further below.

Another important example for the influence of basement structure on the Devonian petroleum system is the Peace River Arch, which separates the Alberta Basin into a northern and a southern part (Figure 5), each having a different tectonic, depositional, geothermal, and hydrogeologic history. The Peace River Arch is a southwest–northeast oriented flexure in the Precambrian basement and formed an emergent landmass throughout much of the Paleozoic until it "collapsed" and was covered by marine strata in the Mississippian (e.g., Ross, 1990). The Peace River Arch controlled the deposition of an Upper Devonian fringing reef complex (Figure 7) and several other distinct sedimentologic and diagenetic-hydrothermal phenomena (e.g., Dix, 1993).

Detached from the orogenies that affected the entire region, the Peace River Arch area had its own tectonic and geothermal dynamics (Stephenson et al., 1989; Ross, 1990). Even today heat flow is elevated relative to normal and significantly higher north than south of the arch, which can be traced back to the Paleozoic, when regional easterly flow of formation fluids probably formed MVT mineralization at Pine Point (Qing and Mountjoy, 1992, 1994). There is no evidence for a corresponding regional flow system south of the Peach River Arch, and no comparable Pb-Zn sulfide mineralization. Furthermore, the general burial and geothermal histories appear to have differed significantly north and south of the Peace River Arch. North of the arch, and especially in the Liard (sub-) Basin area, a regional event of high heat flow with maximum temperatures between the end of the Paleozoic to the Jurassic caused oil maturation and migration in the Late Devonian to Carboniferous (Morrow et al., 1993). There is also circumstantial evidence for late Paleozoic large-scale hydrothermal fluid convection via faults up and through the Devonian carbonates when they were at very shallow burial depths (Morrow et al., 1993; Morrow and Aulstead, 1995). There probably were six distinct fluid events in the northernmost part of the Western Canada Sedimentary Basin (Morris and Nesbitt, 1998); that is, in the southern MacKenzie Mountains (see Figures 2 and 3). However, these events cannot be identified in or correlated with time-equivalent events south of the Peace River Arch, although there are some similarities. Rather, geothermal gradients in the Alberta Basin appear to have been near normal from the late Paleozoic to the Laramide orogeny, and then again after the orogeny, which led to oil and gas maturation and migration in the Late

Figure 5. Major structural elements that influenced Woodbend (D3) and Winterburn (D2) Group deposition. For orientation of map, locate contour of Alberta in Figure 3. Inset frame: see Figure 6. Map is modified from Switzer et al. (1994).

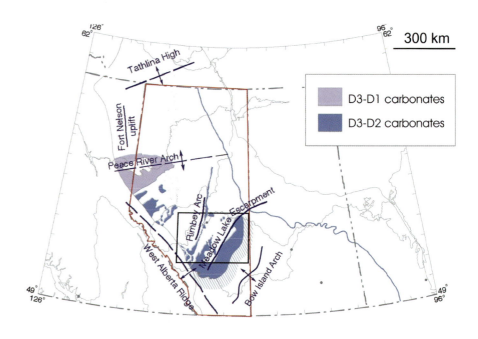

Chapter 9: *The Devonian Petroleum System of the Western Canada Sedimentary Basin*

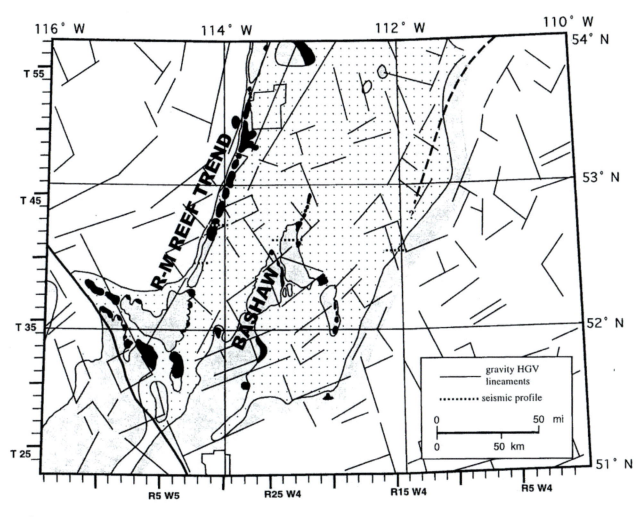

Figure 6. Basement gravity map of southeast Alberta (see inset frame in Figure 5 for location) showing lineaments and fault systems that influenced Woodbend (D3) and Winterburn (D2) Group deposition. The Rimbey-Meadowbrook (R-M) Reef trend and the Bashaw Reef Complex contain D3 and D2 oil and gas reservoirs (black). Map is modified from Edwards and Brown (1995).

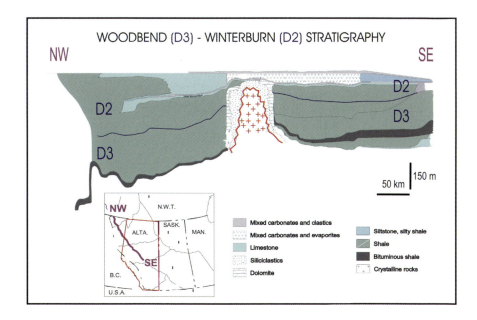

Figure 7. Woodbend and Winterburn Group stratigraphy in a northwest–southeast cross section across the northwestern parts of the Alberta basin (see inset map for location). This particular section crosses the crystalline Peace River Arch (see Figure 5), which effectively divided the Alberta Basin into a northern and a southern part until the early Carboniferous. Map is modified from Switzer et al. (1994).

Cretaceous to early Tertiary (Hacquebard, 1977; Stoakes and Creaney, 1984; Dawson and Kalkreuth, 1994).

Sedimentation and Facies

Deposition of the Devonian strata in the Western Canada Sedimentary Basin was cyclical and developed into four stratigraphic levels, commonly referred to as D1, D2, D3, and D4 (Figure 8). Each level contains carbonate aquifers (including reefal reservoir rocks), and shale, marl, or evaporite aquitards (including source rocks).

The oldest of these four stratigraphic levels contains the Middle Devonian Swan Hills reservoirs. The other three levels are Upper Devonian in age; that is, D3 = Woodbend Group (including reservoirs of the Leduc Formation), D2 = Winterburn Group (including reservoirs of the Nisku Formation), and D1 = Wabamun Group. On a basin-wide scale, most oil and gas are contained in the D2 and D3. Naturally, the overall fourfold cyclicity is also developed in the basinal strata, which contain four source rock intervals.

The depositional cyclicity was controlled mainly by eustatic sea level fluctuations. Thereby, the division of D1–D4 represents third-order eustatic sea level fluctuations, superimposed upon which are several fourth-order cycles that resulted in significant shifts of facies types and lithologic boundaries (Figures 8 and 9). As a result, reservoir levels such as the D3 are internally structured in many parts of the basin (Figure 10).

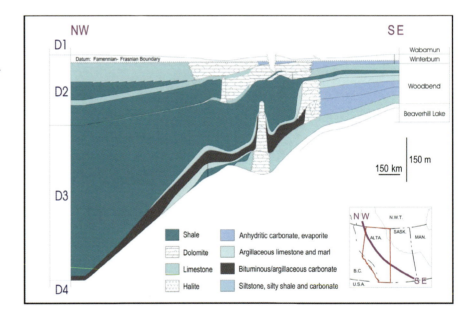

Figure 8. Devonian stratigraphy in a northwest–southeast cross section across almost the entire Western Canada Sedimentary Basin, bypassing the Peace River Arch (see inset map for location). This section shows the overall fourfold cyclicity of the Devonian system, with details in the D3 and D2. Map is modified from Switzer et al. (1994).

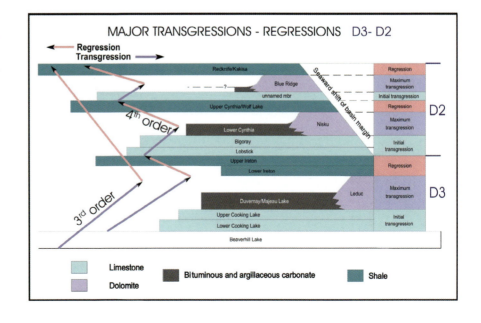

Figure 9. Major eustatic sea level changes and resulting stratigraphy of the Woodbend (D3) and Winterburn (D2) Groups in the Western Canada Sedimentary Basin. The D3–D2 represents a third-order cycle, superimposed upon which are several fourth-order cycles. Diagram is modified from Switzer et al. (1994).

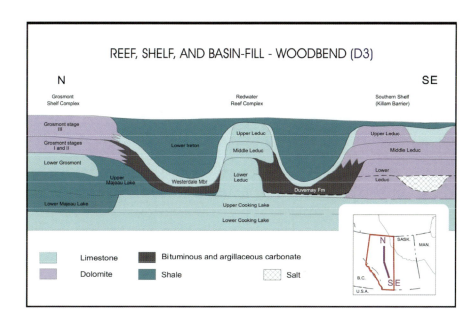

Figure 10. Stratigraphy and lithologies of the Woodbend Group in a schematic north–southeast cross section across eastern Alberta (see inset), showing differentiation within the D3 as a result of fourth-order sea level fluctuations. Diagram is modified from Switzer et al. (1994).

Many authors have studied the facies types and developments of the Devonian reservoir rocks and associated strata, and several facies models have been published. The last comprehensive model for Devonian reef facies deposition is that of Machel and Hunter (1994), which shows that the Devonian carbonates are shallow marine, tropical carbonates that commonly were deposited in two major geologic settings; that is, shallow shelf with shelf margin reef, and gently sloping ramp that grades shoreward into peritidal to supratidal and, in some places (especially the Williston Basin), into evaporitic facies.

Facies types correlate weakly with reservoir properties. Generally, reef and fore reef facies that had relatively large primary porosities and permeabilities form good reservoir rocks (e.g., Amthor et al., 1994). However, diagenesis, especially replacive dolomitization, significantly overprinted most reefs in the basin. Consequently, limestones that are still dominated by their primary porosity and permeability characteristics are relatively rare. Where present, limestones show moderate amounts of marine cements, in a few cases also minor amounts of vadose cements (Figure 11), and petroleum is often contained in remnant primary pores.

Diagenetic Evolution of Carbonates

The Devonian carbonates of the Western Canada Sedimentary Basin underwent a complex diagenetic evolution. Reef rocks commonly display evidence of 10–20 diagenetic processes that affected them from zero depth syndepositionally to burial depths of several kilometers and some 250–350 million years later (Figures 12 and 13). Although some of these diagenetic events are unique to certain locations, there is an astounding uniformity across the basin in the diagenetic evolution, which can be grouped into five major stages (Figure 13). With minor variations, these stages are valid for all Devonian platform and reef carbonates in the Alberta Basin south of the Peace River Arch. The most notable exceptions are reflux dolomitization and karstification, which affected only the easternmost and shallowest Devonian carbonates, including the Grosmont platform.

Stage 1A is syndepositional to shallow burial "early" diagenesis. This stage, which is best recognizable in the few carbonates that are undolomitized, formed various types of calcite cements (Figure 11b; Machel, 1990; Walls and Burrows, 1990). Stage 1 diagenesis generally degraded the reservoir rock quality (i.e., porosity and permeability).

Stage 1B is contemporary with Stage 1A but affected only a few very shallow to supratidal, slightly to moderately evaporative (up to gypsum saturation) environments. The most important example is the Grosmont platform near the northeastern limit of the basin (Huebscher, 1996), which was affected by reflux dolomitization.

Stage 2 is pervasive replacive dolomitization that affected approximately 85% of all of the Devonian carbonates in the Western Canada Sedimentary Basin. In this regard, the Western Canada Sedimentary Basin is unusual because most other Devonian carbonate provinces in the world contain significantly less and/or hardly any dolomite. Pervasive replacive dolomitization typically was matrix-selective, such that larger biochems and allochems remained undolomitized and/or were dissolved to leave molds and vugs (Figure 11c and 11d). Consequently, the resulting dolostones are generally the best reservoir rocks in the Western Canada Sedimentary Basin, although secondary anhydrite often plugs some of the dissolution porosity (Figure 11d). On the basis of textures, spatial relationships to facies and structure, relationships to

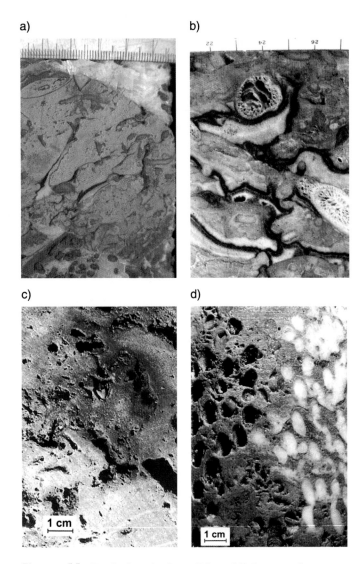

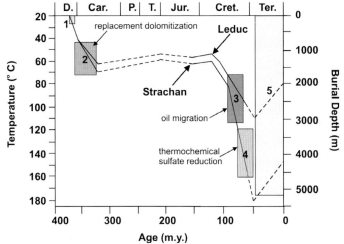

Figure 11. Typical rocks from D2 and D3 reservoirs. (a) Numerous tabular stromatoporoids deposited in carbonate mud, with considerable primary porosity in the matrix. (b) Corals (*Thamnopora*) and dendroid stromatoporoids (*Stachyodes*), and almost no primary mud. The interparticle pores are almost completely cemented with early diagenetic calcites; that is, isopachous marine cements (top half), banded vadose cements (center), and white, blocky cements (center bottom) that formed at relatively shallow burial. (c) Moldic-vuggy dolostone. Molds and vugs originated from dissolution of bioclasts (stromatoporoids, corals), with solution-enlargement of some molds. (d) Moldic dolostone with partial anhydrite cementation. The molds originated from dissolution of dendroid stromatoporoids, such as those shown in (b). Anhydrite cementation took place at depths of approximately 600–1000 m, postdating formation of the molds.

limestone diagenesis and stylolitization, stable isotopes, trace elements, $^{87}Sr/^{86}Sr$ ratios, associated sulfates, recrystallization, and fluid inclusion data, these matrix dolomites are interpreted to have formed during burial at depths of approximately 300–1500 m during the Late

Figure 12. Typical paragenetic sequence for Leduc (D3) reefs in the central to southern parts of the Rimbey-Meadowbrook reef trend.

Figure 13. Burial history diagram for the Rimbey-Meadowbrook reef trend, with an assumed geothermal gradient of 30°C/km. The two burial curves represent the shallow, northeast Leduc reef/pool and the deep, southwest Strachan reef/pool of the reef trend. Numbers 1–5 denote diagenetic stages. Especially noteworthy are the depth and time intervals of replacement dolomitization, oil migration, and TSR. Note that the Leduc reef was never buried deeply enough for TSR. Figure based on Amthor et al. (1993), Machel et al. (1994), and Mountjoy et al. (1999).

Devonian to early Carboniferous (Figure 13). Pervasive matrix dolomitization appears to have been a basin-wide phenomenon, and the dolomitizing solutions were probably diagenetically altered Devonian seawater (Amthor et al., 1993; Machel et al., 1994; Mountjoy et al., 1999).

The major events during Stage 3 were oil maturation and migration during the Late Cretaceous, when the Devonian source rocks underwent burial into and through the liquid oil window (Figure 13). The source rocks are further discussed in the following section.

Stage 4 took place during relatively rapid burial to maximum depths and was marked by thermochemical sulfate reduction (TSR) and associated diagenetic events in the deeper parts of the basin. Depending on several factors, the onset of TSR is constrained to minimum temperatures of approximately 100°C–140°C in general, and took place around 125°C–160°C in the Western Canada Sedimentary Basin (Figure 13). The major reaction products were sour gas, saddle dolomite, and sparry calcite cements, as well as several distinctive geochemical characteristics (Machel et al., 1995a, b; Machel, 1998). In the deepest parts of the basin, close to the limit of the disturbed belt, metamorphic-hydrothermal fluids were injected into some sour gas reservoirs during this stage (Machel and Cavell, 1999).

Stage 5 encompasses the period after maximum burial to the present (Figure 13). This stage left hardly any recognizable traces in most parts of the basin, except widely scattered and usually insignificant amounts of limpid calcite or dolomite cements. However, the Grosmont platform was pervasively karstified during this stage, and its oil was heavily to extremely biodegraded because of its contact with meteoric groundwater at the updip end of the basin (Huebscher and Machel, 1997a). However, the timing of this karstification is uncertain. It could have happened during Stage 5 or, given the location of the Grosmont platform at the northeastern margin of the basin, between Stages 3 and 5 (Huebscher, 1996; Huebscher and Machel, 1997a).

Source Rocks

Source rock facies have been identified at five stratigraphic levels in the Devonian petroleum system of the Western Canada Sedimentary Basin with the aid of stratigraphy and organic geochemistry (Creaney et al., 1994), i.e., one each in the D4 (called Keg River in northern Alberta and Brightholme in southern Saskatchewan), the D3 (Duvernay and/or Majeau Lake), and the D1 (Exshaw-Bakken/Lodgepole); and two in the D2 (Cynthia and an unnamed unit at the stratigraphic level of the Blue Ridge). The stratigraphic levels of the source rock facies of the D3 and D2 are shown in Figures 9 and 10.

All source rocks are marine shales and marls. On a basin-wide scale, the Duvernay/Majeau Lake of the D3 is by far the most prolific and the principal source for most of the Devonian oil and gas in the Western Canada Sedimentary Basin. The Keg River and Brightholme source rock facies of the D4, although richer than the Duvernay/Majeau Lake in parts of the basin, is of lesser importance in the basin as a whole because the D4 source rocks are restricted to two geographically relatively small areas. The remaining three source rocks of the Devonian system contributed overall fairly little to the oil and gas in the basin. For this reason, the following discussion centers on the D4 and the D3 source rocks.

The source rock facies of the D4 level is developed in a long band stretching across Alberta and Saskatchewan. However, this source rock facies is mature (equivalent to the liquid oil window) only in the northwesternmost corner of Alberta, where it is called the Keg River, and in the southeasternmost corner of Saskatchewan, where it is called the Brightholme (Figure 14). Total organic carbon (TOC) in the Keg River reaches up to 15%, and in the Brightholme up to 46%. Oil pools in D4 reservoirs sourced from these source rocks are spatially limited to the two areas of oil maturity, indicating short migration distances and local sourcing. Gas chromatography (GC) and GC-mass spectroscopy (MS) characteristics of these oils show them to be of highly anoxic and, in the case of the Rainbow pools, of hypersaline origin and nonbiodegraded (Creaney et al., 1994). Indeed, many of the Rainbow and Zama reservoirs in northwestern Alberta are surrounded and/or overlain by evaporites.

The Duvernay at the D3 level was deposited across almost the entire Western Canada Sedimentary Basin, with thicknesses ranging from approximately 20 to 160 m (Figure 15). TOC ranges up to 20%. The Duvernay is mature in a wide, north–south stretching band that crosses several large carbonate platforms and reef complexes (Figure 16). Not surprisingly, these platforms and reefs contain Duvernay-sourced oil and/or gas. GC and GC-MS characteristics of Duvernay-sourced oils show them to be of normal marine, anoxic origin, and nonbiodegraded (Creaney et al., 1994). However, in the northeasternmost corner of Alberta, where D3 and D2 carbonates of the Grosmont Formation subcrop or crop out, the Duvernay-sourced oil is biodegraded to heavy oil and/or tar (solid bitumen). Biodegradation is so severe that the source rock(s) cannot be unequivocally identified. Thus, is it open to speculation just how much of the heavy oil now pooled in the Grosmont Formation is derived from Devonian source rocks and/or from younger source rocks, especially the Mississippian Exshaw Formation which is the most prolific source rock in the entire basin (Creaney et al., 1994).

Hydrology and Migration

Migration patterns for aqueous and petroliferous formation fluids in the basin are and were complex. Although

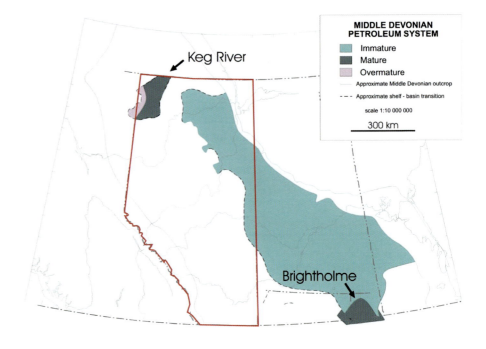

Figure 14. Map of geographic distribution and maturity of the Keg River (D4) source rock facies. Modified from Creaney et al. (1994).

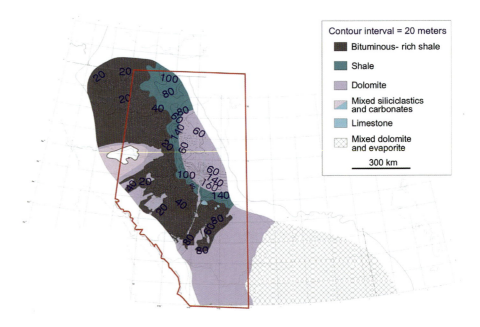

Figure 15. Duvernay (D3) isopach and lithofacies map. Modified from Switzer et al. (1994).

the Devonian appears to be largely closed hydrologically, there are several scenarios of migration and remigration.

The Devonian petroleum system is "closed" hydrologically in that the Devonian reservoirs contain almost exclusively oil and gas derived from Devonian source rocks. A notable exception is the heavy-oil Grosmont Formation (discussed later). There are examples of short-distance migration (i.e., <100 km) and of long-distance migration (i.e., >100 km) (Creaney et al., 1994). For example the Rainbow and Zama pinnacles (i.e., small reefs) in northern Alberta, located in an area where the stratigraphically equivalent Keg River source rock is mature (Figure 17), received their oil from the Keg River source rock via short-distance migration. On the other hand, pools at Red Earth or Nipisi, or in the upper, northeastern part of the Rimbey-Meadowbrook reef trend in the Redwater area, all of which are located at depths where the stratigraphically correlative source rocks are immature, received their oil via long-distance updip migration (Figure 17) that, in some places, took place stratigraphically downward before moving laterally and then updip northeastward. Similarly, there is clear evidence for updip gas migration. Several D3 pools located in the mature "band" of the Duvernay/Majeau Lake contain gas that was derived from the overmature "band" (Figure 17).

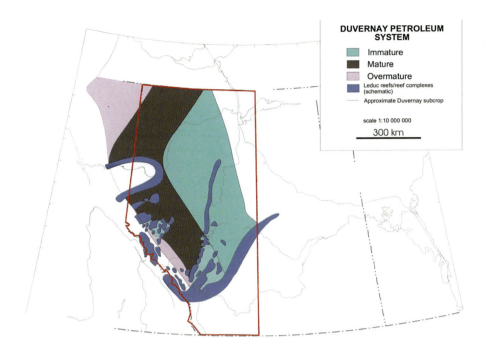

Figure 16. Map of geographic distribution and maturity of the Duvernay source rock facies. Modified from Creaney et al. (1994).

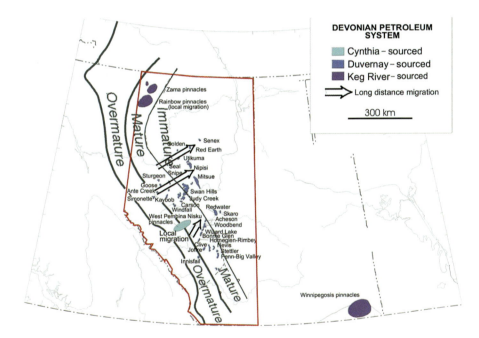

Figure 17. Maturity of the three major Devonian source rock intervals relative to the major oil pools. Most pools contain hydrocarbons equivalent to the maturity of the surrounding Duvernay source rock (i.e., pools located in the overmature band contain gas and gas condensate, whereas pools located in the mature band contain oil). These pools received their oil from short-distance migration. There is also long-distance migration in two general directions (i.e., eastward in central Alberta and northeastward [up the Rimbey-Meadowbrook reef trend] in southern Alberta), which resulted in oil pools in the immature region. Map is modified from Creaney et al. (1994).

The overall closed hydrologic character of the Devonian petroleum system is apparent even in the present hydrogeology of the basin. The Devonian, together with the uppermost Paleozoic and lower Mesozoic strata, forms one of four megahydrostratigraphic units in the basin (Bachu, 1995, 1999: number 4 in Figure 18). This unit appears to be hydrologically separated from the overlying Middle Mesozoic to recent megahydrostratigraphic units (numbers 1 and 2 in Figure 18 — unit 3 is not intersected in this particular cross section) by a basin-wide shaly aquitard, with hardly any communication across (Bachu, 1995, 1999).

Within each of the four Devonian stratigraphic levels, aqueous and petroliferous fluid migration generally occurred through the porous and permeable carbonate aquifers (platforms) into the reef traps that are stacked upon them (Figure 10), whereas the aquitards that separate the carbonate aquifers generally prohibited cross-formational hydrocarbon migration. In some cases, most notably the Rimbey-Meadowbrook reef trend and the underlying Cooking Lake

Figure 18. Present megahydrostratigraphic systems in south–central Alberta (see Figure 20 for location of this section). Flow systems 1, 2, and 4 are taken from Bachu (1995, 1999). Bachu's flow system 3 is not present and/or not identifiable in this part of the basin. The general flow patterns most likely were similar during the Laramide orogeny, except that flow system 2 was absent. A major regional aquitard (effectively coinciding with the Colorado Group and/or Clearwater Formation) separates flow systems 1 and 2 from 4. The four (hydro-)stratigraphic units D1, D2, D3, and D4 each also contain regional aquitards, all of which pinch out near the limit of the disturbed belt. Section is modified from Machel and Cavell (1999).

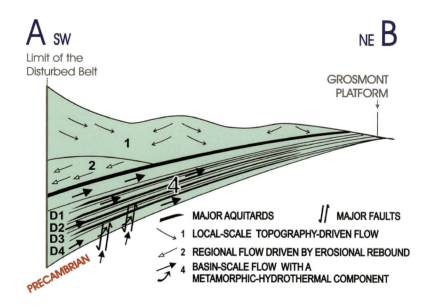

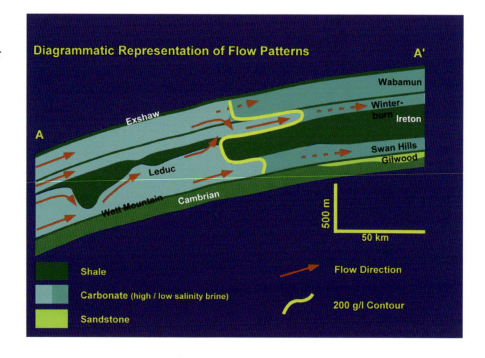

Figure 19. Brine distribution and flow pattern inferred from pressure data in the deepest part of the Devonian aquifer system. Section A-A′ is equivalent to the deepest part of section A-B in Figure 18; that is, between the left margin and the large digit 4 below the regional aquitard. Modified from Michael et al. (2003).

platform (between Innisfail and Redwater; see Figure 17), only the platform margin had significant porosity and permeability and acted as an almost two-dimensional "pipeline" for oil and brines (Stoakes and Creaney, 1984).

On the other hand, in some parts of the basin, the aquitards that normally separate the four Devonian intervals are thin, relatively permeable, and/or missing altogether. This has been identified in the deepest parts of the basin along the front of the fold and thrust belt of the Rocky Mountains, which shows several interesting hydrological and hydrogeochemical phenomena. Right next to the deformation front, the Devonian formations are saturated with a brine of approximately 150–200 g/L TDS (total dissolved solids) that gives way to a much more saline brine of more than 200 g/L in the updip direction (Michael et al., 2003; Figures 18 and 19). Thus, a "light" brine is downdip from a "heavy" brine, which is a gravitationally unstable situation. The inferred overall updip fluid movement (Figures 19 and 20) requires a hydrological drive strong enough to overcome the downdip density drive of the heavy brine. The reasons for this unusual situation are not clear (normally the heavier brines are in the deeper parts of the basin), but one possible explanation is that the brines have been pushed

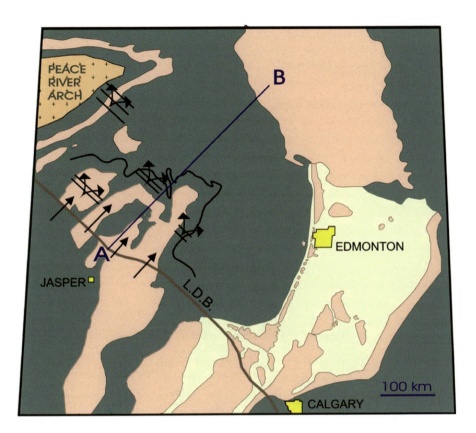

Figure 20. Devonian platform and reef carbonates of the Late Devonian Woodbend Group (D3) in the Alberta Basin. Yellow pattern: undifferentiated Cooking Lake and/or Leduc Formations and their outcrop equivalents; pink pattern: reef complexes; green: basinal areas. Fat double lines denote fault systems; black arrows mark inferred paleofluid flow, as explained in the text. L.D.B. = limit of the disturbed belt. Cross section A-B: see Figure 18.

updip by tectonic loading during the Laramide orogeny (Michael et al., 2003; Machel and Buschkuehle, 2008). There is multiple evidence for a tectonic influence on fluid movement in the deepest part of the basin. First, faults have been identified as fluid conduits in several deep reefs and carbonate platforms relatively close to the Rocky Mountains (Green, 1999), schematically shown in Figures 18 and 20. Secondly, radiogenic strontium expelled from clastic and metamorphic rocks has been identified in calcite cements in carbonates up to 150 km eastward of the deformation front. It thus appears that aqueous fluids entered the basin in its deepest parts and migrated updip through the Devonian carbonates and/or the underlying Cambrian clastic aquifer and faults to mix with brines in several reservoirs within approximately 100–150 km of the deformation front (Figures 18–20; Machel et al., 1996; Machel and Cavell, 1999; Machel and Buschkuehle, 2008).

Crossformational hydrologic connection has also been found in shallower parts of the basin. In such places, hydrocarbons migrated from a stratigraphically lower to a stratigraphically higher reservoir level. The best researched case of this type is the Bashaw Reef Complex, located in southern Alberta (Figure 6), where Duvernay-sourced oil and gas are trapped in D3 and/or D2 reefs, with clear evidence of breaching of the intervening aquitards in several parts of the complex (Hearn et al., 1996; Rostron et al., 1997; Figures 21 and 22). Even the aquitards capping the Devonian section are breached (probably by faulting) in the Bashaw area, and some of the Devonian oil remigrated into clastic traps within the Cretaceous section farther updip, as shown by organic-geochemical data (Riediger et al., 1999).

Reservoirs

On the basis of trapped hydrocarbons, there are five "classes" of hydrocarbon reservoirs in the Western Canada Sedimentary Basin: (a) dry, biogenic gas; (b) heavy oil; (c) sweet, light crude oil; (d) sweet gas and gas condensate; and (e) sour gas and gas condensate. These reservoir classes occur in three regions that correlate with depth: (a and b) shallow (outcrop and subcrop); (c) intermediate (approximately 300–3500 m); and (d and e) deep (>3500 m) (Figure 23). Most traps are reefs that form stratigraphic-diagenetic combination traps, in some cases with a structural component.

Within and crossing these groups, there is an immense diversity of reservoir characteristics. Some reservoir rocks consist of limestone reefs or platforms, but most reservoirs consist of dolostones that have higher porosities and permeabilities than their limestone counterparts. The variable reservoir characteristics reflect the stratigraphic and/or structural location within the basin to some degree. For example, the Grosmont platform in the region labeled "heavy oil" (Figure 23) is distinguished by three major aspects that are not developed elsewhere in the Devonian petroleum system of the basin; that is,

Figure 21. Map of the Bashaw Reef Complex in south–central Alberta (see Figure 6 for location), showing areas where the aquitard between D3 (Leduc) and D2 (Nisku) reservoir intervals was breached by hydrocarbons. G-H marks cross section of Figure 22. Map is modified from Hearn (1996).

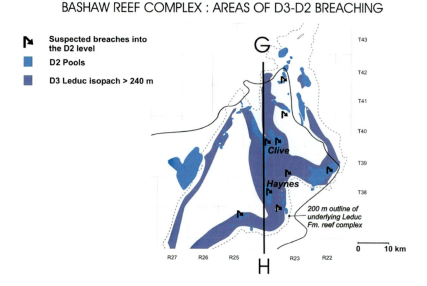

Figure 22. Cross section G-H through the Bashaw Reef Complex in south-central Alberta (see Figure 21 for location), showing the Clive and Haynes pools. The aquitard between the D3 and D2 levels is so thin and permeable that oil migrated from the lower to the upper reservoir level. In the case of Clive, both levels contain oil. In the case of Haynes, the D3 level has completely emptied into the D2. Elsewhere in the reef complex, where the Ireton aquitard is thicker, oil is trapped only in D3 reservoirs. Section is modified from Hearn (1996).

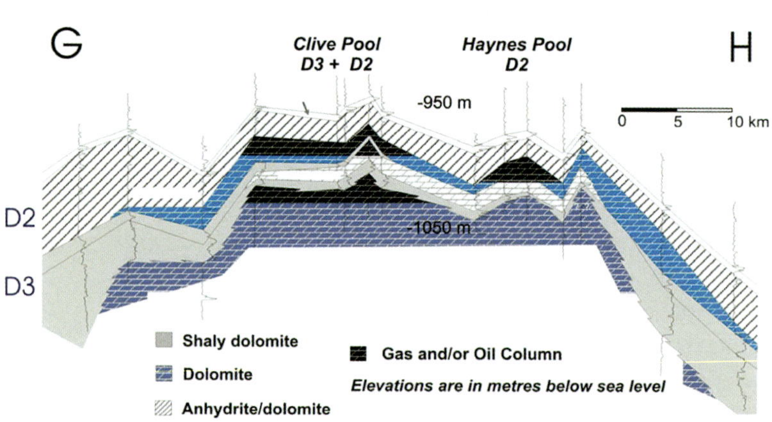

evaporative reflux, karstification, and biodegradation (discussed below). The adjacent Rimbey-Meadowbrook reef trend, which extends southward passing under Edmonton (Figure 23), is peculiar in that it contains numerous light crude oil and sweet gas reservoirs that are connected through an underlying "pipeline," the porous and permeable Cooking Lake Platform, which served as the major migration pathway for brines and hydrocarbons. In many parts of the basin, the Devonian contains oil and/or gas at several Devonian stratigraphic levels; that is, D1, D2, D3, and the D4. Sour gas occurs almost exclusively in the deepest parts of the basin. This is because here the temperatures were high enough for sour gas formation via thermochemical sulfate reduction (Machel, 1998, 2001).

Dry gas reservoirs

Dry gas reservoirs occur in the shallowest part of the basin near the erosional edge of the Canadian Shield (Figure 23). Most of these reservoirs are located in Cretaceous clastics, but some gas is pooled in Devonian carbonates. They are here considered together because genetically they form one group.

Several aspects indicate that the gas is biogenic and generated from the oil that now forms the heavy oil (tar, bitumen) reservoirs nearby in the Cretaceous sands and in the Devonian carbonates: The gas consists almost exclusively of methane, the carbon isotope composition is very low and typical for microbial methane, and organic-geochemical and circumstantial evidence (Head et al., 2003; Jones et al., 2008). In addition, the pressure in these reservoirs, some of which have been producing gas since the 1970s, albeit very low, is sustained over years and even decades of production, indicating that methane is being replenished at about the same rate as it is being extracted. Either the gas is being replenished by microbial action, or it is already "there" and diffuses out of the bitumen and/or formation waters.

Heavy-oil reservoirs

Most heavy oil is pooled in the shallow parts of the Alberta Basin in Lower Cretaceous sands, although some

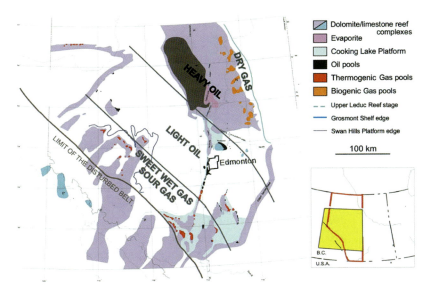

Figure 23. Woodbend (D3) carbonate complexes and hydrocarbon fillings in central Alberta (see inset for location). The contours separating areas of sweet + sour gas versus oil versus heavy oil + biogenic gas roughly parallel the maturity contours of the Duvernay/Majeau Lake shown in Figure 17. Map is modified from Switzer et al. (1994).

smaller deposits are found across the north–central region of Alberta (Figures 23 and 24). For reasons explained in the introduction, this chapter will not discuss the Cretaceous oil sands further but concentrate on the heavy-oil carbonate reservoir(s).

Heavy-oil carbonate reservoirs occur in the so-called "Carbonate Triangle" that underlaps the Cretaceous oil sand deposits (Figure 24). This region contains heavy oil in carbonates at four stratigraphic levels: Grosmont, Nisku (both D2), Shunda, and Debolt (both Mississippian) (AEUB, 2007). The D2, commonly referred to as the "Grosmont reservoir" or "Grosmont Formation," with an estimated $50–71 \times 10^9$ m^3 of original bitumen in place, is touted to be the single largest heavy-oil carbonate reservoir in the world (Alvarez et al., 2008).

A handful of steam injection pilot sites were operated for a few years in the 1980s, but then these sites were abandoned and the Grosmont reservoir has not been exploited and is not under production at present. However, because of the sheer size of the reservoir, the high and generally rising oil price, the increasing shortage of light-oil reserves worldwide, and recent technological advances, the Grosmont is under consideration for several recovery schemes (Alvarez et al., 2008), and one may speculate that the Grosmont reservoir will become an economically viable resource within a decade. The most detailed reservoir studies date from the 1980s and 1990s. Notable accounts are available in Dembicki (1994), Dembicki et al. (1994), Switzer et al. (1994), Huebscher and Machel (1995a, b; 1997a, b), Dembicki and Machel (1996), and Huebscher (1996), and the following description is summarized from these sources, except where indicated otherwise.

The Grosmont platform is distinguished by three major aspects that are not developed elsewhere in the Devonian petroleum system of the basin: (a) the Grosmont was the

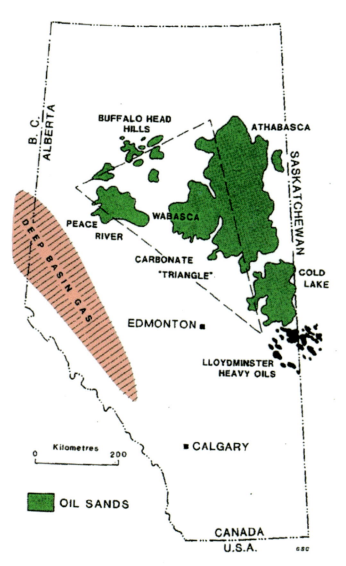

Figure 24. Distribution of heavy-oil (tar, bitumen) reservoirs in Alberta.

only Devonian carbonate complex affected by massive evaporative reflux because of its proximity to a Devonian land mass farther northeast, (b) the Grosmont is the only Devonian carbonate complex with extensive karstification because of its location at the updip end of the basin, and (c) the Grosmont is the only Devonian carbonate reservoir in the basin with extensive biodegradation.

The Grosmont forms a very large carbonate platform in northeastern Alberta (Figures 23 and 25). This platform consists of a series of southward prograding, stacked ramps that were deposited on top of the Cooking Lake (D3) platform during the upper part of the D3 and the D2 (Figure 25). These ramps and the resulting internal stratigraphic subdivisions into Lower Grosmont (LGM), and Upper Grosmont 1, 2, and 3 (UGM 1, 2, 3) are the result of and reflect the eustatic sea level fluctuations in the basin (Figure 9). Thus, the LGM is part of the D3, whereas the UGM 1, 2, and 3 constitute the D2. Lithologically, the Grosmont consists of seven lithofacies types that range from basinal mudstone/shale facies to peritidal carbonates and, in a relatively small area, gypsum and halite. Thin mudstone beds serve as stratigraphic markers to subdivide the Grosmont reservoir into its four stratigraphic intervals (Figures 25 and 26).

Hydrostratigraphic, petrologic, and geochemical data have been used to identify and delineate paleofluid flow and areas of crossformational fluid flow in the Grosmont. The Cooking Lake platform/aquifer, similar to the Leduc reefs sitting on this platform (Figure 26), is completely dolomitized, as are the UGM 2 and 3, whereas the LGM and UGM 1 are only partially dolomitized and contain significant amounts of limestone. Furthermore, the dolomites in the UGM 2 and 3 differ petrologically and geochemically from those below. Those in the UGM

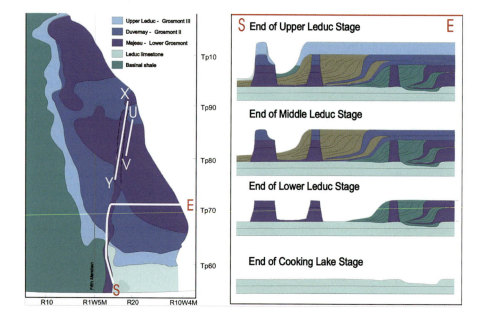

Figure 25. Depositional evolution of the Grosmont platform in northeastern Alberta. The map on the left coincides with the dark area labeled "HEAVY OIL" in Figure 23. Map is modified from Switzer et al. (1994).
Section X-Y: see Figure 26.
Section U-V: see Figure 27.

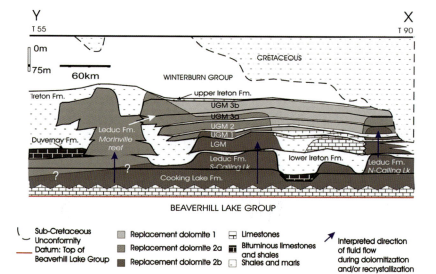

Figure 26. North–south cross section X-Y (see Figure 25 for location), showing the internal architecture of the Grosmont platform, the dolomite types, and areas of upward fluid flow. See text for further explanation. Diagram is modified from Huebscher (1996).

(replacement dolomite 1; Figure 26) resemble fine-crystalline reflux dolomites, whereas those in the Cooking Lake and Leduc are medium- to coarse-crystalline dolomites (replacement dolomites 2a, 2b) and are similar to the burial dolomites developed elsewhere in the basin (Figures 11 and 12). In some places (indicated by black, upward-pointing arrows in Figure 26), replacement types 2a and 2b occur as "halos" within the UGM, replacing replacement dolomite 1. The $\delta^{13}C$ values of the Cooking Lake and Leduc dolomites range from approximately -0.5 to $+5.0‰$ PDB (Peedee Belemnite Formation standard), and their $\delta^{18}O$ values range from -4.4 to $-7.4‰$ PDB. Replacive dolomites 1 in the UGM have $\delta^{13}C$ values of approximately -4.0 to $+3.5‰$ PDB and $\delta^{18}O$ values of -2.6 to $-6.5‰$ PDB, and $^{87}Sr/^{86}Sr$-ratios of approximately 0.7082. In the halos, the dolomites at the base of the Grosmont section have the same isotopic values as Cooking Lake/Leduc dolomites 2a or 2b and then increase upward over a thickness of 20–200 m toward values measured elsewhere in the UGM.

These data taken together suggest the following diagenetic evolution: penecontemporaneous reflux dolomitization (Figure 13, stage 1), followed by pervasive replacive dolomitization by ascending saline formation fluids (Figure 13, stage 2) that penetrated the aquitards separating the Cooking Lake, LGM, and UGM in some places, thereby recrystallizing the reflux dolomites in these locations and forming the halos. Fluid-rock interaction decreased upward, as suggested by increased buffering toward the UGM background isotope values. Dolomite 2a is interpreted to reflect dolomitization by altered seawater at intermediate burial depths, whereas dolomite 2b resulted from recrystallization of dolomite 1. These findings have important implications for oil migration (discussed further below).

The last major diagenetic event that affected the Grosmont before oil migration and emplacement was karstification (Figure 13, stage 5). Karst is developed throughout the top part of the reservoir across the area and appears enhanced (larger cavities) along two discrete levels that crosscut stratigraphic boundaries (Figure 27) that may represent relatively stagnant paleowater tables. Influx of meteoric water probably led to widespread dolomite and evaporite dissolution, resulting in the development of fractures, caves, and collapse structures.

The reservoir rock conditions have thus been shaped by three major factors: primary porosity and permeability distribution controlled by deposition (facies), dolomitization, and karstification. As a result, the Grosmont reservoir now consists of three major reservoir zones mainly controlled by the extent of diagenesis. Limestones that have low average porosities (1–5%) and low permeabilities (commonly 1–100 md) dominate zone I, which encompasses most of the LGM and much of the UGM 1. Dolostones with enhanced porosity (average of 10 to 20%, in places up to 45%) and higher permeabilities (commonly 1 to 1000 md) dominate zone II, which encompasses much of the UGM 2 and parts of the UGM 3. Zone III is the karstified, upper part of the reservoir that contains meter-sized caverns and channels. The best pay generally is in zone II.

Even within these zones, the reservoir properties are highly heterogeneous. Hand specimen, thin section, ultraviolet and scanning electron microscopy petrography, as well as grading scales, mercury capillary pressure curve analysis, and statistics, have been used for characterization of reservoir heterogeneity (Luo et al., 1994; Luo and Machel, 1995). This investigation led to a new pore size classification for carbonate reservoirs with four categories: microporosity (pore diameters <1 m), mesoporosity (pore diameters 1–1000 m), macroporosity (pore diameters 1–256 mm), and megaporosity

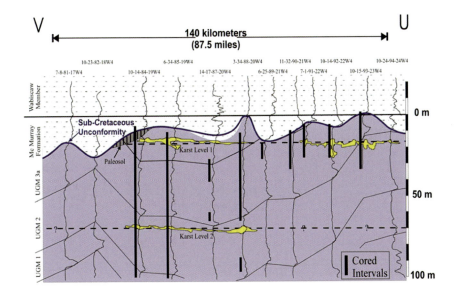

Figure 27. North–south cross section U-V (see Figure 25 for location), showing the two major karst levels within the Grosmont platform. Diagram is modified from Huebscher (1996).

(pore diameters >256 mm). A combination of microscopic observations and capillary pressure curve characteristics led to the recognition of four pore-throat texture types on the microporosity scale and to five types on the mesoporosity scale. Microporosity includes (a) intracrystal dissolution porosity, (b) pervasive intercrystal and intracrystal dissolution porosity, (c) intergranular and/or intercrystal porosity in grainstones, and (d) primary or solution microporosity in mud matrix (only in limestones). Mesoporosity includes (a) intercrystal porosity, (b) solution-enhanced intercrystal porosity, (c) oversized porosity, (d) intragranular solution porosity, and (e) intergranular solution porosity. Some of these types are homogeneous (e.g., nonfabric-selective dissolution porosity and intercrystal primary porosity), whereas others are heterogeneous. Generally, hydrocarbon recovery efficiency is good in the homogeneous pore-throat types but poor in the heterogenous.

Reservoir heterogeneity has further been characterized by comparison and statistical analysis of plug and full diameter core (FDC) data (Luo and Machel, 1995), which indicate that (a) plug porosities are higher on average than FDC porosities; (b) plug permeabilities are, on average, higher than FDC permeabilities; (c) porosity and permeability of plugs show relatively good linear relationships in certain core intervals, but FDC data do not; and (d) generally, the more homogeneous reservoir rocks have relatively high matrix porosity. Heterogeneity is common in Grosmont dolostone reservoir rocks, and it affects overall reservoir quality. At the megascopic scale, the reservoir properties can be classified as homogeneous, dual-porosity, and multilayered. The dual-porosity model has three subcategories: fractured, channeled, and pressure-solution-derived.

The findings regarding dolomitization (Figure 26) provide important clues regarding hydrocarbon migration because the halos also are the most likely locations of hydrocarbon migration from downdip source rocks into the Grosmont reservoir. The downdip Devonian Duvernay Formation (Figure 10) has been identified as the principal source rock for the Grosmont on the basis of biomarker studies (Creaney et al., 1994). However, a long-lasting debate is still ongoing as to which source rocks contributed how much to the combined package of Devonian carbonates + Cretaceous sands, and younger source rocks, especially of Mississippian and Jurassic age, have also been invoked as major source rocks for the heavy-oil accumulations (see summary in Ranger and Gingras, 2006). Identification of source rocks is hampered by advanced biodegradation, which is most severe in the Grosmont (Creaney et al., 1994).

Light crude, sweet, and sour gas reservoirs

Light crude reservoirs occur in a belt crossing the center of Alberta, whereas sweet and sour gases are located in a parallel belt that is located deeper and extends to the deformation front (Figure 23). A thorough discussion would require these belts to be discussed separately. However, in this chapter, these belts are discussed together using the Upper Devonian Nisku West-Pembina play, which straddles the boundary between these two belts and thus encompasses all relevant discriminating characteristics.

The West Pembina Nisku (D2) play is located in central Alberta (Figure 17 — oval area labeled "West Pembina Nisku Pinnacles"). Generally, the Nisku contains sweet oil in its shallower parts, sweet condensate at intermediate depths, and sour condensate in its downdip parts (Figure 28). However, this distribution is only valid for

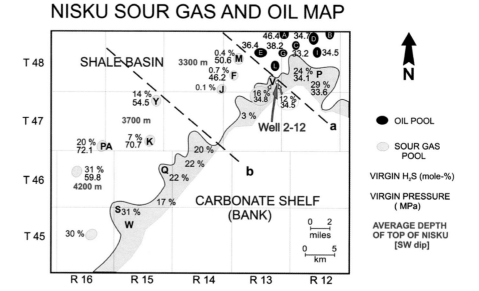

Figure 28. Nisku pools in the West Pembina area, showing distribution of oil versus gas, virgin pressures, H_2S concentrations (mol %), and average depths of the top of the pools. Line a separates oil from gas pools, line b is the "sour gas line" or "TSR line." Figure is taken from Machel et al. (1995b).

the small reef reservoirs that are encased in off-reef marls northwest of the platform margin (such as the PA, K, A, and D pools; see Figure 28). The nearby carbonate platform edge contains sour gas throughout and acted as an updip migration pathway, such that sour gas also occurs as far as 120 km updip in the carbonate platform, juxtaposed against sweet oil pools northwest of the platform margin. As such, the Nisku is the shallowest of the many sour gas plays in Devonian carbonates of the Western Canada Sedimentary Basin, almost all of which are located in the deepest part of the basin where the source rocks are overmature (Figure 17) and where paleotemperatures were high enough for sour gas formation (Machel, 1998, 2001). Present subsurface depths of the Nisku play in West Pembina range from approximately 3000 (northeast) to 4500 (southwest) m. One remarkable feature of the Nisku play is that hydrocarbons and sour gas are contained in numerous closely spaced pools that have been essentially isolated hydrodynamically from one another since hydrocarbon entrapment, which renders them ideal natural laboratories to study depth/temperature-dependent trends. The following characteristics are extracted from Machel et al. (1995b), Riciputi et al. (1996), Machel et al. (1997), and Manzano et al. (1997).

Nisku reefs that form the reservoir facies occur as an elongate shelf margin complex and as a series of small reefs (so-called pinnacles) with diameters of 1–2 km and thicknesses of 50–100 m, located on the slope of the Devonian depositional basin in the northwest (Figure 28). These reefs/pinnacles are surrounded by relatively impermeable calcareous shales and appear to be hydrologically isolated from one another at present, as shown by their virgin pressures (original reservoir pressures prior to production) that vary drastically, even for pools that are located very close to one another (Figure 28). The shallowest pools have pressures close to hydrostatic, whereas the deepest pools have pressures far in excess of hydrostatic. Also, there is no pressure drawdown from pool to pool during production.

Virgin hydrogen sulfide (H_2S) contents (i.e., H_2S contents at discovery and before production of the pools) also vary drastically. In the isolated pools, virgin H_2S contents generally increase downdip, whereas the virgin H_2S contents increase upward in the upper of the two closures (subpools) in the elongate shelf margin reef complex (Figure 28). On the slope toward the basin, the boundary between oil pools updip and gas pools downdip is sharp (line a in Figure 28), but significant (>1%) concentrations of H_2S occur only downdip of line b (Figure 28). The elongate shelf margin reef complex straddles lines a and b and contains H_2S updip and downdip of these lines.

Like most other Devonian carbonates in the Western Canada Sedimentary Basin, the Nisku reefs in the West Pembina area underwent a complicated diagenetic history, which forms a paragenetic sequence almost identical to that shown in Figure 12. The burial history of the Nisku also resembles that of most other Devonian reefs in the Western Canada Sedimentary Basin (Figure 13), with maximum burial depths of approximately 4000–4500 m (i.e., approximately 1500 m greater than the present depths).

Drill cores and fluids extracted from the Nisku contain all petrographic and geochemical features expected to result from thermochemical sulfate reduction, the chief process responsible for sour gas formation (Figure 29), including systematic changes in the composition of the pooled hydrocarbons in the gasoline range, H_2S and elemental sulfur with distinctive $\delta^{34}S$ values, (some) Fe-sulfide, saddle dolomite and blocky calcite with strongly depleted $\delta^{13}C$ values, corroded anhydrite, and solid bitumen. One well (2-12-48-13W5, shown as "2-12" in

REACTIONS AND PRODUCTS OF THERMOCHEMICAL SULFATE REDUCTION

(1) crude oil → light crude oil + H_2S (+ PS) + C_1 + C_{2+}

(2 - 4) $4\,R\text{-}CH_3 + 3SO_4^{2-} + 6H^+ \rightarrow 4\,R\text{-}COOH + 4H_2O + 3H_2S$
(gasoline range:
branched n-alkanes > n-alkanes > cyclic > monoaromatic)

(5,6) $3H_2S + SO_4^{2-} + 2H^+ \rightarrow 4S^\circ + 4H_2O$

(7) $4S^\circ + 1.33\,(\text{-}CH_2\text{-}) + 2.66\,H_2O + 1.33\,OH^- \rightarrow 4H_2S + 1.33\,HCO_3^-$

(8) $hcs + SO_4^{2-} \rightarrow$ altered hcs + solid bitumen
$+ HCO_3^-\,(CO_2) + H_2S\,(HS^-) + H_2O + \text{heat?}$

Figure 29. Simplified reaction scheme for TSR. Reaction 8 represents the net mass and charge balance of TSR. Figure is taken from Machel et al. (1995a).

Figure 28) has sour gas condensate and corroded anhydrite in its upper half, water and saddle dolomite in its lower half, and elemental sulfur and saddle dolomite in the gas-water transition zone from approximately 3210 to 3230 m.

The stable isotope values of saddle dolomite are lowest near the top of the water-saturated zone and increase downward, whereby $\delta^{18}O$ ranges from approximately -8 to -6 ‰ PDB and $\delta^{13}C$ ranges from approximately -12 to $+2$ ‰ PDB. In contrast, the $\delta^{18}O$ and $\delta^{13}C$ values of the gray matrix dolomite samples are invariant with depth and are approximately -5 and $+2$ ‰ PDB, respectively. Late sparry calcite, present in a few vugs, has $\delta^{18}\tilde{O}-8$ to -6‰ PDB and $\delta^{13}\tilde{C}$ -18 to -12‰ PDB. The homogenization temperatures of primary fluid inclusions in saddle dolomite range from 110°C to 180°C, with two median values of approximately 125°C and 145°C (not pressure corrected); freezing point depressions of these inclusions scatter between approximately -5 to $-25°C$. The freezing point depressions of saddle dolomite from well 2-12 correspond to salinities of approximately five times the salinity of seawater. Ion microprobe analyses of anhydrites in well 2-12 yielded $\delta^{34}S$ values of approximately $+24$ to $+28$‰ CDT. Ion microprobe $\delta^{34}S$ data of disseminated pyrite crystals from the syndepositionally well oxygenated reef facies in well 2-12 display two populations — one biogenic and one thermogenic. Powder samples of two crystals of elemental sulfur from the sulfur-bearing interval in this well yielded $\delta^{34}S$ of approximately $+24$‰ CDT.

Stable isotope data of the sour gas components are also distinctive. $\delta^{34}S$ values of H_2S are approximately $+20$‰ CDT and approach those of Nisku anhydrites; $\delta^{13}C$ values of methane (CH_4), ethane (C_2H_6), and propane (C_3H_8) are approximately -41, -29, and -27‰ PDB, respectively; $\delta^{13}C$ values of carbon dioxide (CO_2) are approximately -7‰ PDB. Nisku formation waters tend to be brines with relatively heavy $\delta^{18}O$ and δD values. With increasing degree/extent of TSR downdip from pinnacle pool to pinnacle pool, there is a decrease in the saturate/aromatic ratio, an increase in the relative abundance of organo-sulfur compounds (especially benzothiophenes), the $\delta^{34}S$ values approach that of the solid anhydrite, the $\delta^{13}C$ values of the saturate fraction increase, and the H_2S concentration increases (Figure 30). However, pools P and V in the upper part of the elongate reef structure that fringes the carbonate shelf fall off these trends (Figure 30).

The available data suggest the following scenario: (a) oil migration into the pools once the area entered and passed the liquid oil window (Figure 13, stage 3); (b) in situ thermal maturation and cracking to gas condensate and solid bitumen, with concurrent hydrocarbon expansion and pressure increase, such that the gas-water contact/transition zone was progressively displaced downward; (c) TSR was initiated during advanced stages of gas condensate formation, generating H_2S and further pushing the gas-water zone downward, with dissolution of anhydrite and/or partial replacement by calcite in the upper part of the pools; (d) cessation of TSR, forming elemental sulfur where the gas-water zone reached its present position; and (e) saddle dolomite formation largely after the gas-water zone reached its present position. These processes took place while the study area underwent burial from approximately 2000 to 5500 m during the Late Cretaceous to Early Tertiary (Figures 12 and 13, stage 4). The study area was then partially uplifted and approximately 1500 m of overburden were removed, resulting in the present subsurface depths of the Nisku pools. During and after uplift, the pressures in these pools partially reequilibrated to hydrostatic values commensurate with the present burial depths, as

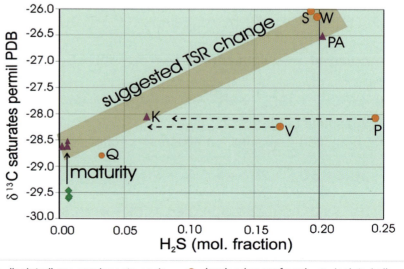

Figure 30. Changes in $\delta^{13}C$ of the saturates as a function of H_2S concentration in the gas phase in Nisku pools in the West Pembina area. Figure taken from Manzano et al. (1997).

indicated by the lowest pressured pools having hydrostatic pressures. Those pools with overpressures are apparently so well sealed that they maintain some or all of the overpressures they had near maximum burial.

The relatively high $\delta^{34}S$ values of H_2S in pools F, M, and V (i.e., well 2-12) and of elemental sulfur in well 2-12 further suggest that these phases were formed by TSR. However, the H_2S contents in pools F and M (0.7% and 0.4%, respectively) and the adjacent pool J (0.1% H_2S) are so low that they do not require TSR, suggesting that TSR did not occur updip of line b (Figure 28), the "TSR line," in the isolated pools. The small amounts of H_2S in pools J, F, and M probably originated from thermal cracking of NSO compounds, and the high $\delta^{34}S$ values of their H_2S probably resulted from Rayleigh fractionation. The $\delta^{13}C$ values of CH_4, C_2H_6, C_3H_8, and carbon dioxide (CO_2) in well 2-12 are similar to those of pools F and M. This phenomenon is best explained by updip migration through the elongated shelf margin reef complex of gases derived from much deeper in the section, presumably adding ^{13}C-enriched CO_2 and ^{13}C-depleted hydrocarbon gases to the system.

Conclusions

1) The Devonian is one of two important petroleum systems in Canada. The Devonian system contains many conventional oil and gas fields, and a giant unconventional heavy-oil reservoir, the Grosmont.
2) Tectonics and structure controlled basin evolution, sedimentation, and petroleum migration. Several Precambrian lineaments and fault systems appear to have been active during deposition of the Devonian section and as pathways for late-diagenetic fluid flow.
3) The Devonian system consists mainly of reefal carbonates and off-reef marls, and to a minor extent evaporites. Deposition was cyclical and controlled mainly by third- and fourth-order eustatic sea level fluctuations.
4) The burial and diagenetic evolution was complex and can be grouped into five major stages.
5) The Devonian petroleum system contains several source rock intervals. The system is closed in that the Devonian source rocks generated nearly all of the petroleum presently pooled in Devonian reservoirs. An exception may be the Grosmont formation, which was probably also sourced from younger source rocks. There are examples of short-distance (<100 km) and long-distance (>100 km) migration. Some Devonian petroleum has migrated into overlying Cretaceous traps.
6) The Devonian section contains four reservoir intervals, D1–D4, and five distinct hydrocarbon plays that are controlled mainly by subsurface depth: heavy oil in the shallowest part of the basin; biogenic gas associated with heavy oil; light oil, sweet gas, and gas condensate; and sour gas in the deepest part of the basin. The traps are mainly stratigraphic-diagenetic combination traps.
7) The Western Canada Sedimentary Basin today contains four megahydrostratigraphic units. The Devonian makes up the major part of the lowest of these units, which appear to have been established during and/or after the Laramide orogeny. Petroleum migration took place in the Late Cretaceous and Early Tertiary through various migration routes. In the deepest part of the basin, tectonic expulsion of formation fluids during the Laramide orogeny may have been involved in petroleum migration.

Acknowledgments

The Natural Science and Engineering Research Council of Canada (NSERC), several oil companies (including Amoco, Crestar, Chevron, Esso, Home Oil, Husky Oil, Mobil, Norcen, Numac, Pan Canadian, Petro-Canada, and Shell), and the Alberta Research Council have funded my research on various aspects of the Devonian petroleum system of the Western Canada Sedimentary Basin. Numerous colleagues have sharpened my understanding of the problems at hand in various ways. Special thanks go to Eric Mountjoy, who introduced me to the Western Canada Sedimentary Basin in 1980 and has been a scientific collaborator and inspiration ever since. Eric passed away earlier this year. This article is dedicated to his memory.

References

Alberta Energy and Utilities Board, 2007, Alberta's energy reserves 2006 and supply/demand outlook 2007-216: ST98-2007.

Alvarez, J. M., R. P. Sawatzky, L. M. Forster, and R. M. Coats, 2008, Alberta's bitumen carbonate reservoirs — moving forward with advances R&D: Presented at the World Heavy Oil Congress, Paper 2008-467.

Amthor, J. E., E. W. Mountjoy, and H. G. Machel, 1993, Subsurface dolomites in Upper Devonian Leduc Formation buildups, central part of Rimbey-Meadowbrook reef trend, Alberta, Canada: Bulletin of Canadian Petroleum Geology, **41**, 164–185.

Amthor, J. E., E. W. Mountjoy, and H. G. Machel, 1994, Regional-scale porosity and permeability variations in Upper Devonian Leduc buildups: implications for reservoir development and prediction in carbonates: AAPG Bulletin, **78**, 1541–1559.

Bachu, S., 1995, Synthesis and model of formation-water flow, Alberta Basin, Canada: AAPG Bulletin, **79**, 1159–1178.

Bachu, S., 1999, Flow systems in the Alberta Basin: patterns, types and driving mechanisms: Bulletin Canadian Petroleum Geology, **47**, 455–474.

Barrett, K. R., J. C. Hopkins, K. N. Wilde, and M. E. Connelly, 2008, The origin of matrix and fracture mega-porosity in a carbonate bitumen reservoir, Grosmont Formation, Saleski, Alberta: Presented at the World Heavy Oil Congress, Paper 2008-344.

Creaney, S., J. Allen, K. S. Cole, M. G. Fowler, P. W. Brooks, K. G. Osadetz, R. W. Macqueen, L. R. Snowdon, and C. L. Riediger, 1994, Petroleum generation and migration in the Western Canada Sedimentary Basin, *in* G. D. Mossop and I. Shetsen, eds., Geologic atlas of the Western Canada Sedimentary Basin: Canadian Society of Petroleum Geologists and Alberta Research Council, 455–468.

Dawson, F. M., and W. Kalkreuth, 1994, Coal rank and coalbed methane potential of Cretaceous/Tertiary coals in the Canadian Rocky Mountain foothills and adjacent foreland: 1. Hinton and Grande Cache areas, Alberta: Bulletin of Canadian Petroleum Geology, **42**, 544–561.

Dembicki, E. A., 1994, The Upper Devonian Grosmont Formation: well log evaluation and regional mapping of a heavy oil carbonate reservoir in northeastern Alberta: M.S. thesis, University of Alberta.

Dembicki, E. A., and H. G. Machel, 1996, Recognition and delineation of Paleokarst zones by the use of wireline logs in the bitumen-saturated Upper Devonian Grosmont Formation of northeastern Alberta, Canada: AAPG Bulletin, **80**, 695–712.

Dembicki, E. A., H. G. Machel, and H. Huebscher, 1994, Reservoir characteristics of a heavy-oil carbonate reservoir: the Grosmont Formation, Alberta, Canada: Presented at the CSPG-CSEG Joint Annual Meeting, Book of Abstracts, 375–376.

Dix, G. R., 1993, Patterns of burial- and tectonically controlled dolomitization in an Upper Devonian fringing-reef complex: Leduc Formation, Peace River Arch area, Alberta, Canada: Journal of Sedimentary Petrology, **63**, 628–640.

Edwards, D. J., and R. J. Brown, 1995, A geophysical perspective on the question of basement involvement with the distribution of Upper Devonian carbonates in central Alberta, *in* G. M. Ross, ed., Alberta Basement Transects Workshop, Lithoprobe Report 51, Lithoprobe Secretariat, 225–233.

Green, D. G., 1999, Dolomitization and deep burial of the Devonian of west-central Alberta deep basin: Kaybob South and Fox Creek (Swan Hills Formation) and Pine Creek fields (Leduc and Wabamun Formations): Ph.D. thesis, McGill University.

Hacquebard, P. A., 1977, *in* G. Deroo, T. G. Powell, B. M. Tissot, R. G. McCrossan, and P. A. Hacquebard, eds., The origin and migration of petroleum in the Western Canada Sedimentary Basin: Geological Survey of Canada Bulletin, **262**, 11–22.

Hay, P. W., 1994, Oil and gas resources in the Western Canada Sedimentary Basin, *in* G. D. Mossop and I. Shetsen, eds., Geologic atlas of the Western Canada Sedimentary Basin: Canadian Society of Petroleum Geologists and Alberta Research Council, 469–470.

Head., I. M., M. Jones, and S. R. Larter, 2003, Biological activity in the deep subsurface and the origin of heavy oil: Nature, **426**, 344–352.

Hearn, M. R., 1996, Stratigraphic and diagenetic controls on aquitard integrity and hydrocarbon entrapment, Bashaw Reef Complex, Alberta, Canada: M.S. thesis, University of Alberta.

Hearn, M. R., H. G. Machel, and B. J. Rostron, 1996, Stratigraphic and diagenetic controls on petroleum entrapment and re-migration in Devonian reefs, Bashaw Area, Alberta: Presented at the CSPG Annual Conference.

Hein, F. J., and R. A. Marsh, 2008, Regional geologic framework, depositional models and resource estimates of the oil sands of Alberta, Canada: World Heavy Oil Congress, Paper 2008-320.

Huebscher, H., 1996, Regional controls on the stratigraphic and diagenetic evolution of Woodbend Group carbonates, north-central Alberta: Ph.D. thesis, University of Alberta.

Huebscher, H., and H. G. Machel, 1995a, Cross-formational fluid flow in Devonian dolostones: Presented at the 1st SEPM Congress on Sedimentary Geology, Congress Program and Abstracts.

Huebscher, H., and H. G. Machel, 1995b, Seal quality related to facies distribution and diagenesis, an example from the Woodbend Group, north-central Alberta: Presented at the CSPG/Canadian Well Logging Society Joint Annual Convention, Program and Abstracts.

Huebscher, H., and H. G. Machel, 1997a, Paleokarst in the Grosmont Formation, northeastern Alberta, *in* J. Wood, and B. Martindale, eds., CSPG-Society for Sedimentary Geology Joint Convention, Core Conference, 129–151.

Huebscher, H., and H. G. Machel, 1997b, Stratigraphic and facies architecture of the Upper Devonian Woodbend Group, north-central Alberta; CSPG-SEPM Joint Convention, Program with Abstracts, 135.

Jones, D. M., I. M. Head, N. D. Gray, J. J. Adams, A. K. Rowan, C. M. Aitken, B. Bennett, H. Huang, A. Brown, B. F. J. Bowler, T. Oldenburg, M. Erdman, and S. R. Larter, 2008, Crude-oil biodegradation via methanogenesis in subsurface petroleum reservoirs: Nature, **451**, 176–180.

Luo, P., H. G. Machel, and J. Shaw, 1994, Petrophysical properties of matrix blocks of a heterogeneous dolostone reservoir — the Upper Devonian Grosmont Formation, Alberta, Canada: Bulletin of Canadian Petroleum Geology, 42, 465–481.

Luo, P., and H. G. Machel, 1995, Pore size and pore-throat types in a heterogeneous Dolostone reservoir, Devonian Grosmont Formation, Western Canada Sedimentary Basin: AAPG Bulletin, 79, 1698–1720.

Machel, H. G., 1990, Burial diagenesis, porosity and permeability development in carbonates, in G. R. Bloy, and M. G. Hadley, eds., The development of porosity in carbonate reservoirs: CSPG Short Course Notes, 2-1–2-18.

Machel, H. G., 1998, Gas souring by thermochemical sulfate reduction at 140°C: Discussion: AAPG Bulletin, 82, 1870–1873.

Machel, H. G., 2001 Bacterial and thermochemical sulfate reduction in diagenetic settings: Sedimentary Geology, 140, 143–175.

Machel, H. G., and B. E. Buschkuehle, 2008, Diagenesis of the Devonian Southesk-Cairn Carbonate Complex, Alberta, Canada: marine cementation, burial dolomitization, thermochemical sulfate reduction, anhydritization, and squeegee fluid flow: Journal of Sedimentary Research, 78, 366–389.

Machel, H. G., and P. A. Cavell, 1999, Low-flux, tectonically induced squeegee fluid flow ("hot flash") into the Rocky Mountain Foreland Basin: Bulletin Canadian Petroleum Geology, 47, 510–533.

Machel, H. G., and I. G. Hunter, 1994, Facies models for Middle to Late Devonian shallow-marine carbonates, with comparisons to modern reefs — a guide for facies analysis: Facies, 30, 155–176.

Machel, H. G., H. R. Krouse, and R. Sassen, 1995a, Products and distinguishing criteria of bacterial and thermochemical sulfate reduction: Applied Geochemistry, 10, 373–389.

Machel, H. G., H. R. Krouse, L. R. Riciputi, and D. R. Cole, 1995b, Devonian Nisku sour gas play, Canada: a unique natural laboratory for study of thermochemical sulfate reduction, in M. A. Vairavamurthy and M. A. A. Schoonen, eds., Geochemical transformations of sedimentary sulfur: American Chemical Society Symposium Series, 612, 439–454.

Machel, H. G., P. A. Cavell, and K. S. Patey, 1996, Isotopic evidence for carbonate cementation and recrystallization, and for tectonic expulsion of fluids into the Western Canada Sedimentary Basin: Geological Society of America Bulletin, 108, 1108–1119.

Machel, H. G., L. R. Riciputi, and D. R. Cole, 1997, Ion microprobe investigation of diagenetic carbonates and sulfides in the Devonian Nisku Formation, Alberta, Canada, in I. P. Montañez, J. M. Gregg, and K. L. Shelton, eds., Basin-wide diagenetic patterns: integrated petrologic, geochemical, and hydrologic considerations, Society for Sedimentary Geology Special Publication, 57, 157–165.

Manzano, B. K., M. G. Fowler, and H. G. Machel, 1997, The influence of thermochemical sulfate reduction on hydrocarbon composition in Nisku reservoirs, Brazeau River area, Alberta, Canada: Organic Geochemistry, 27, 507–521.

Michael, K., H. G. Machel, and S. Bachu, 2003, New insights into the origin and migration of brines in deep Devonian aquifers, Alberta, Canada: Journal of Geochemical Exploration, 80, 193–219.

Morris, G. A., and B. E. Nesbitt, 1998, Geology and timing of paleohydrogeological events in the MacKenzie Mountains, Northwest Territories, Canada, in J. Parnell, ed., Dating and duration of fluid flow and fluid-rock interaction: Geological Society (London), Special Publication 144, 161–172.

Morrow, D. W., and K. L. Aulstead, 1995, The Manetoe Dolomite — a Cretaceous-Tertiary or a Paleozoic event? Fluid inclusion and isotopic evidence: Bulletin of Canadian Petroleum Geology, 43, 267–280.

Morrow, D. W., J. Potter, B. Richards, and F. Goodarzi, 1993, Paleozoic burial and organic maturation in the Liard Basin region, northern Canada: Bulletin of Canadian Petroleum Geology, 41, 17–31.

Mountjoy, E. W., H. G. Machel, D. Green, J. Duggan, and A. E. Williams-Jones, 1999, Devonian matrix dolomites and deep burial carbonate cements: a comparison between the Rimbey-Meadowbrook reef trend and the deep basin of west-central Alberta: Bulletin Canadian Petroleum Geology, 47, 487–509.

Price, R. N., 1994, Cordilleran tectonics and the evolution of the Western Canada Sedimentary Basin, in G. D. Mossop and I. Shetsen, eds., Geologic atlas of the Western Canada Sedimentary Basin, Calgary: CSPG and Alberta Research Council, 13–24.

Qing, H., and E. W. Mountjoy, 1992, Large-scale fluid flow in the Middle Devonian Presqúile barrier, Western Canada Sedimentary Basin: Geology, 20, 903–906.

Qing, H., and E. W. Mountjoy, 1994, Formation of coarsely crystalline, hydrothermal dolomite reservoirs in the Presqúile Barrier, Western Canada Sedimentary Basin: AAPG Bulletin, 78, 55–77.

Ranger, M. J., and M. K. Gingras, 2006, Geology of the Athabasca oil sands. Field guide and overview, 5th ed.

Riciputi, L. R., D. R. Cole, and H. G. Machel, 1996, Sulfide formation in reservoir carbonates of the Devonian Nisku Formation, Alberta, Canada: Geochimica Cosmochimica Acta 60, 325–336.

Riediger, C. L., M. G. Fowler, L. R. Snowdon, R. MacDonald, and M. D. Sherwin, 1999, Origin and alteration of

Lower Cretaceous Mannville Group oils from the Provost oil field, east central Alberta, Canada: Bulletin Canadian Petroleum Geology, **47**, 43–62.

Ross, R. M., 1990, Deep crust and basement structure of the Peace River Arch region: constraints on mechanisms and formation, *in* S. C. O'Connel and J. S. Bell, eds., Geology of the Peace River Arch: Bulletin of Canadian Petroleum Geology, **38**, 25–35.

Rostron, B. J., J. Tóth, and H. G. Machel, 1997, Fluid flow, hydrochemistry, and petroleum entrapment in Devonian reef complexes, south-central Alberta, Canada, *in* I. P. Montañez, J. M. Gregg, and K. L. Shelton, eds., Basin-wide diagenetic patterns: integrated petrologic, geochemical, and hydrologic considerations: Society for Sedimentary Geology Special Publication No. 57, 139–155.

Stanton, M. S., 2004, Origin of the Lower Cretaceous heavy oils ("tar sands") or Alberta: Search and Discovery Article 10067, http://www.searchanddiscovery.net/documents/2004/stanton/images/stanton.pdf.

Stephenson, R. A., C. A. Zelt, Z. Jajnal, P. Morel-à-l'Huissier, R. F. Mereu, D. J. Northey, G. F. West, and E. R. Kanasewich, 1989, Crust and upper mantle structure and the origin of the Peace River Arch: Bulletin of Canadian Petroleum Geology, **37**, 224–235.

Stoakes, F. A., and S. Creaney, 1984, Sedimentology of a carbonate source rock: the Duvernay Formation of Central Alberta, *in* L. Eliuk, ed., Carbonates in subsurface and outcrop: CSPG Core Conference, 132–147.

Switzer, S. B., W. G. Holland, D. S. Christie, G. C. Graf, A. S. Hedinger, R. J. McAuley, R. A. Wierzbicki, and J. J. Packard, 1994, Devonian Woodbend-Winterburn strata of the Western Canada Sedimentary Basin, *in* G. D. Mossop and I. Shetsen, eds., Geologic atlas of the Western Canada Sedimentary Basin, Calgary: CSPG and Alberta Research Council, 165–195.

Walls, R. A., and G. Burrowes, 1990, Diagenesis and reservoir development in Devonian limestone and dolostone reefs of Western Canada: CSPG, Short Course Notes, Section 5, 5.1–5.17.

Chapter 10

Review of Geology of a Giant Carbonate Bitumen Reservoir, Grosmont Formation, Saleski, Alberta

Kent R. Barrett[1] and J. C. Hopkins[2]

Introduction

The Grosmont Formation is an Upper Devonian carbonate succession that is present in northeastern Alberta. It contains 318 billion barrels of bitumen on the basis of Alberta government estimates (Alberta Energy Resources Conservation Board, 1996). Figure 1 is a map of the interpreted bitumen resource on the basis of Energy Resources Conservation Board mapping. This map also shows the location of Laricina Energy's Saleski land block in the heart of the bitumen accumulation. During the winters of 2006–2007 and 2007–2008, Laricina Energy drilled 21 vertical wells for the purposes of bitumen resource delineation. In addition, one horizontal well was drilled as part of a proposed steam-assisted gravity drainage (SAGD) pilot.

The Grosmont Formation is a 120-m thick carbonate succession that is sandwiched between shales of the Upper and Lower Ireton Formation within the Woodbend Group. It has been subdivided chronologically into the A, B, C, and D units. These subdivisions correspond to Cutler's (1983) units LG, G1, G2, and G3, respectively. The lowermost three units were deposited during shallowing-upward depositional cycles. The uppermost unit, the Grosmont D, is an aggradational depositional unit. Grosmont strata dip gently to the southwest. The Grosmont has been bevelled by erosion in an easterly direction.

The upper two Grosmont units, Grosmont C and D, contain the bulk of the bitumen resource. At Saleski, there is up to 45 m of bitumen pay exceeding 12% porosity. This resource is characterized by very high porosities and bitumen saturations. Laricina Energy's Saleski land holdings are located downdip from the regional Grosmont gas cap and updip from the regional water leg.

The Grosmont Formation has had a complex geological history that culminated in the accumulation of the bitumen deposit.

- The Grosmont Formation is a carbonate ramp succession deposited in a shallow marine setting on the eastern edge of the Ireton Shale Basin.

- Grosmont limestones were dolomitized during the Late Devonian or Early Mississippian.

- The Grosmont was uplifted, tilted, and eroded during the Late Jurassic to Early Cretaceous Larimide Orogeny. The Grosmont subcrop was exposed to the influences of meteoric water. Fresh surface water percolated down into the Grosmont carbonates, preferentially leaching more soluble strata—largely calcite and selected dolomite grains. This karsting episode substantially enlarged the existing pore system, producing regional zones of dissolution termed "megaporosity zones."

- Sometime after reservoir development, oil migrated into the Grosmont Formation. This preserved the reservoir from the deleterious effects of subsequent porosity reduction. This oil was later biodegraded to bitumen.

Reservoir geology of the Grosmont C and D

At Saleski, the bitumen resource is confined to the Grosmont C and D units. The gross interval is approximately 65 m. Individual reservoir units within the Grosmont C and D are stratigraphically continuous within the Saleski area. Porosity is largely secondary, but its distribution is controlled by depositional facies that are crudely layer cake

---

[1]Laricina Energy, Calgary, Alberta, Canada
[2]University of Calgary, Calgary, Alberta, Canada (retired)

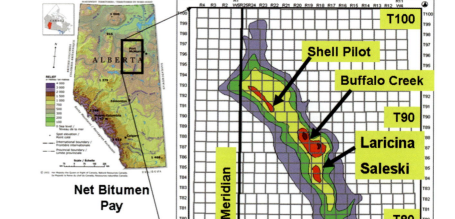

Figure 1. Location map showing the Alberta Energy Resources Conservation Board's interpretation of bitumen pay thickness in northeast Alberta. The location of Laricina's Saleski project is also shown.

in their distribution. The Grosmont C has been subdivided into three units: the Lower Argillaceous Dolomite, Vuggy, and Upper units from bottom to top (Figure 2).

The Grosmont C Lower Argillaceous Dolomite unit (381–394.3 m in well 7-26-85-19W4) is a dolomitic wackestone with numerous wispy shale partings. The lower 8.5 m is nonporous, more argillaceous, and contains an abundance of subvertical lined burrows and brachiopod-rich intervals (see Figure 3a and b). The upper 5 m is also highly bioturbated and contains scattered vugs but is generally of low reservoir quality. The Lower Grosmont C Argillaceous unit was deposited in an open marine setting.

The Grosmont C Vuggy unit (368–381 m in well 7-26-85-19W4) is a poorly bedded dolo-wackestone. This unit contains ubiquitous, irregular, 0.5–1.5-cm diameter vugs commonly connected by short, subvertical fractures. Fracturing is often so intense that a mosaic breccia texture with high intraclast porosity results.

In areas where leaching is less severe, it is possible to observe subvertical burrows 0.4–1.0 cm in width (Figure 4b). A computed tomography (CT) scan of core from 7-26-85-19W4 (Figure 4a) revealed possible horizontal Thalassanoides burrows. Burrowing generally develops from biological activity in less agitated depositional conditions typical of a subtidal depositional environment.

The Upper Grosmont C unit (361–368 m in well 7-26-85-19W4) is a clean, very fine to fine-grained, laminated, dolomitic grainstone. Fenestral fabric and stromatolitic laminations are common. It is interpreted to be a tidal flat

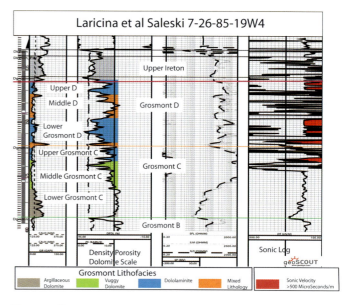

Figure 2. Well logs from Laricina Saleski 7-26-85-19W4 showing the petrophysical expression of the Grosmont C and D bitumen bearing intervals. The gamma-ray and density logs are labeled to show the distribution of Grosmont lithofacies and tripartite subdivisions of the Grosmont C and D. Porosity >12% is colored. Portions of the sonic log are shaded red. These are zones of very low sonic velocity, reflecting high bitumen content of the associated zones.

deposit. It has good intraparticle (because of leaching of carbonate grains or clasts) and intercrystalline (pore space between dolomite crystals) porosity. It commonly has intervals of chaotic solution breccias.

Figure 3. Core samples form the Lower Grosmont C Argillaceous Facies. (a) Subvertical lined burrows in a bioturbated dolomite from well 10-22-85-19W4 (409 m). (b) Two brachiopod-rich layers from well 10-22-85-19W4 (415.25 m).

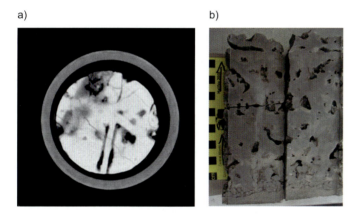

Figure 4. Core samples from the Grosmont C Vuggy Facies. (a) Image of a CT scan of a sample of core showing horizontal burrows from well 7-26-85-19W4. (b) Leached Thalassanoides burrows in a muddy dolomite matrix from well 10-17-84-19W4 (443 m).

The Grosmont C-D Marl (359.5–361 m in well 7-26-85-19W4) separates the Grosmont C and D. It consists of a white dolomudstone with irregular wisps of shale that is capped by a 0.5 to 1 m thick interlaminated siliciclastic green shale and fine dolomitic grainstone interval (see Figure 5a and b). The C-D Marl is thought to have been deposited in quiet shallow waters that were subjected to frequent subaerial exposure on the basis of the presence of mudcracks. It is lithologically similar to the Middle Grosmont D unit.

The Grosmont D has been subdivided into the Lower, Middle, and Upper units (see Figure 2).

The Lower Grosmont D unit (347.5–359.5 m in well 7-26-85-19W4) is a bitumen-saturated megaporosity zone.

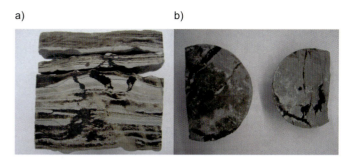

Figure 5. Core samples from the C-D Marl. (a) Interlaminated pale gray shale and light- to dark-brown bitumen stained porous dolomite layers. Irregular breaks in the layering are mudcracks indicating exposure during deposition. (b) Top view of the same mudcracks from well 7-26-85-19W4 (361.1 m).

In the Saleski area, it consists of angular clasts of white dolomite encased in a matrix of bitumen-saturated fine disaggregated dolomite sediment. There are sporadic intervals displaying relict fenestral fabric and laminations with variable grain size that support a tidal flat origin.

The Middle Grosmont D unit (338.3–347.5 m in well 7-26-85-19W4) has a mixed lithology consisting of wispy-laminated dolomudstone overlain by amphipora-floatstone capped by a thin unit consisting of green siliciclastic shale interlaminated with dolomite mudstone. Helminthopsis burrows have been observed in the dolomudstone.

The Middle Grosmont D unit has a moderate amount of interparticle and intercrystalline porosity throughout. The amphipora-floatstone facies has vugular porosity because of leaching of stromatoporoids.

The Upper Grosmont D is a relatively clay-free, bitumen-saturated, laminated dolomite grainstone that corresponds to the 331 to 338.3 m interval in well 7-26-85-19W4. The unit is very fine to fine-grained with some coarser grained interbeds. Fenestral texture and stromatolitic lamination are common. The Upper Grosmont D unit is interpreted to have been deposited in an intertidal to supratidal environment. The primary porosity types are interparticle and intercrystalline porosity. Solution breccia zones up to 3 m thick are frequently present.

Fracturing

The Grosmont C and D units are heavily fractured. Fractures are typically short (generally <10 cm in length), nonplanar, subvertical, and open or occluded by bitumen. Rotation of fractured blocks is widespread. Widening of fractures by dissolution is commonly observed, especially in the Grosmont C Vuggy unit. Fractures are devoid of cement. They compensate for their short length by their large quantity and contribute greatly to vertical permeability.

Their short length, irregular morphology, and apparently random orientation suggest a nontectonic origin for Grosmont fracturing. They are undoubtedly a by-product of karst collapse associated with the Early Cretaceous freshwater dissolution event.

Megaporosity zones

One remarkable aspect of the Grosmont geology in the Saleski area is the presence of megaporosity zones within the C and D cycles of the Grosmont Formation. These zones are defined as having greater than 25% porosity on the basis of neutron-density log measurements. Their occurrence is mainly confined to the Grosmont C Vuggy, Upper Grosmont C, and Upper and Lower Grosmont D units. They are up to 12 m thick and can be correlated over a considerable distance in the Saleski area on the basis of well logs. Figure 6 shows the occurrence of megaporosity zones in well 7-26-85-19W4.

There are two types of megaporosity zones: solution breccia zones and conventional dolomite reservoirs. In core, the solution breccia zones have a distinctly black appearance because of their high bitumen content. Dolomite clasts, when present, are angular, unimodal, and may also be bitumen stained. Breccias vary from mosaic breccias (clast-supported) to matrix/bitumen-supported, depending on the ratio of clasts to matrix (Loucks, 1999) (Figure 7a–c). Interbedded intervals with conformable or tilted bedding and relict sedimentary textures are common.

Matrix/bitumen-dominated solution breccia zones defy scrutiny by conventional core analysis procedures because of their unconsolidated nature. Most of the porosity is intergranular within the breccia matrix. Porosity is often greater than 40%, and permeability is commonly estimated to exceed 10 Darcies.

A subordinate proportion of megaporosity zones are conventional dolomite reservoirs with high matrix porosity. Porosity types include vugular, intraparticle, and fenestral. These units are commonly interbedded with or laterally adjacent to solution breccia zones and are important reservoir units.

Grosmont megaporosity zones at Saleski have a distinct petrophysical signature. In addition to their high log porosity measurements, their resistivity usually exceeds 100 Ω-m and commonly is greater than 2000 Ω-m. Sonic log velocities of megaporosity zones commonly contain intervals of very low sonic velocity (see Figure 3), typically less than 1250–2000 m/s (>500–800 µs/m). Other authors

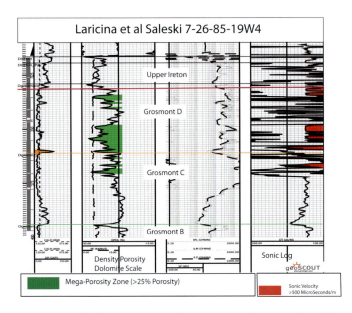

Figure 6. Well logs for Laricina et al., well 7-26-85-19W4, illustrating the distribution of megaporosity zones in the Grosmont C and D.

Figure 7. Core photos of bitumen-saturated breccia facies ranging from clast-supported to matrix-supported fabrics.

have suggested that these readings are spurious and have attributed them to washed out wellbore conditions (Dembicki and Machel, 1996). Such is not the case at Saleski where the caliper log readings through the breccia zones are typically in-gauge. Also the corresponding density log readings, which are notoriously more sensitive to wellbore washout, show no signs of bad hole conditions.

It is our interpretation that these sonic velocity readings are valid but are not representative of porous dolomite velocities. The velocities are close to the expected value for bitumen (Mochinaga et al., 2006). It is speculated that sonic velocities through bitumen-saturated, matrix-supported breccia zones would be closer to pure bitumen velocities than that of dolomite.

Origin of Breccia Zones

Two periods of diagenesis have played a major role in Grosmont reservoir development. Dolomitization of the Grosmont occurred in one or more stages but is believed to have been complete in early Mississippian time (Huebscher and Machel, 2006). Some porosity development probably occurred at that time, but not of the magnitude that is found here today.

The second phase of reservoir development occurred during the pre-McMurray formation erosional event. The Devonian subcrop was exposed to fresh surface water, which percolated into the Grosmont along the subcrop margin and also downward along faults and fractures (Figure 8). Preexisting reservoir development would have facilitated the flow of groundwater within the Grosmont.

Limestone and dolomite are most soluble at low temperatures. Secondly, the solubility of limestone and dolomite are inversely proportional to water salinity. Consequently, the near-surface environment above sea level offers the optimum setting for carbonate dissolution. Rainwater is a particularly effective agent for dissolving limestone and dolomite because it is weakly acidic by nature.

Most nonbreccia related porosity development in the Grosmont Formation is the result of selective leaching. Intraparticle and vugular porosity are by products of preferential removal of skeletal grains, fossil material, and burrows (see Figure 3b and Figure 9). There is a strong bias toward the preferential dissolution of coarser grains. Clearly the material that has been selectively leached was more soluble than the remnant material.

What was the source of the solubility contrast?

The theory of this chapter is that the chemistry of the affected grains is the key. Dolomite is much less soluble than calcite. Our hypothesis is that carbonate grains that contained an elevated calcium carbonate content were

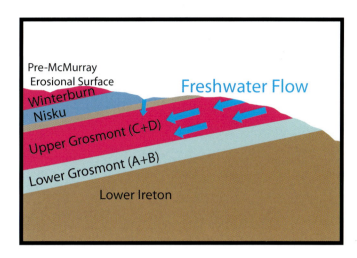

Figure 8. Model for the development of porosity along the Grosmont subcrop during Cretaceous age exposure to freshwater leaching.

Figure 9. Intraparticle porosity caused by selective leaching of individual grains in a carbonate grainstone from well 10-17-84-19W4 (437 m).

preferentially dissolved because they were less stable in the presence of freshwater.

There are several lines of evidence that calcite that survived dolomitization has been dissolved during meteoric diagenesis.

- *Rock property data:* The Grosmont C and D units have very low calcite content. Normally there is some relict calcite after dolomitization. Measured rock densities from Grosmont log measurements and core analyses are close to the expected values for a pure dolomite.

- *Thin section data:* Figure 10 is a photomicrograph of a thin section taken from the Grosmont C Vuggy unit in well 10-26-85-19W4. Two types of dolomite crystal developments are present. There are large irregular

Figure 10. Photomicrograph of a thin section showing individual dolomite rhombic crystals and irregular clumps of subhedral dolomite crystals. The sample was impregnated with a blue epoxy to highlight the porosity. Porosity was formed when freshwater dissolved calcite that had been preserved during the earlier dolomitization event. Taken from well 10-26-85-19W4, 367.69 m, top of Grosmont C Vuggy.

Figure 11. Scanning electron image of dolomite rhombs from the Grosmont D from well 6-34-85-19W4 (350.16 m).

masses of crystals (Idiotopic-S texture of Greg and Sibley, 1984) and discrete dolomite rhombs that barely touch one another (Idiotopic-E texture, Greg and Sibley, 1984). Dolomite crystals consumed the preexisting calcite as the rock was dolomitized. The Idiotopic-E texture forms when dolomite crystal growth is interrupted, leaving the regions between crystals undolomitized. Later dissolution removed the matrix calcite, leaving the more resistant dolomite crystals surrounded by pore space.

- *Scanning electron microscopy and X-ray diffraction data:* Five samples were collected within two wells (10-26-85-19W4 and 6-34-85-19W4) to examine the nature of the internal sediment within solution breccia zones. Examination of this sediment by scanning electron microscopy reveals that the main constituent of this sediment is a dolomite dust consisting of well formed dolomite crystals (Figure 12). X-ray diffraction (XRD) analysis of these samples reveals that they contain a high dolomite content (see Figure 6). Calcite is virtually absent from the analyses. Jakucs (1977, pp 71–76) described this phenomenon from European examples of karsted dolomitic terrain. The origin of this fine dolomite sediment was also attributed to preferential dissolution of calcite leaving a residue of dolomite crystals.

It is clear that dolomite was not totally immune from dissolution during the freshwater dissolution event. It is evident from Figure 13 that some corrosion of dolomite crystals has taken place that is most likely attributed to freshwater leaching. Creation of thick breccia zones in intervals that are largely dolomite requires indiscriminate dissolution of dolomite and calcite. This would only be accomplished by the flow of large quantities of freshwater through the megaporosity interval. At Saleski, this was a very efficient process because there is virtually no carbonate cement in the Upper Grosmont. The dissolved material was discharged at the surface or it was precipitated deeper in the subsurface as carbonate cement.

The process that has been described here has all of the attributes of karst drainage dynamics. Chemical denudation of a carbonate terrain by fresh meteoric water resulted in the development of a subterranean drainage system.

One aspect of karst drainage is the development of cave systems. Caves commonly develop when subterranean flow is diverted into a single path along regional joint systems. There are several negative assumptions about caves and their effect on reservoir quality. One is that caves are thought to be narrow, widely spaced flow systems. They are generally assumed to contain a significant volume of argillaceous sediment derived from the surface mixed with breccia material spalled off of the cave roof and walls. It is assumed that cave systems are inherently unstable and prone to collapse after a minimal amount of burial, and that the host rock in which caves formed has low reservoir quality.

These assumptions are largely true for karsted limestone terrains but do not appear to be true for the effects of groundwater leaching of a dolomitic terrain. At Saleski, the Upper Grosmont has stratigraphically controlled porosity development (Figure 14) with the highest porosity confined largely to the tidal facies deposits. Breccia development was an overprint on the porosity system because of high groundwater flow through the porous beds. Unconsolidated, porous disaggregated dolomite sediment can be found in the cavernous porosity. Cavern collapse has not been recognized at Saleski on the basis of seismic data and well control.

X-Ray Diffraction Data

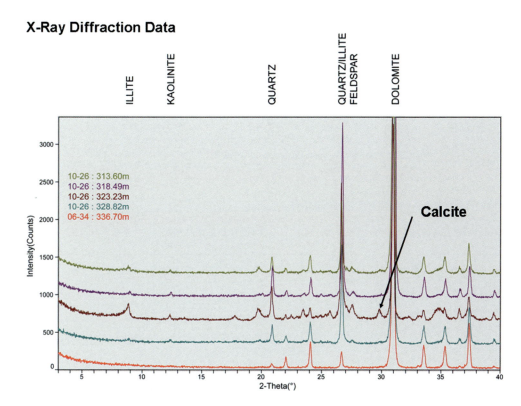

Figure 12. The results of six XRD analyses from Grosmont samples at Saleski. Dolomite, quartz, and clays dominate the samples. The calcite content is almost nil.

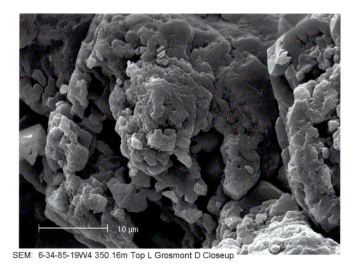

Figure 13. Closeup of Figure 10 showing evidence of corrosion of dolomite crystals during freshwater dissolution.

Exploitation of the Bitumen Resource

Laricina Energy is considering several exploitation strategies to gain bitumen production from the Grosmont Formation at Saleski. One promising strategy is the injection of steam to mobilize the bitumen. The most commonly used practice in Western Canada is the SAGD (Steam Assisted Gravity Drainage) method in which two stacked horizontal wells are drilled approximately 5 m apart. The upper well is used to inject steam into the reservoir. As the steam rises, the bitumen bearing strata heats up. As bitumen's temperature increases, its viscosity decreases until it becomes mobile and drains by gravity to the lower producing horizontal well.

Why is the Grosmont an attractive SAGD candidate?

There are at least five reasons why SAGD is appropriate for extraction of bitumen from the Grosmont Formation:

1) *Size:* A large resource is needed to justify the high capital costs of building a SAGD facility. The Grosmont has large reserves and a thick bitumen column.

2) *Predictability of reservoir units:* Before embarking on a steam injection project, certainty is required in terms of continuity of reservoir units. Geological cross sections demonstrate that reservoir units extend over large distances and that they maintain their reservoir quality.

3) *Good permeability:* High reservoir permeability is essential. A commercially viable bitumen zone must be able to accept large quantities of steam. The Grosmont bitumen reservoirs have high porosity and permeability. Vertical permeability is particularly important for a successful SAGD process. Steam needs to be free to rise within the reservoir. Once the heat from the steam has mobilized the bitumen it needs to be able to flow downward to the producing well. The Grosmont reservoirs have excellent vertical permeability because of their high fracture density.

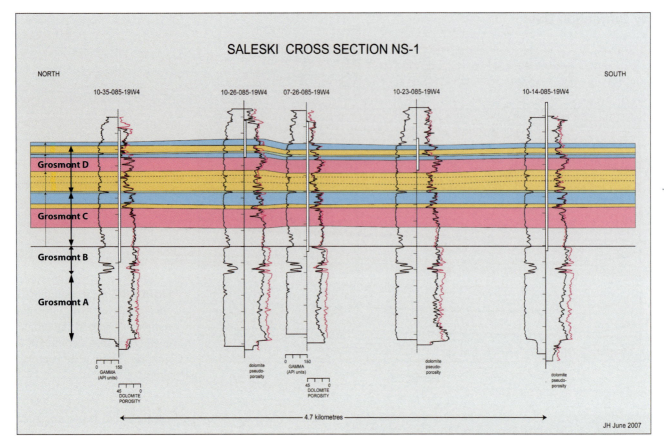

Figure 14. North-south trending cross section demonstrating the continuity of reservoir units within the Upper Grosmont formation.

They also do not contain shale beds that would provide effective barriers to steam flow.

4) *Good top seal to steam flow:* The Grosmont is overlain by the Wabiskaw shale, which is a competent but ductile unit that will prevent the flow of steam upward into the overlying Cretaceous strata.

5) *Lack of bottom water and thief zones:* For the region around the proposed Saleski Pilot and much of the updip portion of the Grosmont trend, there is not an underlying aquifer. Therefore, the only water produced during a SAGD process should be condensed steam. The Grosmont at Saleski also is free of a gas cap or lean bitumen zones that could diminish the effectiveness of the steam process.

Conclusions

The main conclusions of our studies are as follows:

- The Grosmont C and D units have each been subdivided into three subdivisions on the basis of log response and depositional environment.
- Reservoir quality within the Grosmont C and D is strongly controlled by depositional environment. Tidal facies sediments were the preferred strata for the best reservoir development.
- Intervals with greater than 25% porosity have been arbitrarily defined as megaporosity zones on the basis of neutron-density open hole logs. These zones occur preferentially but not exclusively within tidal flat deposits.
- Selective dissolution of calcite-rich grains and strata by freshwater flow during the pre-McMurray formation erosional event was responsible for most of the porosity development. Breccias formed when high groundwater flow dissolved and excavated dolomitic strata and calcite-rich beds.
- The Grosmont is a good candidate for SAGD exploitation because of its large bitumen reserves, high reservoir quality and vertical permeability, predictability of reservoir units, lack of bottom water and thief zones, and good top seal.

Acknowledgments

The authors thank the management of Laricina Energy Limited and OSUM Oil Sands Corporation for permission and support in preparing this paper.

References

Alberta Energy and Resource Board, 1996, Crude bitumen reserves atlas: Statistical series 96-38.

Cutler, W. G., 1983, Stratigraphy and sedimentology of the Upper Devonian Grosmont Formation, Northern Alberta: Bulletin Canadian Petroleum Geology, **31**, 282–325.

Dembicki, E. A., and H. G. Machel, 1996, Recognition and delineation of paleokarst zones by use of wireline logs in the bitumen-saturated Upper Devonian Grosmont Formation of northeastern Alberta, Canada: AAPG Bulletin, **80**, 695–712.

Gregg, J. M., and D. F. Sibley, 1984, Epigenetic dolomitization and the origin of xenotopic dolomite texture: Journal Sedimentary Petrography, **54**, 908–931.

Huebscher, H., and H. G. Machel, 2006, Reflux and burial dolomitization in the Upper Devonian Woodbend Group of north-central Alberta, Canada: CSPG Dolomite Conference, Calgary, Alberta.

Jakucs, L., 1977, Morphogenics of Karst regions, variants of Karst evolution: Akademiani Kiado.

Loucks, R. G., 1999, Paleocave carbonate reservoirs: origins, burial-depth modifications, spatial complexity, and reservoir implications: AAPG Bulletin, **83**, 1795–1834.

Mochinaga, H. S., S. Onozuka, T. Kono, A. Ogawa, A. Takahashi, and T. Jogmec, 2006, Properties of oil sands and bitumen in Athabasca: 2006 CSPG-Canadian Society of Exploration Geophysicists-Canadian Well Logging Society Annual Convention abstract.

Chapter 11

Deterministic Mapping of Reservoir Heterogeneity in Athabasca Oil Sands Using Surface Seismic Data

Yong Xu[1] and Satinder Chopra[2]

Bitumen reserves in oil sands in Alberta, Canada represent one of the biggest such deposits in the world. The Athabasca region contains the bulk of this resource, and the Lower Cretaceous McMurray formation contains the most significant target interval. Inclined heterolithic strata and associated sand accumulations comprise most of the formation. However, the distribution of bitumen in the formation varies because of the high degree of facies heterogeneity throughout the deposit. This lithological heterogeneity causes difficulties in interpreting geology and estimating the bitumen distribution.

Surface seismic data could play an important role in characterizing the subsurface heterogeneity because they provide lateral and vertical coverage and a link to rock physics through amplitude variation with offset (AVO). However, most (with notable recent exceptions using deterministic lambda-mu-rho, shown by Bellman, 2007 and Evans and Hua, 2008) applications of surface seismic in Athabasca have been to provide attributes for statistical and neural network predictions (Tonn, 2002; Anderson et al., 2005). The relationships between seismic data and lithology are determined at the well control points by multivariate analysis or neural networks and then the lithology between wells is predicted from these relationships. However, interpreters often find these relationships less straightforward than conventional techniques.

In this chapter, we describe a two-step approach to understanding the heterogeneity of Athabasca oil-sands reservoirs. The first step involves a rock physics study to understand the relationship among lithology and the related rock parameters and pick lithology-sensitive rock parameters that can be seismically derived. The second step is deriving the chosen parameters from the seismic data.

We demonstrate this method with a case history that begins with rock physics analysis of an Athabasca reservoir zone using well log data, and then uses seismic inversion to derive the lithology-sensitive parameters from a 2D profile. For the case study presented here, the derived results are encouraging as they calibrate with the available log curves, and a blind well test confirmed the accuracy of the calibration.

Rock Physics Analysis

We begin the rock physics analysis by crossplotting (Figure 1) different pairs of parameters for the McMurray formation reservoir which is at a depth of about 100 m. Figure 1a shows a strong linear correlation between bulk density and gamma ray. Clean sand samples have average densities of 2.075 g/cc with average gamma-ray values of API 25°, whereas 100% shale has an average density of 2.24 g/cc with average gamma-ray value of API 85°. If a linear relationship between gamma ray and shale volume (Vshale) is assumed, then Vshale can be estimated from density by using the relation Vshale = (density − 2.075)/0.165. Figure 1b reveals a weak correlation between gamma ray (shale volume) and V_P/V_S. Because P impedance can be accurately derived from seismic data, it is always desirable to look for any strong correlation between impedance and another rock parameter of interest. However, as seen in Figure 1c and d, P impedance is unable to indicate lithology variation, because in this case, shale and sandstone have similar P-wave impedance values as shown by the uncorrelated gamma ray and density scatter.

In the Athabasca region, the depth of the McMurray formation varies laterally from very shallow to over 600 m. Such a large variation in depth means the rock parameters in different areas exhibit different behavior due to

[1]Formerly at Arcis Corporation, Calgary; now with Imperial Oil, Calgary, Alberta, Canada.
[2]Arcis Corporation, Calgary, Alberta, Canada
This paper appeared in the September 2008 issue of THE LEADING EDGE and has been edited for inclusion in this volume.

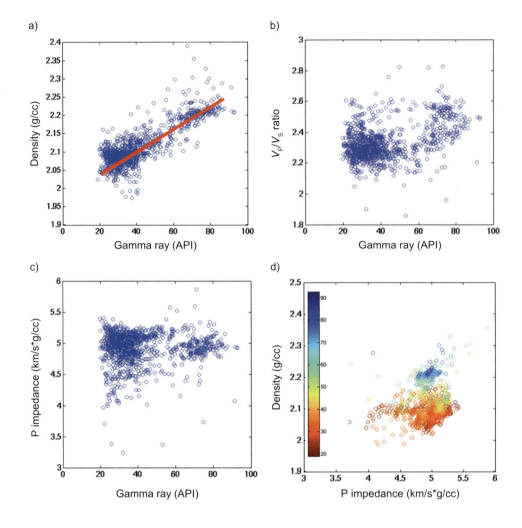

Figure 1. Crossplots of (a) density versus gamma ray, (b) V_P/V_S versus gamma ray, (c) P impedance versus gamma ray, and (d) density versus P impedance, with colors coded by gamma-ray values. Data samples come from the McMurray formation in four wells in the Athabasca oil sands. Reservoir depth is 100 m. A linear relationship between density and gamma ray (red line in Figure 1a) can be used to estimate Vshale (or pseudo-gamma ray) from density.

varying overburden compaction. Figure 2 shows crossplots from a McMurray reservoir at a depth of 400 m in an area different from the one under study. These crossplots suggest good correlation between P-impedance and density. Consequently, the relationship between these parameters could be a lithology indicator, though density as such is still a good indicator. Thus, rock physics analysis in a given area is important for, first, determination of those rock parameters that may exhibit some useful lithology-correlated relationship and, then, using this relationship to estimate such attributes from seismic data.

Workflow for Mapping Reservoir Heterogeneity

Figure 3 shows our workflow for mapping reservoir heterogeneity in the study area. This workflow is based mainly on conventional P-wave surface seismic, although we believe it can be extended to incorporate multicomponent data. As stated earlier, because of the heterogeneity within the formation and weak correlation between seismic (P impedance or reflectivity) and lithology, "normal" attempts at geologic interpretation usually prove futile. We address this problem by using AVO attributes from surface data.

Because the reservoir is shallow and seismic data usually have sufficiently high resolution in shallow zones, it was expected that reasonably convincing estimates of reservoir heterogeneity could be obtained. Note that our approach should not in any way discourage the application of statistical methods for the same goal. When more than a couple of seismic attributes are available, neural network approaches could determine reservoir properties within the interval of interest. However, we emphasize that the approach in this work for estimating the lithology-sensitive density reflectivity attribute provides good quality control and validation with well ties.

Improved Three-term AVO inversion

Linearized three-term AVO inversion is commonly used to extract P-, S-, and density reflectivity from prestack seismic data; however, straightforward application on a sample-by-sample basis can generate unreliable solutions. This is because of the ill-posed nature of the

Chapter 11: Deterministic Mapping of Reservoir Heterogeneity in Athabasca Oil Sands

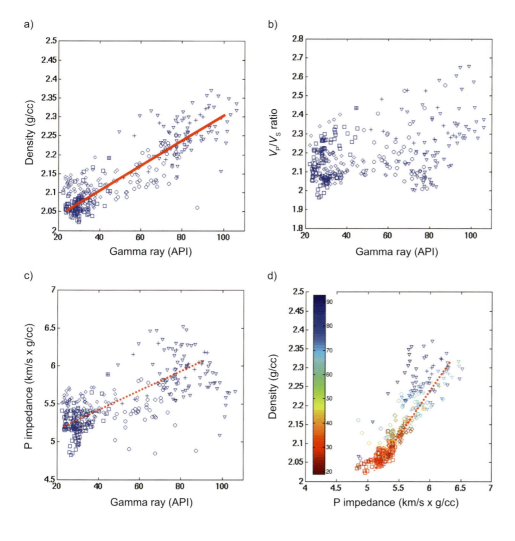

Figure 2. Crossplots of rock physics parameters using samples from an Athabasca oil-sands well in a different area than the study area. Crossplots are arranged in the same way as in Figure 1. The McMurray reservoir is at a depth of approximately 400 m. Although the correlation between the density and gamma ray is strong, the correlation between P-impedance and gamma ray is much stronger compared with Figure 1. This can be seen in c and d.

inverse problem necessitating the use of certain constraints for stabilizing the inversion. Furthermore, large incident angles are required for reliable results from three-term inversion (Downton and Lines, 2001; Roy et al., 2006). In addition to this, the use of statistical constraints in the inversion yields a solution that usually exhibits a reasonable variance but underestimates the lithology anomalies of interest.

Our method improves the three-term linear AVO inversion by (1) using a windowed approach instead of a sample-by-sample basis, (2) applying error-based weights in the frequency domain, (3) reducing the uncertainty of the inverse problem, (4) accounting for the strong reflection from the McMurray–Devonian interface, and (5) reducing the distortion due to normal moveout (NMO) stretch and the related offset-dependent tuning. This results in a more reliable inversion process and also relaxes the requirement for large angles in the inversion.

Figure 4 compares the density reflectivity derived by the application of different AVO inversion methods on a synthetic gather that has an angle range of 0–40°. The true density reflectivity used to evaluate the inversion is calculated using the density log. The results clearly demonstrate that the density reflectivity derived by the improved three-term inversion is closer to the true reflectivity than the density reflectivities derived by other methods.

Application to Real Data

A 2D seismic profile running through 11 wells in the study area was taken through an amplitude-preserved AVO processing flow. Data quality was reasonably good, and the usual noise problems in terms of ground roll and other wave modes were skillfully tackled using adaptive and iterative noise-attenuation schemes. The surface elevation variation along the profile is approximately 45 m, and this is of the same order as the reservoir depth variation of 55–90 m. This elevation variation was a significant factor regarding the amplitude recovery and the stability of the AVO inversion at the reservoir level. Care was exercised for amplitude recovery, and superbinning was part of the data conditioning for AVO inversion.

Figure 3. Workflow of deterministic interpretation of reservoir heterogeneity of oil sands using seismic data. This approach combines rock physics with a reliable inversion and quality control including synthetic and well tie calibration with colored seismic attributes.

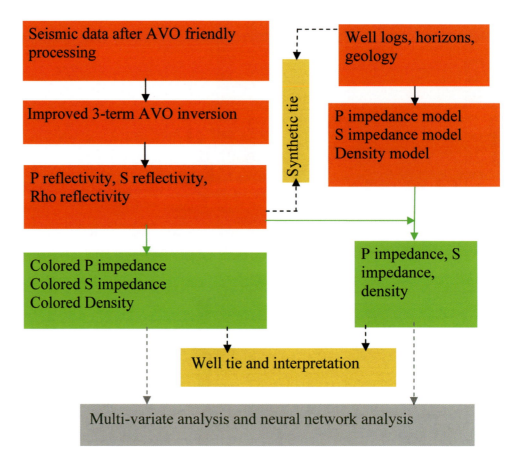

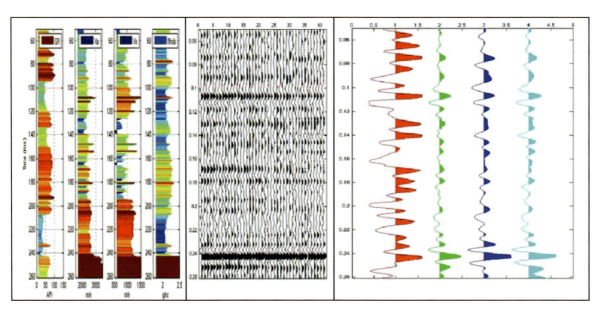

Figure 4. Comparison of density reflectivity derived by application of different AVO inversion methods on a synthetic gather. Well logs to generate synthetic gather are shown in the left panel. The synthetic gather in the middle panel (inclusive of noise) is sorted with respect to angle and the range of angles used is from 0° (left) to 40° (right). In the right panel, the red trace is density reflectivity obtained from the application of least-squares inversion and scaled down three times for display, the green trace is obtained after application of Bayesian-constrained AVO inversion, the blue trace is the true density reflectivity calculated from the density log, and the cyan trace on the extreme right is obtained by the application of the improved three-term AVO inversion. Clearly, the three-term AVO estimate shown in the cyan trace is very close to the true density reflectivity and is also superior to reflectivities derived by other methods.

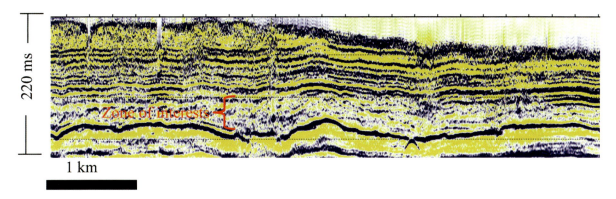

Figure 5. Stack section showing the zone of interest.

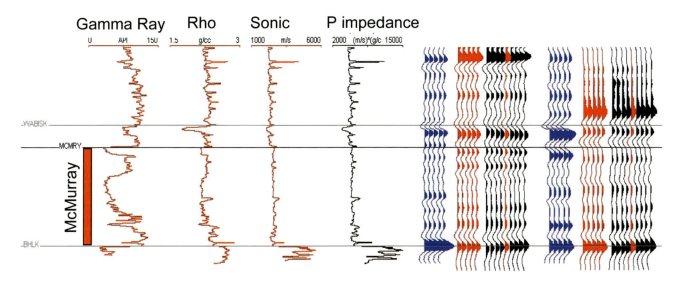

Figure 6. Synthetic tie with AVO derived reflectivities. Blue traces are synthetic reflectivity, red traces are inverted reflectivity at the well location, and black traces are a portion of the inverted reflectivity section at the well location. The left group of traces represents P reflectivity, and the right group of traces represents density reflectivity.

Figure 5 shows a stacked section for the seismic profile with the zone of interest indicated.

Figure 6 shows log curves and synthetics for a typical well and their correlation with the derived P impedance and density reflectivities from seismic data. The correlation between the two pairs of reflectivities is reasonably good and encouraging. Figure 7 shows the results of different AVO attributes derived as per the workflow in Figure 3. The density reflectivity derived after AVO inversion is shown in Figure 7a.

Colored density was derived from density reflectivity after simple trace integration without using density logs from wells, and this result (Figure 7b) indicates the richest sand areas (dark green) are in the middle of the McMurray formation with a good seal cap in the Upper McMurray around wells 5 and 6. These are verified by the overlain gamma-ray log curves.

Next, the density logs from all of the wells except wells 3 and 7 (not available at the time) were used to generate a density model. Model-based poststack inversion was performed on density reflectivity utilizing the density model to generate a density section (Figure 7c). This section has higher resolution than the colored density and better matches the log curves. The linear relationship indicated on the crossplot between the density and gamma ray (Figure 1a) was used to transform the derived density section into a Vshale section (Figure 7d). Two recently drilled wells (3 and 7) were used in a blind test; well 3 is mainly shaley within the McMurray, and the density inversion result verifies this. Well 7 indicates good sand in the middle McMurray but a sandy cap in the Upper McMurray. These results are clearly confirmed on the inverted density (Figure 7c) and the derived Vshale sections (Figure 7d). The same results are seen on the colored density sections.

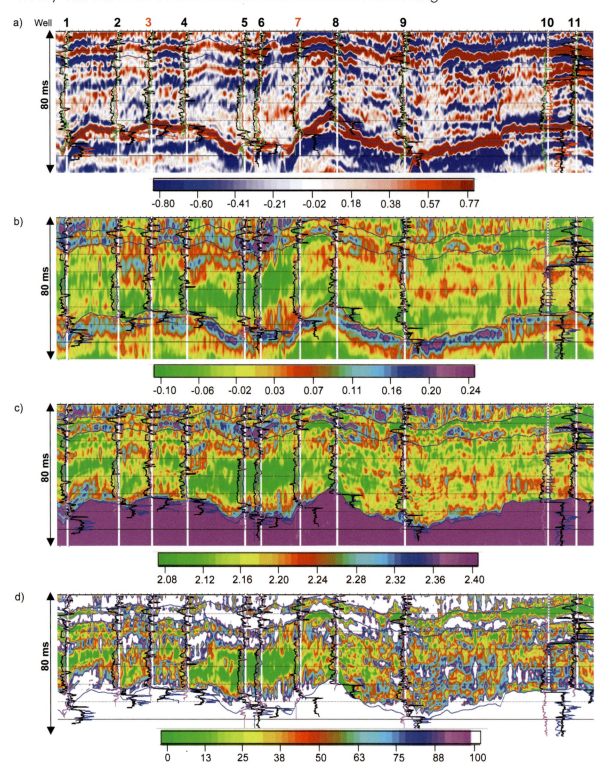

Figure 7. (a) Density reflectivity, (b) colored density (the trace-integration version of density reflectivity), and (c) the density section from model-based inversion. (d) Vshale transformed from density in (c) using the linear relationship between density and gamma ray shown in Figure 1a. Log curves are overlaid on the section. In (b, c, and d), the black curves are density logs, the purple curves are gamma-ray logs, and the blue curves are impedance logs. No density logs are used in the derivation of (b), and density logs from all wells except wells 3 and 7 and horizons are used to generate a density model to derive (c) from a using model-based poststack inversion. The middle McMurray is usually the reservoir, whereas the Upper McMurray is cap rock. The richest sand areas (dark green) within the middle McMurray are around wells 5 and 6, with good shaley cap rocks in the Upper McMurray; these are verified by gamma-ray logs of both wells. Recently drilled wells 3 and 7 serve as blind well tests. Well 3 is mainly shaley within McMurray, and the density inversion result verifies this. Well 7 is drilled at the edge of the richest sand zone, and its reservoir also matches the inversion results. In addition, the sandy cap rock within the Upper McMurray in well 7 is convincingly predicted by the inversion.

All four derived density estimate sections in Figure 7 yield encouraging confirmation with well logs and exhibit believable lateral variation in reservoir heterogeneity within the target zone.

Conclusions

A new workflow using rock physics analysis and an improved three-term AVO inversion to map reservoir heterogeneity has been demonstrated on a case study from the Athabasca oil sands in Alberta. Rock physics analysis helps find a relationship between lithology and seismically driven elastic attributes and picks out lithology-sensitive parameter(s). In the study presented here, density is closely correlated with lithology. However, with increasing depth of burial, other rock physics parameters (e.g., acoustic impedance) may also correlate with lithology. The density reflectivity is reliably derived from 2D data in the area using the improved three-term AVO inversion. Other attributes derived from density reflectivity, confirmed with calibration to existing well log data, have further provided convincing calibration to the two recently drilled wells.

The discussed workflow has successfully demonstrated a methodology for mapping heterogeneity in oil-sands reservoirs. Considering the importance of the characterization of oil-sands reservoirs in Alberta and other places around the world, this methodology could have very promising applications.

Acknowledgments

We thank an anonymous company for allowing us to show the data example in this case study. We thank Arcis Corporation for encouraging the development of this work and for permission to present these results.

References

Anderson, P. F., L. Chabot, and F. D. Gray, 2005, A proposed workflow for reservoir characterization using multicomponent seismic data: 75th Annual International Meeting, SEG, Expanded Abstracts, 991–994.

Bellman, L. W., 2007, Oil sands reservoir characterization: a case study at Nexen/Opti Long Lake: CSEG, Expanded Abstracts, 640–641.

Downton, J. E., and L. R. Lines, 2001, Constrained three-parameter AVO inversion and uncertainty analysis: 71st Annual International Meeting, SEG, Expanded Abstracts, 251–254.

Evans, J., and Y. Hua, 2008, Early project seismic application at the Ellis River heavy oil project-maximizing the value of VSP and well data: CSEG, Expanded Abstracts.

Roy, B., P. Anno, and M. Gurch, 2005, Wide-angle inversion for density: tests for heavy-oil reservoir characterization: 75th Annual International Meeting, SEG, Expanded Abstracts, 1660–1664.

Tonn, R., 2002, Neural network seismic reservoir characterization in a heavy oil reservoir: The Leading Edge, **12**, 309–312.

Chapter 12

Imaging Oil-sands Reservoir Heterogeneities Using Wide-angle Prestack Seismic Inversion

Baishali Roy,[1] Phil Anno,[1] and Michael Gurch[1]

Introduction

Mass density, because of its linear relationship with porosity, has long been recognized as a potential seismic indicator of fluid saturation. Given its dependence on mineral composition, density can also be diagnostic for lithology. In this chapter, we discuss some key aspects of a wide-angle processing and density inversion workflow and apply it to a bitumen reservoir in Canada for imaging reservoir heterogeneities (e.g., shales) that can potentially act as permeability baffles. In this field, intrareservoir shales typically have higher densities than surrounding reservoir sands. This wide-angle workflow yields stable density estimates, from reflected P-waves alone, at a resolution suitable for mapping the intrareservoir shales.

This study is based on data from the Surmont bitumen reservoir approximately 60 km southeast of Fort McMurray, Alberta, Canada, in the Lower Cretaceous McMurray formation. The oil is too deep (400 m) to mine. Steam-assisted gravity drainage (SAGD) technology is being used to inject steam into the reservoir and heat the oil so that it can be produced. Shale heterogeneities within the reservoir (Figure 1) thicker than 3 m could have an impact on steam chamber development and affect SAGD performance. Predicting the areal extent and the thickness of these bodies would lead to better reservoir management.

Petrophysical analysis of well log data points to density as the elastic property with the most consistent shale-to-reservoir contrast in the McMurray formation. The intrareservoir shale bodies typically have higher density than surrounding reservoir sediments. Figure 2a shows logged densities against gamma-ray log data (a lithology indicator). Nonreservoir rocks have slightly higher densities on average than bitumen sands. Although inversion for impedance and Poisson's ratio is almost routine today, Surmont shale prediction from these two properties alone suffers from large uncertainties. In Figure 2b, the plot of velocity versus gamma-ray data shows very little correlation to lithology. These observations provided the original motivation for improving the density estimates produced through seismic inversion. Although uncertainty is not eliminated entirely, Surmont oil sands are much better discriminated from shale by density.

However, estimation of density from seismic is notorious for its difficulty (Debski and Tarantola, 1995). AVO modeling of reflectivity demonstrates the need for a wide aperture of reflection angles to constrain density estimates from P-wave inversion. For this oil-sands reservoir, we demonstrate a stabilizing constraint on density estimates through additional reflection data acquired and processed beyond 40°. In this case, we achieved reflection data up to 60° at the reservoir depth. Khare et al. (2007) have also mentioned the value of using large-angle P-P data to achieve stability in inversion.

Amplitude variation with offset (AVO) analysis and inversion routinely utilize reflection angles only up to 35–40°. These more typical angle limits essentially eliminate seismic information on density, even for data with only a small amount of noise. Given these typical data limitations, a density estimate is usually not attempted from inversion of P-waves alone. Instead, two-parameter inversions are carried out for, say, intercept and gradient or impedance and Poisson's ratio, leaving the density term as an unknown third independent parameter.

Gray (2005) sought to reduce the burdens introduced by wide-angle acquisition and processing. He instead inverted P-S data, in which density information is imprinted in reflection amplitudes at smaller angles. Of course a key tradeoff is the requirement for multicomponent acquisition and

[1]ConocoPhillips, Houston, Texas, U.S.A.
This paper appeared in the September 2008 issue of THE LEADING EDGE and has been edited for inclusion in this volume.

processing. And it remains to be seen whether modes with an S-wave component can retain the resolution required for mapping intrareservoir shale bodies. However, multicomponent recording contains full wavefield information that can be used in polarization filtering to attenuate surface-wave noise (Chiu et al., 2007) and increase the fidelity of near offsets.

Our field test indicates that data from an additional 15–20° of reflection-angle aperture, beyond the conventional 40° far-angle limit, act as an effective constraint on the inversion density solution. However, these wide (60°) reflection angles cannot be successfully processed and inverted with conventional workflows. Some key wide-angle considerations include anisotropic prestack imaging, a wavelet stretch correction, compensation for inelastic losses, and regularization of the inversion solution.

Additional requirements are set on the bandwidth of the P-P reflection data to resolve shale bodies with thickness of 3–5 m. Fortunately, the P-wave data acquired with a dynamite source contain a broad band of frequencies that can be processed and preserved for the target depth (Figure 3).

Angle Requirements and Difficulties of Wide-angle Inversion

Angle requirements for robust wide-angle inversion can be qualitatively understood through forward modeling. Angle requirements are higher when the density contrast of the adjacent lithologies is small (approximately 10%), and the angle aperture requirement diminishes if the contrast is higher (>20%). Figure 4a and b shows the contributions from the density term of the Aki and Richards approximation (equation 1) of P-P reflectivity versus angle

$$r(\theta) \approx c_1(\theta; \overline{\frac{V_S}{V_P}}) \frac{\delta Z_p}{Z_p} + c_2(\theta; \overline{\frac{V_S}{V_P}}) \delta\sigma + c_3(\theta; \overline{\frac{V_S}{V_P}}) \frac{\delta\rho}{\rho} \quad (1)$$

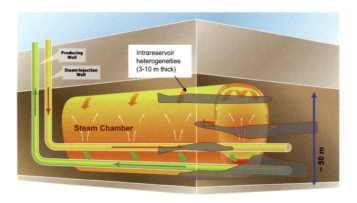

Figure 1. Intrareservoir heterogeneities (e.g., shale lenses thicker than 3 m) can act as baffles during steam chamber growth.

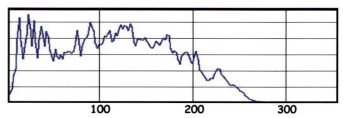

Figure 3. Broad bandwidth of P-P seismic reflection data (20–200 Hz) helps to resolve thin shale layers.

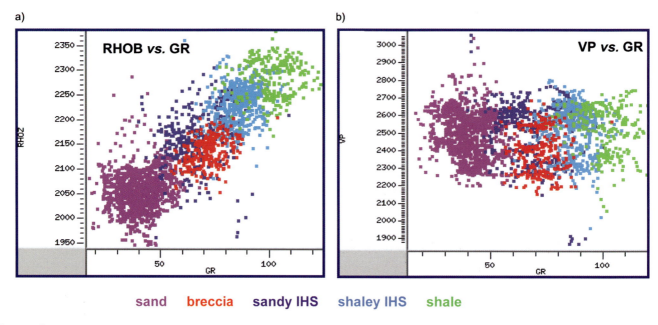

Figure 2. (a) Density log data from the area show a high correlation with gamma ray. (b) Sonic data show a weak correlation, implying density is the better lithology indicator.

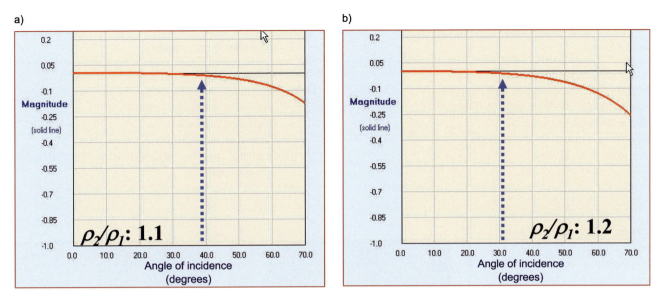

Figure 4. (a) Reflectivity versus angle curve for a 10% density contrast between sand and shales keeping other parameters constant. This shows the requirement for angles >50° in the seismic data to observe density effects above the background noise level. The requirement of larger angles is less with larger density contrasts (20%) such as shown in (b).

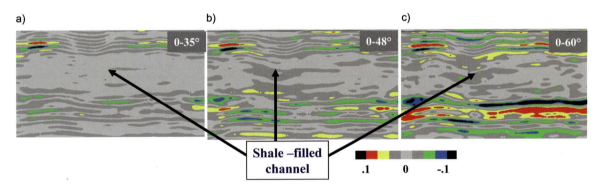

Figure 5. Band-limited density cross sections obtained from maximum reflection angles of (a) 35°, (b) 48°, and (c) 60° reflection data. The data clearly indicate the value of the additional wide-angle data for reservoir characterization as shown by the shale-filled channel.

where $r(\theta)$ is the P-wave reflectivity as a function of the reflection angle θ parameterized in terms of the impedance, Poisson's ratio, and density perturbations. c_1, c_2, and c_3 are constants that include the geometrical terms and the background V_S/V_P ratio. Impedance and Poisson's ratios are kept unchanged at a sand-shale boundary. Figure 4a shows that any detectable density signal for a 10% density contrast is beyond the 50° reflection angle; for a 20% contrast, 45° may contain the necessary density signal. Most measured sand-shale density contrasts in the bitumen reservoir at Surmont are approximately 10%. The value of the larger angles for density inversion is also demonstrated by the real data examples shown in Figure 5. This figure shows the added characteristics in the reservoir architecture, including the channel structure shown by the arrows. Adequately designed forward modeling exercises show in this case that the elastic property contrasts between the sand and shales are small enough that linearized approximations of the Zoeppritz equations are sufficiently accurate for large reflection angles.

Difficulty in density inversion can be attributed to the difficulties (Table 1) in processing and preserving amplitudes of large-offset data. It requires additional processing effort to preserve signal-to-noise ratio (S/N) and relative amplitudes and frequency from the near angles that are dominated by multiples and ground roll to mid angles that are relatively clean to the far angles dominated by stretch, anisotropy, and overall degradation of signal quality.

Wide-angle processing and inversion for density

A wide aperture of reflection angles — including angles well beyond 40° — produces a significant constraint

Table 1. Difficulties in wide-angle processing and its effect on image quality.

Physical cause	Effect on image quality
Signal-to-noise ratio	Degrades image quality and inversion
Nonhyperbolic moveout from long raypaths	Imaging/positioning errors
Anisotropy (such as VTI from overburden)	Imaging/positioning errors
Wavelet stretch due to imaging	Decreases resolution
Inelastic losses (effective attenuation)	Decreases resolution
Density inversion: ill posed	Degrades inversion quality and stability, introduces nonuniqueness

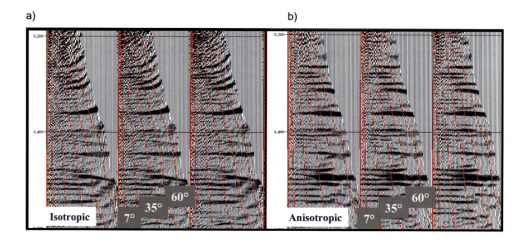

Figure 6. (a) Isotropic imaging of prestack gathers. (b) Anisotropic imaging of the same gathers. Notice the improvement in flattening the reflections at larger offsets.

on the P-wave estimates of density contrasts within the reservoir. However, without appropriate preprocessing, the wide-angle prestack data will degrade rather than constrain inversion. This workflow is designed to preserve wide-angle reflection amplitudes consistent with the small-angle data. Ground-roll suppression using polarization filters on the 3-C data (Chiu et al., 2007) was key to improving the S/N of the nears on the P-P data. Anisotropic imaging and velocity analysis are required to image large-angle reflections to their corresponding small-angle image time. A wavelet stretch correction minimizes distortion introduced through imaging as a function of reflection angle. Inversion regularization through very general prior information takes over where the data leave off, providing an additional constraint on an otherwise nonunique density solution.

Anisotropic imaging

Figure 6 compares isotropic and anisotropic imaging with a Kirchhoff prestack time migration. Anisotropic imaging produces a superior result beyond 40°. The same small-angle, hyperbolic velocity field was used in both cases. The anisotropic result also relies on an additional parameter field that characterizes moveout at large offsets. Both parameter fields are updated after imaging with a residual analysis for each gather. A secondary benefit is that anisotropic imaging also improves Radon demultiple on these data. Residual moveout from isotropic imaging reduced the distinction in the Radon domain between primaries and multiples.

Wavelet stretch correction

Wavelet stretch due to imaging significantly alters wide-angle reflectivity parameters. Roy et al. (2005) introduced an analytic correction for wavelet stretch. We implement this correction as a stationary deconvolution operator on traces formed at a constant reflection angle. Wavelet stretch is given in terms of reflection angle θ as

$$\left.\frac{\partial t}{\partial \tau}\right|_x = \cos\theta. \qquad (2)$$

Here $\partial \tau$ is a differential time at zero offset that reduces to t at offset x according to the moveout prescribed by ray theory; that is, equation 2 is derived from the parametric ray equations, exact for any stratified medium. This correction is therefore an isotropic approximation of wavelet stretch. According to equation 2, wavelet stretch increases nonlinearly from approximately 6% at a reflection angle of 20° to 50% at a 60° reflection angle. Figure 7 shows a schematic diagram of the convergence of moveout curves with increasing offset and consequent stretch after imaging.

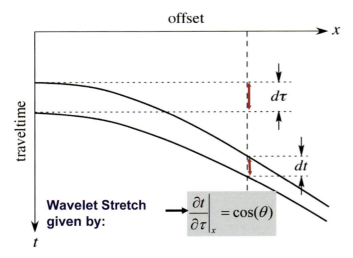

Figure 7. Schematic diagram indicating the convergence of moveout curves with increasing offset and consequent stretch after imaging. The stretch at any offset is related to the cosine of the reflection angle.

Figure 8 shows the result of inversion on synthetic angle gathers, with and without the stretch correction. Figure 8a plots a shale volume curve and upscaled elastic logs from a well in the bitumen reservoir. Figure 8b and d shows, respectively, prestack data without and with the stretch correction. Figure 8c and e highlights the densities estimated by inversion and their association with intrareservoir shale. Uncorrected wavelet stretch significantly degrades the density estimates from inversion.

Inelastic losses

Long raypaths through shallow unconsolidated geology and target reservoirs, saturated with viscoelastic fluids such as bitumen, cause the wave energy to attenuate significantly. Figure 9 compares the seismic bandwidth over a 200-ms window around the target from a 10° stack (Figure 9a), a 45° stack (Figure 9b), and an average reflectivity spectrum computed using well log data (Figure 9c) that shows the true nature of earth's reflectivity. This apparent attenuation (loss of frequencies) is a combined effect of wavelet stretch due to imaging and scattering and intrinsic losses. High-frequency laboratory measurements of the bitumen fluid and the rocks have

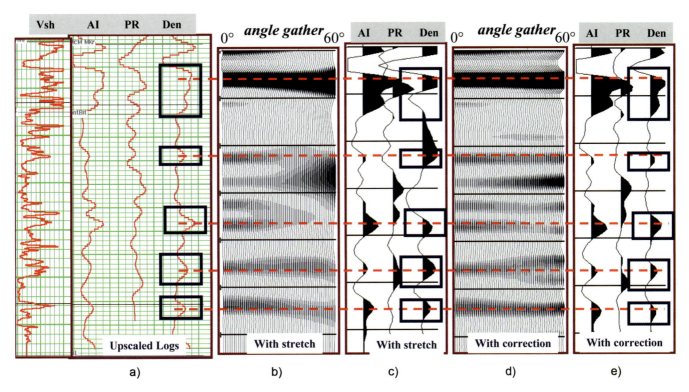

Figure 8. (a) Shale volume curve and upscaled (filtered to seismic bandwidth) AI, Poisson's ratio, and density logs through an oil-sands reservoir. (b) Imaged synthetic angle gather showing significant wavelet stretch at large angles. Synthetic data were created from unfiltered logs. (c) Prestack inversion of synthetic in (b). AI is unaffected by wavelet stretch. The density estimate suffers greatly due to stretch. (d) The same synthetic as in (b) after wavelet stretch correction. The far-angle corrected data show much greater temporal resolution, leading to greater resolution on density and intrareservoir shales. (e) Prestack inversion of corrected data in (d). Estimated densities from inversion now compare favorably to log densities.

178 Heavy Oils: Reservoir Characterization and Production Monitoring

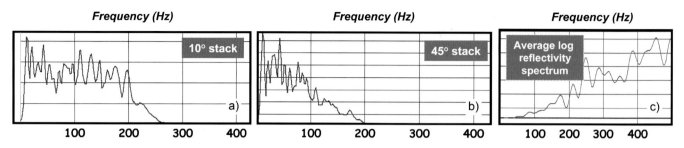

Figure 9. The bandwidth of seismic data decreases significantly from a (a) 10° to (b) 45° angle. This loss is attributed to an effective attenuation from long raypaths in an attenuative medium. (c) Average reflectivity spectrum from well data showing the true nature of earth's reflectivity.

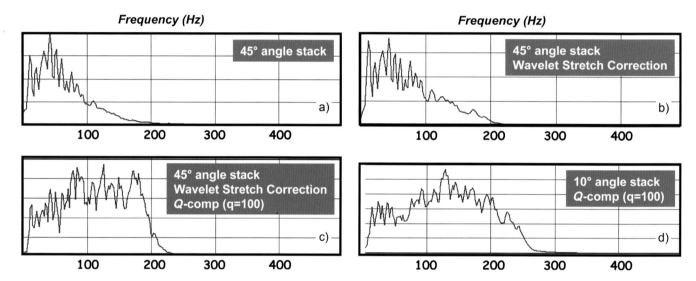

Figure 10. (a) Attenuated bandwidth of 45° angle stack, (b) 45° stack after wavelet stretch correction, (c) Q-compensation to balance the spectrum relative to the near offsets, and (d) 10° angle stack after Q-compensation. The goal of Q-compensation was to balance the spectral bandwidth of the far offsets relative to the near offsets.

shown these to be highly attenuative (Batzle et al., 2006). Wide-angle inversion of 0–60° reflection data requires correction for such losses such that a single wavelet can model the entire aperture. The alternative is to extract a wavelet at each angle to model the data correctly; however, that process does not correct for the frequency losses and hence reduces the resolution of the inversion results. Figure 10 shows the progressive enhancement of the spectral bandwidth with the application of a stretch filter (Figure 10b), Q-compensation using $Q = 100$ (Figure 10c), and a 10° stack with a Q-compensation using $Q = 100$ (Figure 10d), which is our target spectral bandwidth. The goal was to match the spectral shape of the far stacks to the near stacks and be closer to the well-log-based reflectivity bandwidth. Q-correction filters tend to also enhance noise, so the maximum gain was set to 20 dB for stability.

Constrained prestack linear inversion

Inversion estimates of the elastic parameters may be further constrained by incorporating prior information into the statement of the inverse problem. For shale prediction within oil-sands reservoirs, this information should be restricted to a very general, statistical nature. For example, prespecified layer boundaries for model constraints will usually be wrong because of extreme lateral heterogeneity in these reservoirs. For these reservoirs, we chose to constrain our solutions with the statistics of correlations between elastic parameters measured from local well data.

Our statement of an inverse problem with this prior statistical information is given by equation 3.

$$(F^T F + \varepsilon \, C_m^{-1})m = F^T d. \quad (3)$$

F denotes the forward-modeling operator, and ε is a regularization weight that is chosen experimentally on the basis of the S/N of the input prestack data, d. Prior statistical information in equation 3 is conveyed through a model covariance matrix C_m. Generalizing the Downton and Lines (2001) approach, we define the covariance matrix as

$$C_m = \begin{bmatrix} \sigma_1^2 & \sigma_1\sigma_2 r_{12} & \sigma_1\sigma_3 r_{13} \\ \sigma_1\sigma_2 r_{12} & \sigma_2^2 & \sigma_2\sigma_3 r_{23} \\ \sigma_1\sigma_3 r_{13} & \sigma_2\sigma_3 r_{23} & \sigma_3^2 \end{bmatrix}. \quad (4)$$

Symbols σ_1, σ_2, and σ_3 denote standard deviations measured from well logs that have been cast as impedance reflectivity, Poisson's ratio reflectivity, and density reflectivity, respectively. These contrasts also constitute the model (m) in equation 3. Symbols r_{12}, r_{13}, and r_{23} denote crosscorrelation coefficients between the three model parameters. The solution of equation 3 is known formally as a maximum likelihood solution. The solution with $\varepsilon = 0$ is an unconstrained least-squares solution.

Figure 11a and c shows the unconstrained inversion solution for density contrasts in the test area, with the far-angle input limited to 48° (in Figure 11a) and 60° (in Figure 11c). Figure b and d shows maximum likelihood densities from the same input, regularized with the covariance in equation 4. Scaling of the density estimates in Figure 11b and d is much improved. Close inspection also reveals improvements in lateral continuity. Importantly, inclusion of a 60° angle stack as input data minimizes the contribution of the regularization term. However, addition of the 60° angle stack also improved the unconstrained solution indicating the value of information in the far angles. This confirms, as expected from AVO modeling, that properly processed wide-angle data exercise a strong constraint on the density solution.

Results

Figure 12 directly compares density solutions from 5–60°, 5–40°, and a conventional phase-rotated stack

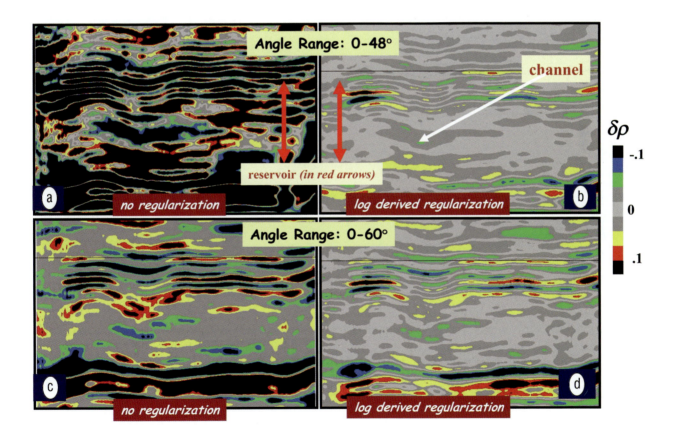

Figure 11. (a) Inverted density section using no regularization for constraining the linear inversion. (b) The same density section inverted with a log-derived covariance matrix for constraining the inversion. Note the improvement in the scaling and continuity of the density amplitudes due to regularization. The channel-shaped feature at the top of the reservoir is more clearly identified in (b). (c) Same as (a); however, an additional 60° angle stack was used for the inversion that improved the solution without regularization. (d) Density using widest aperture and regularization produces the best results.

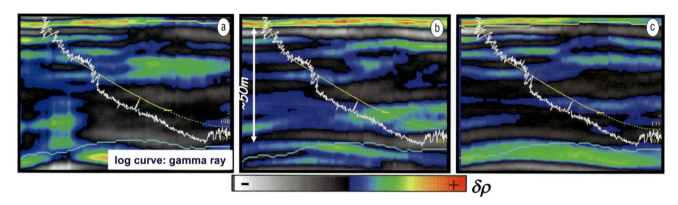

Figure 12. (a) Density cross section from inverting 5–60° reflection data. Inserted gamma-ray curve shows the high net-gross sands in low density overlain by a high-density nonreservoir unit. (b) Density cross section from inverting 5–40° reflection data showing poor imaging of the sands and shales. (c) Conventional 40° phase-rotated stack.

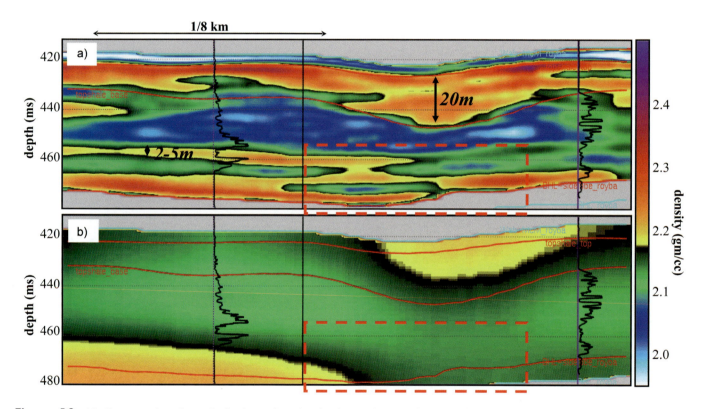

Figure 13. (a) Cross section through the inversion density image intersecting two Surmont wells. A resistivity curve from each well is overlain in black. Bitumen-carrying reservoir facies have high resistivity and are typically low density relative to intrareservoir shale facies. The well on the left shows an intrareservoir shale body at 460 ms. (b) Low-resolution density model interpolated from well data to calibrate the inversion density image in (a). This calibration requires only low-resolution guidance from well control. High-resolution details such as intrareservoir shale bodies are provided through inversion of prestack seismic data.

from 0 to 40° input data. Both inversions were regularized with the same covariance matrix. Well data (gamma-ray curve) through the bitumen interval are overlain on the density estimates. On the basis of the well data, we would interpret higher density units as likely shale bodies overlying a high net-gross reservoir interval. The transition from the overlying shale to the bitumen reservoir back to an intrareservoir shale lens is best imaged in the density inverted from 60° data.

Figure 13a shows a cross section through the Surmont inversion density image. A resistivity curve from each well identifies the oil-bearing reservoir facies. At the well on the left, the reservoir is divided by a high-density shale unit approximately 3-ms thick (3 m for these rocks) at a depth of about 460 ms. The inversion image shows elevated densities for this intrareservoir unit relative to the surrounding reservoir. Additional measured logs in this well confirm that this unit is high-density shale. The 12-m thick bitumen interval above this shale would potentially be cut off from contact with steam injected below the shale. The well on the right in Figure 13a logged one of the few low-density shale facies at approximately 465 ms, also correctly characterized by the inversion.

Figure 12 shows uncalibrated density images, derived solely from seismic data. An underlying, low-resolution geologic model is required to calibrate density data from inversion to measurement units comparable to density logs recorded in wells. Figure 13b shows the cross section through the density calibration model. A comparison of Figure 13a with Figure 13b reveals that the density model imparts a low-frequency trend in the density image. The prestack seismic inversion supplies details of the images (namely, intrareservoir heterogeneities) too detailed to be mapped from well control alone.

Conclusions

Key requirements for robust density estimation from P-waves alone are data from a wide aperture of reflection angles that are preprocessed appropriately for wide-angle inversion. For the intrareservoir density contrasts observed in this field, any data beyond the conventional 40° reflection angle serve to constrain and stabilize the inversion estimate of density. Future studies should investigate whether additional data constraints such as mode-converted waves further improve reservoir characterization without compromising resolution.

Attenuation of surface waves and anisotropic imaging followed by demultiple was critical for preserving P-wave reflections beyond 40° for inversion. However, this wide-angle imaging introduces a great deal of wavelet stretch (approximately 50% stretch in a 60° angle stack). A wavelet stretch deconvolution algorithm corrected the distortion on the 60° data to approximately 13% (equivalent to stretch at 30°). Frequency losses from an effective attenuation were compensated by Q-compensation. This helped to meet the resolution objectives of imaging intrareservoir shale bodies down to 3 m. Regularization of the inversion solution through statistical constraints further improves scaling and continuity of density estimates. We express these constraints through a model covariance matrix derived from local well log data. The net effect of an appropriate wide-angle P-wave workflow is stable density estimates at a resolution suitable for mapping intrareservoir shales in this oil-sands field.

Acknowledgments

We thank Ajay Badachhape, Matt Hall, and Gary Myers for all their help during the project. We also thank ConocoPhillips and Total for permission to publish this paper.

References

Batzle, M., and R. Hofmann, 2006, Heavy oil seismic properties: The Leading Edge, **25**, 750–756.

Chiu, S., N. Whitmore, and M. Gurch, 2007, Polarization filter by eigenimages and adaptive subtraction to attenuate surface-wave noise: CSEG, Expanded Abstracts, 445–449.

Debski, W., and A. Tarantola, 1995, Information on elastic parameters obtained from the amplitudes of reflected waves: Geophysics, **60**, 1426–1436.

Downton, J. E., and L. R. Lines, 2001, Constrained three-parameter AVO inversion and uncertainty analysis: 71st Annual International Meeting, SEG, Expanded Abstracts, 251–254. doi: 10.1190/1.1816583.

Gray, D., 2003, P-S converted-wave AVO: 73rd Annual International Meeting, SEG, Expanded Abstracts, 165–168.

Khare, V., and T. Rape, 2007, Density inversion using joint PP/PS data: Sensitivity to the angle range: 77th Annual International Meeting, SEG, Expanded Abstracts, 965–969.

Roy, B., P. Anno, R. Baumel, and D. Javaid, 2005, Analytic correction for wavelet stretch due to imaging: 75th Annual International Meeting, SEG, Expanded Abstracts, 234–237.

Chapter 13

Characterization of Heavy-oil Reservoir Using V_P/V_S Ratio and Neural Networks Analysis

Carmen C. Dumitrescu[1] and Larry Lines[2]

Introduction

The oil-sands reservoir related to the Long Lake South (LLS) project is contained within the McMurray formation, which is the basal unit of the Lower Cretaceous Mannville Group. The McMurray formation directly overlies the sub-Cretaceous unconformity, which is developed on Paleozoic carbonates of the Beaver Hill Lake Group and is overlain by the Wabiskaw, Clearwater, and Grand Rapids Formations of the Mannville Group.

The study area (Figure 1) is located along the axis of the McMurray Valley system, which was localized by the dissolution of underlying Devonian evaporates, creating the preferred depositional fairway for the Lower Cretaceous McMurray sediments. The most significant bitumen reservoirs within the McMurray formation are found within the multiple channels that represent lowstand system tracts, incised into the regional, prograding parasequence sets that represent highstand system tracts. During sea level rise, these incised channel systems were filled with a transgressive estuarine complex, consisting of sandy to muddy estuarine point bars. In the Long Lake area, the McMurray formation is dominantly composed of these multiple, sand-rich, fluvial, and estuarine channels, which are incised into each other and stacked along a preferred path of deposition. This preferred path is aligned north-northwest to south–southeast in the Long Lake area (Dumitrescu et al., 2009).

A typical seismic line (IL-562) from the prestack time-migrated volume is presented in Figure 2. The reservoir is between the McMurray and Devonian horizons. These high-quality seismic data reveal large-scale depositional elements such as sand-dominated point-bar deposits (#1) and mud-dominated abandoned channel fill deposits (#2).

These depositional elements can be visualized on horizon time slices on the edge detector volume attributes such as semblance and volume curvature (most negative, most positive, and dip curvature) calculated on the prestack time-migrated volume. Figure 3 is an example of a horizon time slice at McMurray +10 ms of the semblance attribute.

Depending on their size and configuration, nonreservoir shale bodies can impede steam chamber growth and fluid drainage within a steam-assisted gravity drainage (SAGD) production process. Distinguishing between reservoir and nonreservoir using a conventional seismic interpretation approach has proved ambiguous. However, petrophysical analysis has determined that the V_P/V_S ratio and density are key discriminators between sand and shale. Therefore, deriving the V_P/V_S ratio and density volumes from seismic data is a useful and important objective.

It is well known and accepted by the industry that inversion is a necessary step in imaging and interpreting the reservoir and there is a continuous struggle to improve the resolution of the inverted volume. Depending on the seismic data and the number of wells available, a V_P/V_S ratio volume can be obtained from (a) traveltime measurements on the vertical and radial components of multicomponent records (Lines et al., 2005), (b) amplitude versus offset (AVO) analysis and prestack (simultaneous) inversion using only the PP component (Dumitrescu and Lines, 2006), or (c) joint inversion of the PP and PS (registered in PP time) poststack seismic data (Dumitrescu and Lines, 2007).

For this project, the processing was designed to preserve prestack amplitudes and involved true amplitude recovery, noise attenuation, statics, and prestack time migration. This was followed by the creation of supergathers (5×5) with a bin size of 50 m. Prestack (simultaneous)

---

[1] Sensor Geophysical Ltd., Calgary, Alberta, Canada
[2] CHORUS, University of Calgary, Calgary, Alberta, Canada

184 Heavy Oils: Reservoir Characterization and Production Monitoring

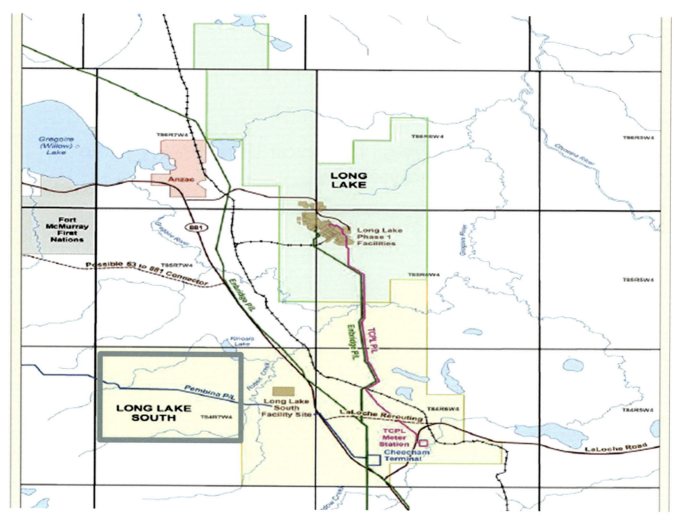

Figure 1. Map showing the LLS area in Alberta, Canada.

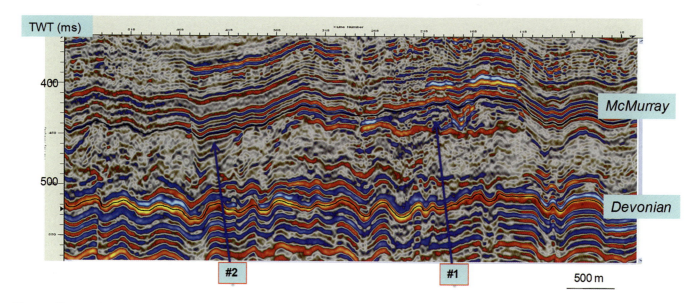

Figure 2. IL-562 showing two features within the McMurray reservoir: sand-dominated point-bar deposit (#1) and mud-dominated abandoned channel fill deposit (#2).

inversion was performed with the computed angle gathers. The output of this deterministic inversion consisted of volumes of P-impedance, S-impedance, V_P/V_S ratio, and density. Neural networks analysis (NNA) was used for estimating a new seismic volume by integrating well information and several seismic volumes. The estimated V_P/V_S ratio volume was used successfully in mapping bitumen sands in heavy-oil reservoirs.

Sercel digital multicomponent recording system. Out of 42 wells (Figure 4) with dipole sonic logs, only 31 were used in the NNA.

Petrophysical analysis was performed on all of the wells to provide a trustworthy set of logs (Figure 5) for inversions and NNA. The analysis included edits and corrections for poor-quality logs. Missing curves (e.g., shear sonic and density) were estimated using the specific mudrock line for the zone of interest or the multiattribute

Method

The LLS Project includes 3D seismic and approximately 50 logged and cored wells over a 50-km^2 surface area. The 3D seismic was acquired in 2005 using the

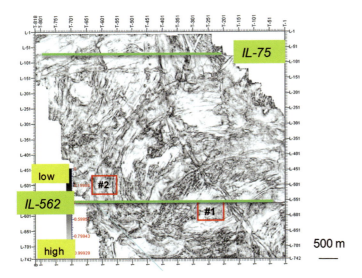

Figure 3. Horizon time slice of the semblance at McMurray $+10$ ms (showing features #1 and #2).

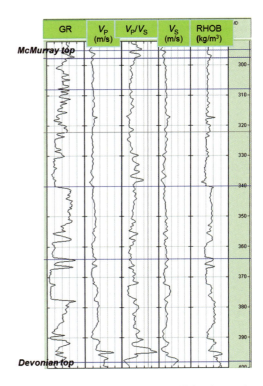

Figure 5. Typical set of logs for a well in the project.

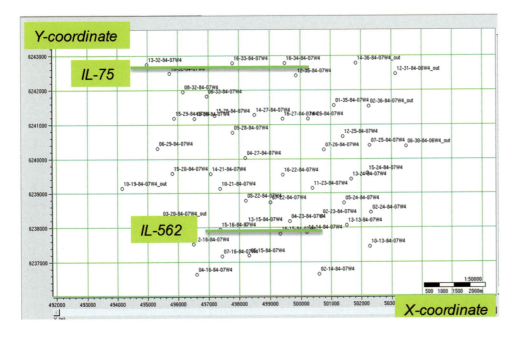

Figure 4. Base map with wells used for NNA and the inlines presented in this chapter (IL-75 and IL-562).

analysis that allows us to estimate logs from existing logs (Hampson et al., 2001). Crossplots of several properties over the target interval showed a good separation between gas sand, bitumen sand, and shale.

The crossplot in Figure 6 is between density (vertical axis) and V_P/V_S (horizontal axis) logs from wells in the study area. The three zones — gas sand, bitumen sand, and shale — are defined based on the associated histograms. The gas sands have low density and the lowest V_P/V_S ratio, the bitumen sands have low density and V_P/V_S ratio values that vary with bitumen quality, and the shales have high values for densities and V_P/V_S ratios.

Some of the inversion attributes (P-impedance (Z_P), S-impedance (Z_S), and V_P/V_S ratio) used in the NNA are results from prestack inversion. Prestack inversion uses the fact that the basic variables Z_P, Z_S, and density are coupled by two relationships that should hold for the background "wet" trend. Figure 7 presents the two crossplots in logarithmic domain between (a) P-impedance and S-impedance and (b) P-impedance and density. The regional rock property trends were derived from log data within the Wabiskaw-to-Devonian interval.

The NNA was performed on wells and several seismic attributes that can be classified in (a) instantaneous attributes, derived from a combination of the input seismic trace and the Hilbert transform of the trace (i.e., trace envelope, instantaneous phase, and instantaneous frequency); (b) recursive attributes, derived by applying a recursive operator along the seismic trace (i.e., the integrated and differentiated seismic trace); (c) band-pass attributes of the seismic trace; (d) AVO attributes derived from prestack seismic data (i.e., P- and S-wave reflectivity

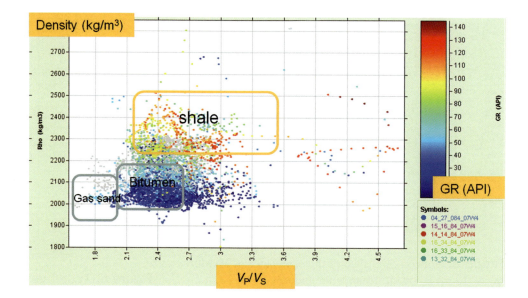

Figure 6. Crossplot of the density logs (vertical axis) and V_P/V_S (horizontal axis), colored by gamma ray count.

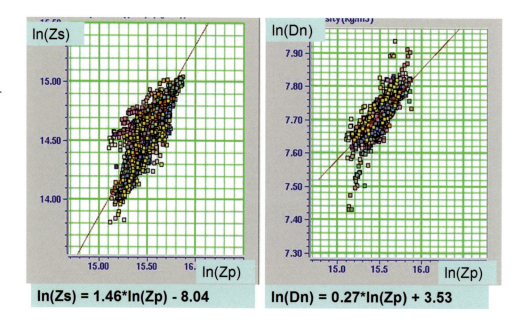

Figure 7. Crossplots of (left) ln(Z_S) versus ln(Z_P) and (right) ln(density) versus ln(Z_P) using data from the Wabiskaw-to-Devonian interval from 20 wells.

ln(Zs) = 1.46*ln(Zp) - 8.04

ln(Dn) = 0.27*ln(Zp) + 3.53

and fluid factor); and (e) attributes derived from prestack inversion (i.e., P- and S-impedance and V_P/V_S ratio) and from previous NNA (density) (Herrera et al., 2006).

Neural networks analysis has four steps: (a) perform a multiattribute step-wise linear regression and its validation, (b) train neural networks to establish the nonlinear relationships between seismic attributes and reservoir properties at well locations, (c) validate results on the wells withheld from the training, and (d) apply trained neural networks to the 3D seismic data volume.

Neural networks can be classified (a) by the type of problem that can be solved (i.e., classification or prediction) and (b) by the type of training used (i.e., supervised or unsupervised). In our approach, we used a supervised prediction type that has the advantage that the output can be interpreted based on the training values. In terms of implementations, there are several different types of neural networks: (a) the multilayer feedforward neural network (MLFN), (b) the probabilistic neural network (PNN), and (c) the radial basis function (RBF). In our approach, we selected the PNN that uses Gaussian weighting functions that fit the seismic attributes to the training samples by a generalized nonlinear regression approach. The key parameter in the PNN method is the sigma factor that controls the width of each Gaussian function and is allowed to vary for each input attribute (Hampson et al., 2001).

Using the ranking process available within the software and after checking the errors, we selected the

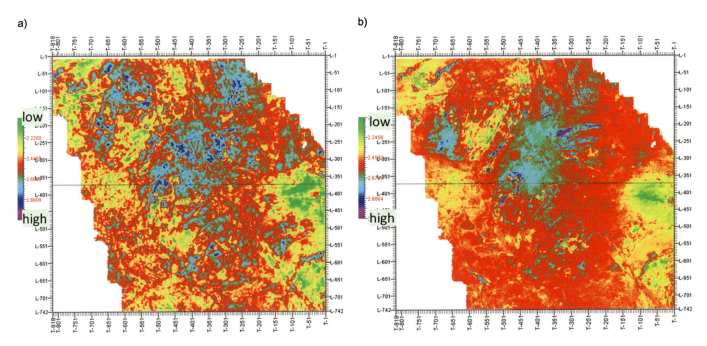

Figure 8. Horizon slice at McMurray +7 ms on V_P/V_S volume obtained from (left) prestack inversion and from (right) NNA.

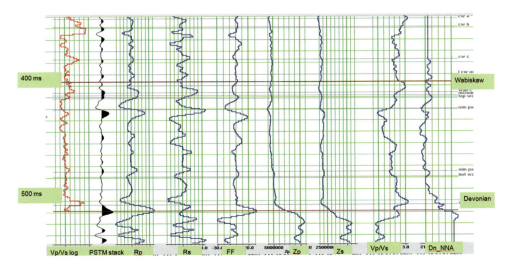

Figure 9. An example (at well 16-34) of the V_P/V_S log and the seismic attributes used in the training procedure. The first track from the left is the V_P/V_S log, the second is the PSTM stack, the third is R_P, the fourth is R_S, the fifth is FF, the sixth is Z_P, the seventh is Z_S, the eighth is V_P/V_S, and the ninth is Dn_NNA.

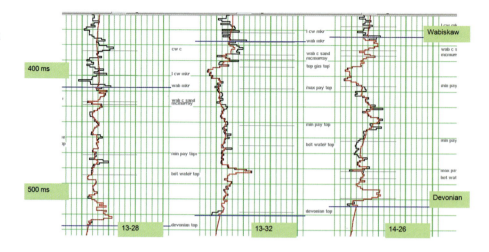

Figure 10. Application of the PNN comparing the predicted V_P/V_S logs (in red) and the real V_P/V_S logs for some of the wells used in the training.

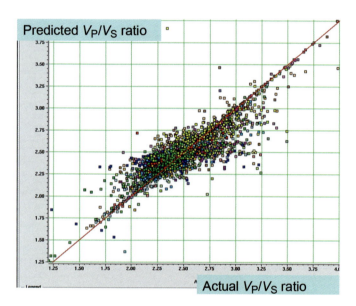

Figure 11. Crossplot of the V_P/V_S predicted by the NNA versus the actual V_P/V_S. Data points are from the analysis zone of all 32 wells.

attributes for training the networks. The neural networks were used in an effort to account for nonlinear relationships between logs and seismic after first testing the linear multiattribute method alone.

Results

In this study, NNA was used to predict the V_P/V_S log throughout the LLS 3D in an attempt to get a better definition of the bitumen sand and to better differentiate between sands and shale in the McMurray formation. Figure 8 shows a comparison between a V_P/V_S volume from deterministic inversion and one from a neural networks analysis. V_P/V_S results obtained from the two inversions are comparable, but it is easy to observe that NNA results are less noisy. Later we will try to connect this lithology indicator attribute with the edge detector attributes to ease the interpretation.

The logs resulting from petrophysical analysis were used in the prestack inversion and in the NNA. Only the well ties with good correlations were used in the analysis. As a result, 31 of 42 wells were used. For each of the wells, the data used in the training procedure were the V_P/V_S curves, along with the extracted and correlated seismic attributes at each well: the PSTM stack, P-wave reflectivity (R_P), S-wave reflectivity (R_S), fluid factor (FF), Z_P, Z_S, V_P/V_S ratio, and density from NNA (Dn_NNA). An example for well 16-34 is presented in Figure 9. For each of the reflectivity traces, we used not only the trace itself but also transforms of this trace (e.g., instantaneous phase, instantaneous frequency, etc.).

As mentioned before, the first step in NNA is to perform a multiattribute step-wise linear regression and its validation. The analysis indicated that the optimum number of attributes to use is nine, and the attributes, ranked on their ability to predict the V_P/V_S logs, are

- V_P/V_S ratio
- integrated FF trace
- integrated R_S trace
- PSTM stack
- density from NNA
- apparent polarity of the FF
- amplitude-weighted frequency of the PSTM stack
- filter 5/10–15/20 of the R_S
- quadrature trace of the R_P

The correlation coefficient between the actual and the predicted result was equal to 0.73, and the prediction error was 0.22. The validation was computed by leaving out one well at a time and predicting its values using the other wells in the training and the defined linear relationship.

The next step was to train the neural networks (using the PNN algorithm) and to establish the nonlinear relationships between seismic attributes and reservoir properties at well locations. A comparison between the real V_P/V_S and the predicted V_P/V_S logs is presented in Figure 10.

After this second step, the correlation coefficient increased to 0.87 and the error dropped to 0.15. A crossplot of the predicted V_P/V_S ratio and the actual V_P/V_S ratio for all of the wells in the study area is presented in Figure 11.

The last step was to apply the trained PNN to the whole 3D seismic volume. Figure 12 shows the estimated V_P/V_S ratio results on inline 75 located in the north part of the 3D and wells no more than 50 m offline. The V_P/V_S volume was estimated within the target interval: from 5 ms above the McMurray to 5 ms below the Devonian. (All of the data outside of the calculation interval are extrapolated end-point values.)

The predicted V_P/V_S volume obtained from the NNA analysis provides meaningful and reliable information about the McMurray reservoir. Density is also an excellent discriminator between gas sand, bitumen sand, and shale. Figure 13 shows the estimated V_P/V_S ratio and the density on a horizon time slice at McMurray + 10 ms. Each of the two volumes is co-rendered with semblance. By displaying these two attributes in this manner, we combined edge information with the variation of a physical property (e.g., V_P/V_S ratio and density).

By integrating all available attributes, we characterized and mapped the reservoir heterogeneities affecting the SAGD operation (i.e., the extent of bitumen sand and gas-saturated and shale zones). Two ways of doing so are presented here: (a) by co-rendering V_P/V_S ratio with semblance (Figure 13) and (b) by crossplotting the rock physics volumes (Figure 14). Figure 14 shows the spatial

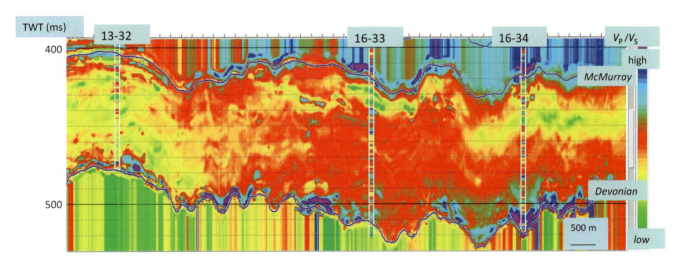

Figure 12. V_P/V_S results at IL-75 with inserted V_P/V_S logs. All of the wells shown tie this line within a 50-m projection distance.

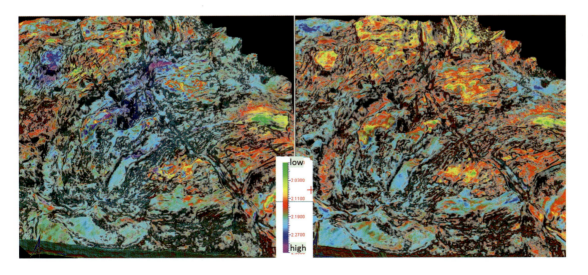

Figure 13. Horizon time slice at McMurray + 10 ms from V_P/V_S (left) and the density volume (right) obtained from NNA and co-rendered with semblance.

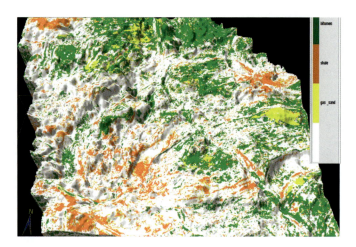

Figure 14. Horizon slice at McMurray $+ 10$ ms of the distribution of gas sand (yellow), bitumen sand (green), and shale (brown) resulting from crossplotting density versus V_P/V_S.

distribution of the three zones created in the crossplot presented in Figure 6.

Conclusions

We presented a case study for improving the resolution of a V_P/V_S volume (obtained from deterministic inversion) by using NNA. The first step in the NNA was multiattribute regression that provides the optimum number and ordering of the attributes. The next step was to use a probabilistic neural network to increase the resolution of the predicted V_P/V_S logs. The attributes that were used are standard seismic attributes as well as attributes from AVO analysis and prestack inversion. The derived neural networks results show a strong correlation with the target logs at training well locations and for the rest of the wells, suggesting that rock properties can be accurately estimated with NNA when deterministic inversion results are used as external attributes in training the networks.

The results of this analysis correlated well with recent drilling, making the NNA part of the workflow for future projects. Utilizing the V_P/V_S volume computed with NNA minimizes the uncertainty in gas sand, bitumen sand, and shale identification, thereby contributing to optimal horizontal well placement. In turn, this has the ultimate effect of increased production and economic efficiency.

Finally, the integration of edge attributes with rock physics attributes allows for a realistic geologic interpretation for oil-sands development purposes.

Acknowledgments

We acknowledge and thank the following organizations for their support: Sensor Geophysical Ltd., Nexen Inc., OPTI Canada Inc., and CHORUS (the Consortium for Heavy Oil Research by University Scientists).

References

Dumitrescu, C. C., and L. Lines, 2006, V_P/V_S ratio of a heavy oil reservoir from Canada: CSPG-Canadian Society of Exploration Geophysicists-Canadian Well Logging Society Convention abstracts, 10–15.

Dumitrescu, C. C., and L. Lines, 2007, Heavy oil reservoir characterization using V_P/V_S ratios from multicomponent data: EAGE 69th Conference and Exhibition, Extended Abstracts, P301.

Dumitrescu, C. C., L. Lines, D. Vanhooren, and D. Hinks, 2009, Characterization of heavy oil reservoir using V_P/V_S ratio and neural networks analysis: CSEG Convention.

Hampson D., T. Todorov, and B. Russell, 2001, Using multi-attribute transforms to predict log properties from seismic data: Exploration Geophysics, **31**, 481–487.

Herrera, V. M., B. Russell, and A. Flores, 2006, Neural networks in reservoir characterization: The Leading Edge, **25**, 402–411.

Lines, L., Y. Zou, A. Zhang, K. Hall, J. Embleton, B. Palmiere, C. Reine, P. Bessette, P. Cary, and D. Secord, 2005, V_P/V_S characterization of a heavy-oil reservoir: 75th Annual International Meeting, SEG, Expanded Abstracts, 1397–1400.

Chapter 14

Multicomponent Processing of Seismic Data at the Jackfish Heavy-oil Project, Alberta

Karen J. Pengelly,[1] Larry R. Lines,[2] and Don C. Lawton[2]

Introduction

This investigation was undertaken to evaluate the processing flows needed to obtain vertical and radial post-stack-migrated seismic sections from a heavy-oil reservoir in eastern Alberta. Converted-wave seismic processing flows have been previously investigated and documented by Harrison (1992) and Isaac (1996). Of particular importance to converted-wave processing is the analysis of receiver statics. Isaac (1996) and Cary and Eaton (1993) showed that S-wave receiver statics can be extremely large and variable compared to P-wave receiver statics. It is not uncommon to have S-wave receiver statics on the order of ±200 ms, whereas P-wave receiver statics are commonly small, typically less than 20 ms.

Velocity analysis is an integral component of converted-wave processing. There has been extensive research relating to nonhyperbolic moveout, valid for weak anisotropy. In many cases, for short to medium offset P-P data, hyperbolic normal moveout (NMO) is an adequate approximation for moveout used in velocity estimations (Al-Chalabi, 1973; Tsvankin and Thomsen, 1994; Alkhalifah, 1997). For P-S data, the hyperbolic NMO correction is valid only for short offsets (Iverson et al., 1989). Furthermore, Castagna and Chen (2000) found that conventional processing software assumes hyperbolic moveout and may produce false structure and false responses below anisotropic regions because of improper removal of NMO. It has been found that the overlying rock in some heavy-oil areas exhibits high values of anisotropy. Newrick and Lawton (2003) found that at Pikes Peak, Saskatchewan, the Thomsen parameters of anisotropy have values of $\epsilon = 0.12 \pm 0.02$ and $\delta = 0.30 \pm 0.06$, from data using a multioffset vertical seismic profile. If the Jackfish area is similar, there is a need to explore the results based on nonhyperbolic NMO as opposed to the standard hyperbolic NMO calculations.

Seismic Survey Acquisition

On 24 October 2002, Devon Canada recorded a three-component (3-C), 2D test line at Jackfish Thermal Heavy-Oil Project, Alberta, a steam-assisted gravity drainage (SAGD) heavy-oil field presently in the development phase and presteam injection. The Jackfish location, as shown in Figure 1, is approximately 170 km south of Fort McMurray. The seismic survey was acquired using dynamite charges of 60, 125, and 250 g in sequence along the seismic line, each with a hole depth of 10 m. The seismic line is oriented north–south. The charge size used for this study is 125 g and spaced at 22.5 m. There were gaps on the seismic line that created areas of lower fold on the resulting stacked seismic section. Multicomponent receivers were used with a receiver spacing of 7.5 m.

When using a P-wave source to acquire multicomponent data, the components that are recorded at the 3-C geophone include the vertical, radial, and transverse components (Figure 2). On Figure 2, the small black and blue arrows indicate the direction of particle motion, whereas the blue arrows indicate the P-SV particle motion. Make note that for P-SV component shot record, there is a change from positive to negative polarity across the shot. The conversion point for the P-P data is referred to as the common depth point (CDP), whereas the P-SV conversion point is called the common conversion point (CCP).

In the Jackfish study area, the geology was initially assumed to be isotropic. As discussed by Tatham and McCormick (1991), the P-wave particle motion is confined to the source-receiver plane; therefore, the mode-converted

[1]Chevron, Houston, Texas, U.S.A.
[2]University of Calgary, Department of Geoscience, Calgary, Alberta, Canada.

Figure 1. Jackfish SAGD location (modified from Multimap and Autodesk Mapguide online mapping software).

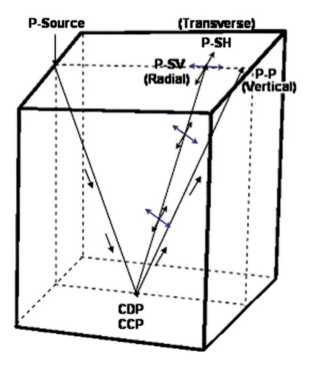

Figure 2. Polarization vectors of a 3-C geophone using a P-source (modified from Tatham and McCormack, 1991).

shear-wave particle motion must also be in this plane. Because the mode-converted P-S data are polarized in this way, each corresponding pair of traces from the horizontal components of the shot gather needed to be rotated into the source-receiver plane (radial) and orthogonal plane (transverse) (Isaac, 1996). However, because this data set consisted of a single 2D line with the radial component of the geophone already oriented along the source-receiver plane, rotation of the horizontal components is not required. The azimuthal orientation of the geophone with respect to magnetic north was not thoroughly investigated and therefore not corrected for.

Vertical Component Data Processing

Geometry

Shown in Figure 3 is the vertical component processing flow for the final migrated section. Before correcting

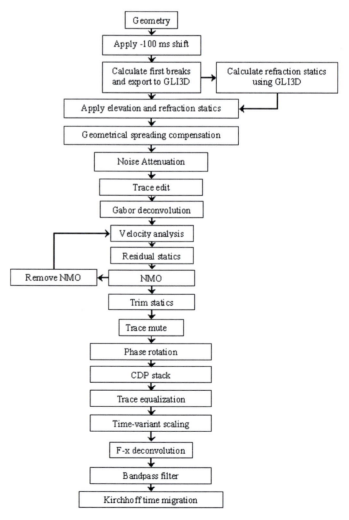

Figure 3. Vertical component processing flow through to migration.

the geometry, the raw shot records were divided into three separate lines according to dynamite charge size. Using the 125-g charge size, a −100 ms header static shift was first applied to compensate for the shift that was applied during VectorSeis acquisition. First breaks were calculated using Promax2D and exported to Hampson Russell GLI3D where the refraction statics were calculated. A two-layer model was built in GLI3D and the refraction statics were imported into Promax2D, where the elevation corrections and refraction statics were applied to the time-shifted data set. Elevation corrections were applied using a datum of 638.0 m and a replacement velocity of 2000 m/s. The elevation profile for the Jackfish test line varies from 609.4 to 637.6 m, resulting in 28.2 m in elevation change along the 2-km seismic line.

Radial filter

The remaining vertical component processing flow was standard once an effective noise attenuation and deconvolution method was found. Various noise attenuation and geometrical spreading methods were investigated, including true amplitude recovery (TAR), surface-wave attenuation, surface consistent amplitudes, and radial filter. However, the radial filter was the only method that did not remove near-surface events or create a "shadow zone" below the first breaks (Henley, 2003). Because the target reservoir is less than 500 m in depth, the radial filter was ideal for removing low-frequency shot-generated linear noise. In Figure 4a, the Devonian event is "masked" by the low-frequency shot-generated noise. By applying a radial fan filter, the noise has been attenuated, leaving a continuous Devonian reflector on the near-offset traces (Figure 4b). The radial fan filter and dip filter were used to attenuate the linear noise. By using a minimum and maximum radial trace velocity of −2000 and 2000 m/s, respectively, velocities below these values are attenuated. From investigating the data that are being removed, there were no reflections being removed at or below and low frequencies of 10–15 Hz, and therefore 15 Hz was used as the low cut for this data set. Similar parameters were used in the radial dip filter. Because there was still noise present at 2000 and 600 m/s, these were removed by using a dip filter. A velocity range of 20 m/s was used on all dip filters and applied in the frequency domain.

Gabor deconvolution

Gabor deconvolution was used because it was superior to conventional deconvolution processes for preserving the near-surface signal. Margrave et al. (2003) have shown Gabor deconvolution to be superior at increasing the resolution of seismic data compared with traditional deconvolution processes in near-surface data. This superiority is because Gabor deconvolution is a time-adaptive

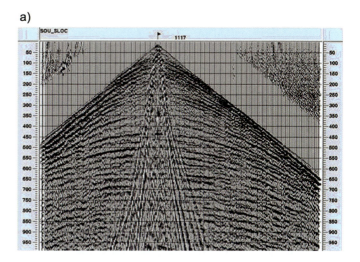

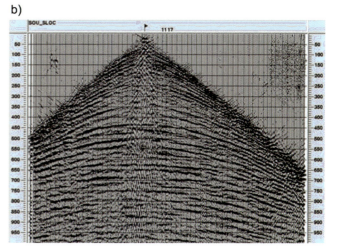

Figure 4. (a) Vertical shot with geometrical spreading compensation, shot and receiver statics, and 5-8-180-240 Hz Ormsby filter. (b) After noise attenuation, with radial filtering.

method as opposed to conventional methods that assume statistical stationarity.

Successful parameters include a half-width analysis window of 100 ms with a window overlap of 2. The Burg spectrum used for wavelet estimation was Burg, with ten coefficients and minimum phase, using a hyperbolic time frequency smoother of 5 Hz-s and window length of 10 Hz. A postdeconvolution filter was not applied during this process. However, a high-pass filter does need to be applied before migration because of Gabor whitening of noise in the input traces.

Near-surface static solution

Residual and trim statics were derived using Promax2D's external model correlation utility, which consists of four stages: model building, correlation computation, statics computation, and statics application. The external modeling utility derives a static by crosscorrelation of the individual traces sorted by CDP with the model. The trim statics were calculated in a similar way; however, the resulting statics were applied directly to the channel static, whereas the residual statics were applied to CDP and shot and receiver static headers. Residual statics and velocity analysis were computed in an iterative fashion in which the final velocities for several CDPs indicated small lateral variations in stacking velocities along the line. The Devonian event is located between 400 and 500 ms.

When stacked without applying the radial filter, events were contaminated with low-frequency noise (Figure 5). The final stack was created using the updated velocity field, although a significant amount of linear noise remained (Figure 6). With the radial filter applied, the low-frequency noise has been attenuated, although in some areas (e.g., the north end of the seismic line above the McMurray formation) the radial filter appears to have removed signal. In Figures 5–7, the McMurray formation lies between 350 and 410 ms on the north end of the lines. An Ormsby filter of 10-15-180-240 Hz, trace equalization, and a poststack f-x deconvolution was applied to the final stack before migration.

Kirchhoff migration

The final stack was migrated with 15° dip aperture (Figure 7) using Kirchhoff time migration. Because the geology is relatively flat, the final stacked and migrated sections look similar except for noise that was removed in the deeper part of the section below the Devonian event and along the edges of the data. The final migration with or without noise attenuation is almost identical except for small areas where signal appears to be lost when using the radial filter. For inversion, the f-x deconvolution was not applied; therefore, only the migration with the radial filter applied was used for inversion of the vertical component data.

Shown in Figures 8 and 9 are the frequency spectra for the raw shots and migrated sections, respectively. From raw shots, the frequency content varies considerably from shot to shot. Figure 8 indicates a frequency content of 10-20-70-90 Hz, whereas the frequency spectrum after migration indicates that the Gabor deconvolution operator is successfully boosting the amplitudes and flattening the frequency spectrum of the data as shown in Figure 9. In Figure 9, the low frequencies are constrained by the radial filter; however, there does appear to be frequencies up to 200 Hz.

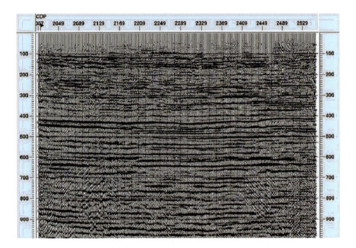

Figure 5. Vertical component final stack with Gabor deconvolution and poststack 10-15-180-240 Hz Ormsby filter applied.

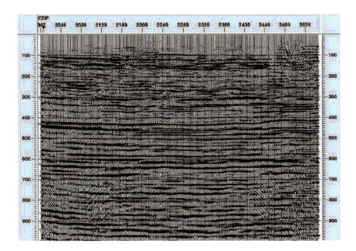

Figure 6. Vertical component final stack with radial filter, Gabor deconvolution, and poststack 10-15-180-240 Hz Ormsby filter applied.

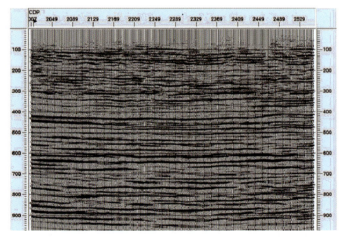

Figure 7. Vertical component Kirchhoff time migration with radial filter, Gabor deconvolution, and 10-15-180-240 Hz Ormsby filter applied before migration.

Chapter 14: *Multicomponent Processing of Seismic Data at the Jackfish Heavy-oil Project, Alberta*

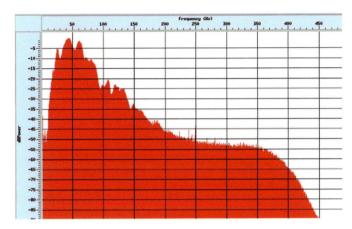

Figure 8. Vertical component frequency spectrum of a typical shot record.

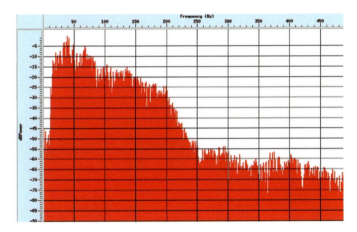

Figure 9. Vertical component frequency spectrum following final migration.

Radial Component Data Processing

As with many converted-wave processing flows, this data set proved to be challenging to process because of large receiver statics and contamination of reflection data with shot-generated noise. The processing flow was based on established multicomponent processing flows created by Harrison (1992), Isaac (1996), and Lu and Hall (2003). The radial component processing flow was similar to the vertical component processing flow with some exceptions, which include P-S asymptotic binning, polarity reversal, the calculation of receiver statics, and the incorporation of a depth-variant stack. The processing flow is shown in Figure 10 from initial geometry through to final migration. As discussed earlier, rotation of the horizontal components was not needed and therefore will not be included as part of the discussion about the radial component processing flow.

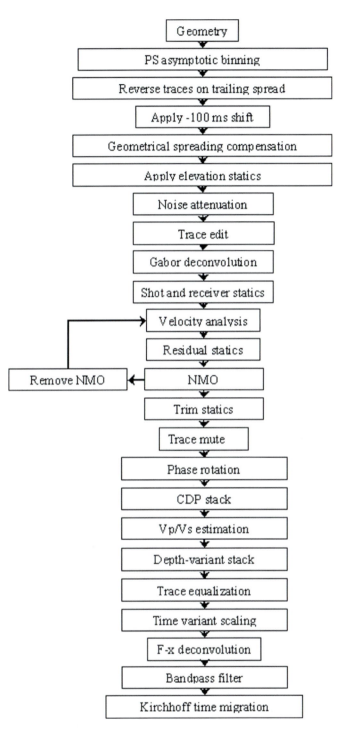

Figure 10. Radial component processing flow through to migration.

Radial component geometry

After the geometry assignment, CCP binning using an asymptotic assumption was used to account for the asymmetry of the mode-converted raypath (Behle and Dohr, 1985; Tessmer and Behle, 1988). By using bins from the vertical component, both data sets will have identical CDP numbers, but this is not the natural coordinate system of

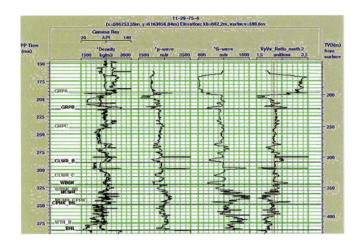

Figure 11. Log display of dipole sonic well 11-29-75-6.

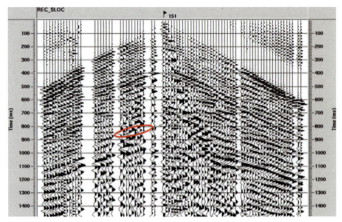

Figure 12. P-S receiver gather with 8-12-50-60 Hz Ormsby filter and 500 ms AGC applied.

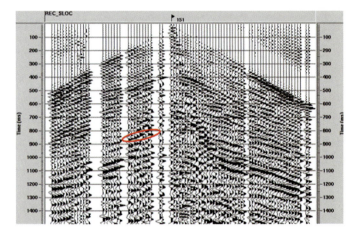

Figure 13. P-S receiver gather with radial filter, 8-12-50-60 Hz Ormsby filter, and 500 ms AGC applied.

P-S data. In this process, a V_P/V_S value needs to be estimated. In this case, a value of 2.2 was derived from well log data (Figure 11).

The polarity of the trailing spread traces was reversed, ensuring that all traces in the gather have identical P-S polarity. A 100-ms bulk shift was also applied to compensate for the shift applied to the VectorSeis data during acquisition. To calculate geometrical spreading compensation, an initial P-S velocity must first be calculated. The velocity table from P-P final stack was converted to P-S velocities using the following equation:

$$V_{PS} = \frac{V_{PP}}{\sqrt{V_P/V_S}} \qquad (1)$$

where V_P = P-P stacking velocity, V_{PS} = P-S stacking velocity, and V_P/V_S is the average velocity ratio.

This result is based on the result from Tessmer and Behle (1988) that the stacking velocity for P-S reflections is given by $\overline{V_{PS}} = \sqrt{V_P V_S}$.

Radial filter

Before calculating the receiver statics, the P-S reflections needed to be identified. To accomplish this, applying a radial filter attenuated linear noise over the shallow part of the section. One of the benefits of using the radial filter, designed by Dave Henley, is its ability to remove linear noise at various velocities, whether originating from one source or from several sources (Henley, 2003). After testing the radial filter, there do not appear to be events under the 8- to 12-Hz low-pass bandwidth. Artifacts were removed in the data from the application of the radial filter by applying an additional iteration of the radial filter to remove backscatter remaining on the shot gathers.

The radial fan filter parameters were similar to those used for the vertical component with the exception of velocity, in which there was a second fan filter of ±400 m/s, with time coordinate of 0.4 s. Additionally, a filter was incorporated to remove backscatter for the first radial filter, which is placed at time coordinate of −1 s. Shown in Figures 12 and 13 are receiver gathers of the raw data before and after noise attenuation. The red oval on both figures shows the removal of low-frequency noise, resulting in the appearance of a coherent event at 800–900 ms.

Gabor deconvolution

To enhance the signal-to-noise ratios and boost the frequency content of the seismic data, Gabor deconvolution was applied. Similar parameters were used as in the vertical component data, with the exception of the number of coefficients for the Burg spectrum, which were five as opposed to ten in the vertical component. Results of Gabor deconvolution were favorable over conventional deconvolution processes at either end of the seismic line, where there is low signal-to-noise ratio and low fold.

Shown in Figures 14 and 15 are the results of applying the radial filter and Gabor deconvolution to prestack data. The brute stack in Figure 14 contains a significant amount of linear noise that is "masking" the reflections. By removing the linear noise and boosting the amplitudes, the Devonian event at 630 ms is now a continuous bright event in Figure 15.

Near-surface statics solution

For the converted-wave data, shot statics were imported from the vertical component database. The shot statics should be identical for the P-P and P-S data sets because the downgoing P-wave is the same for both components. As an initial estimate of receiver statics, the vertical component shot statics were multiplied by the average V_P/V_S of 2.2 and applied to the converted-wave data. However, large statics still remained on the receiver gathers because of the slow upgoing S-wave; therefore, additional static calculations needed to be implemented to calculate receiver statics.

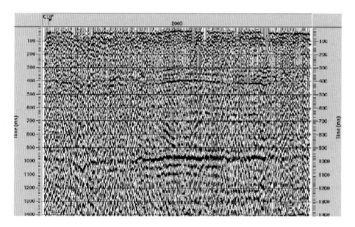

Figure 14. Radial component brute stack with 8-12-50-60 Hz Ormsby filter and 200 ms AGC applied.

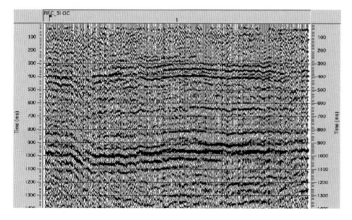

Figure 15. Radial component brute stack with radial filter, 8-12-50-60 Hz Ormsby filter, and 200 ms AGC applied.

The method used to calculate receiver statics includes sorting the P-S data into receiver gathers, applying NMO corrections, and stacking the data to create an initial brute stack. Because the receiver statics were large on the north end of the line, a commonly used method of calculating receiver statics by using autostatics (Cary and Eaton, 1993) was not used. The method of hand statics used by Isaac (1996) was applied and involved picking an event in the common receiver stack domain, flattening it, and incorporating the static shift to the database. This procedure is only valid for flat reflectors. The reflector used in the hand static procedure is that which occurs at approximately 1 s. Unfortunately, the Devonian event is not completely visible on the receiver stack. The remaining receiver static was removed through the residual autostatics procedure.

To calculate residual statics, the same method was used for the radial component as for the vertical component. Shots used for the residual statics had offsets of 100–1700 m to remove the near-offset noise in the shot gathers. The initial crosscorrelation time gate was centered at the predicted Devonian event of 630 ms at 300–960 ms, with an allowable static of 36 ms, and decreased by half with each iteration. The static applied at the third iteration was only ±1–2 ms. However, the results of using the predicted P-S stacking velocities were not ideal, and therefore a stacking velocity using the P-S data needs to be calculated. Because converted-wave data are only hyperbolic for near offsets, hyperbolic and nonhyperbolic NMO equations were also tested. The standard two-term NMO equation was used for the hyperbolic NMO estimation.

$$t^2 = t_0^2 + \frac{x^2}{v_{\text{rms}}^2} \quad (2)$$

where t is traveltime at x; t_0 is traveltime at zero offset; x is source-receiver offset; and V_{rms} is rms velocity, which controls the moveout (Taner and Koehler, 1969).

Tsvankin's long offset equation for nonhyperbolic NMO uses the full three-term equation with Taylor coefficients valid for weak anisotropy.

$$t_A^2 \equiv t_0^2 + A_2 x^2 + \frac{A_4 x^4}{1 + A^* x^2} \quad (3)$$

where t_0 is the traveltime at zero offset, t_A is the traveltime, $A^* = \frac{A_4}{\frac{1}{V_h^2} - A_2}$, and x is the source-receiver offset (Tsvankin and Thomsen, 1994).

Shot gathers using hyperbolic and nonhyperbolic NMO are displayed in Figures 16 and 17, respectively, and there does not appear to be an improvement using the nonhyperbolic NMO approximation. Therefore, the hyperbolic NMO equation was used for the remainder of the processing. Residual static computation and velocity

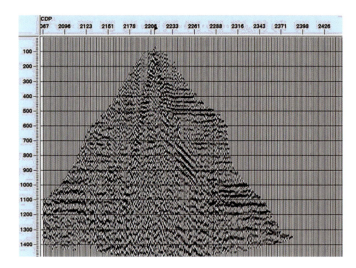

Figure 16. P-S shot gather with hyperbolic NMO and 200 ms AGC applied.

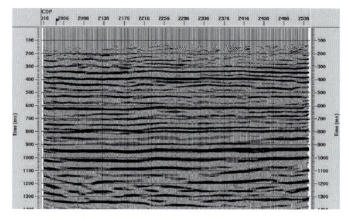

Figure 18. Radial component Kirchhoff time migration from CDP stack, with radial filter, Gabor deconvolution, with 8-12-70-90 Hz Ormsby filter, and 200 ms AGC applied.

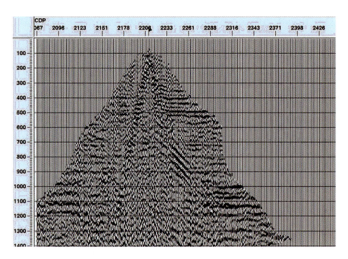

Figure 17. P-S shot gather with nonhyperbolic NMO and 200 ms AGC applied.

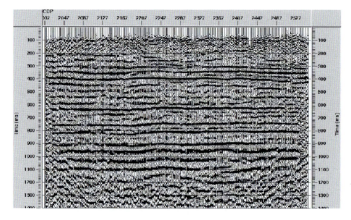

Figure 19. Radial component depth-variant stack, with radial filter, Gabor deconvolution, 8-12-70-90 Hz Ormsby filter, and 200 ms AGC applied.

analysis were computed in an iterative fashion, followed by trim statics.

Depth-variant stack

One problem with velocity analysis is that near-surface velocities are not pickable on the semblance plot. The depth-variant stack was introduced by Eaton et al. (1990) to correctly locate converted-wave reflections and minimize some binning-related artifacts. To complete a depth-variant stack, several V_P/V_S values must be calculated using the same horizons from the vertical and radial components. The V_P/V_S values are then calculated using the following equation from Harrison (1992):

$$\frac{V_P}{V_S} = \frac{2I_S}{I_P} - 1 \qquad (4)$$

where I_S is the time interval between P-S reflections and I_P is the time interval between the P-P reflections.

For near-surface data above 300 ms, V_P/V_S values from logs were used in place of the preceding method. The depth-variant stack takes the calculated V_P/V_S values and the final vertical component velocity cube to create a converted-wave velocity cube. Because only one V_P/V_S value is used, the resulting velocities may not show lateral variations that are present on the radial component data if the same variations are not present in the vertical component data. This can cause a decrease in resolution of the image of laterally varying reservoirs. In Figure 19, the depth-variant stack has successfully resolved the near-surface events. However, there is concern about preserving lateral variations. The final migrated section from the depth-variant stack is shown in Figure 20. The first 100 ms were removed from the top of the data because of removing the first 100 m of offset in my stacks. In

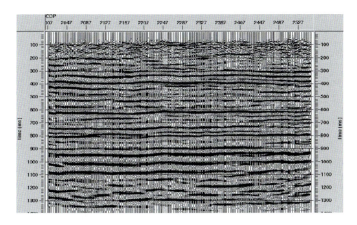

Figure 20. Radial component Kirchhoff time migration from depth-variant stack, with radial filter, Gabor deconvolution, 8-12-70-90 Hz Ormsby filter, and 200 ms AGC applied.

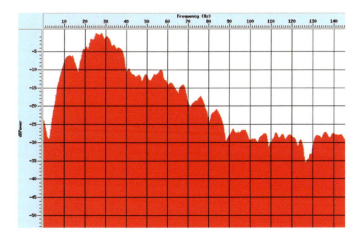

Figure 21. Frequency spectrum for a typical raw shot.

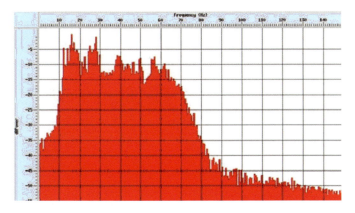

Figure 22. Frequency spectrum for radial data following Kirchhoff migration.

Figures 18–20, the McMurray formation appears in the window 500–600 ms at the north end of the lines.

Because heavy-oil reservoirs have lateral variations in lithology and porosity, both stacks were migrated, although the migrated depth-variant stack was used for inversion. The frequency spectra for the raw data and after migration are shown in Figures 21 and 22, respectively. The bandwidth of the processed data is not as high as was expected (<70 Hz), although the deconvolution has broadened the amplitude spectrum by enhancing the high-frequency content.

Conclusions

The final migrated section for the vertical component was the result of a standard processing flow. Particularly noteworthy is the result from using noise attenuation, and removing noise attenuation filters yielded similar results when using f-x deconvolution before migration. For the radial component, the radial filter was found to remove a significant amount of linear noise in the shot gathers. This noise suppression allowed the reflections to be more clearly identified after radial filtering. The asymptotic, CCP, and depth-variant stacks were migrated. The depth-variant stack may remove lateral variations in the velocity field but has shown to have greater frequency in the near-surface data. The CDP stack was found to have higher frequency and resolution for the deeper events below 700 ms. The enhancement of the spectral bandwidth was advantageous for obtaining enhanced inversions of the data, as shown by Pengelly (2005).

Acknowledgments

We would like to thank Devon Canada for providing the seismic data, the Consortium for Research in Elastic Wave Exploration Seismology (CREWES) for technical support, and Landmark and Hampson Russell for donating software used to complete this project. Karen Pengelly would also like to thank the Alberta Ingenuity Centre for In Situ Energy (AICISE) for funding her thesis research.

References

Al-Chalabi, M., 1973, Series approximation in velocity and traveltime computations: Geophysical Prospecting, **21**, 783–795.

Alkhalifah, T., 1997, Velocity analysis using nonhyperbolic moveout in transversely isotropic media: Geophysics, **62**, 1839–1854.

Behle, A., and G. Dohr, 1985, Converted waves in exploration seismics, *in* G. Dohr, ed., Seismic shear-waves: Part B: Applications: Geophysical Press, 15B.

Cary P., and D. Eaton, 1993, A simple method for resolving large converted-wave (P-SV) statics. Geophysics, **58**, 429–433.

Castagna, J., and H. Chen, 2000, Anisotropy effects on full and partial stacks: Geophysics, **65**, 1028–1031.

Eaton, D. W. S., R. T. Slotboom, R. R. Stewart, and D. C. Lawton, 1990, Depth variant converted wave stacking: 60th Annual International Meeting, SEG, Expanded Abstracts, 1107–1110.

Harrison, M. P., 1992, Processing of P-SV surface seismic data: Anisotropy analysis, dip moveout and migration: Ph.D. thesis, University of Calgary.

Henley, D., 2003, Coherent noise attenuation in the radial trace domain: Geophysics, **68**, 1408–1416.

Isaac, J. H., 1996, Seismic methods for heavy oil reservoir monitoring: Ph.D. thesis, University of Calgary.

Iverson, W., B. Fahmy, and S. Smithson, 1989, V_P/V_S from mode-converted P-S reflections: Geophysics, **54**, 843–852.

Lu, H., and K. Hall, 2003, Tutorial: converted wave (2D PS) processing: CREWES Research Report, **15**.

Margrave, G., L. Dong, P. Gibson, J. Grossman, D. Henley, and M. Lamoureux, 2003, Gabor deconvolution: extending Wiener's method to nonstationarity: CREWES Research Report, **15**.

Newrick, R., and D. Lawton, 2003, Investigation of turning rays in the Western Canada Sedimentary Basin: 2003 CSPG/CSEG Convention-Partners in a New Environment, Expanded Abstracts, CD-ROM.

Pengelly, K., 2005, Processing and interpretation of multicomponent seismic data from the Jackfish Heavy Oil Field, Alberta: M.S. thesis, University of Calgary.

Taner M., and F. Koehler, 1969, Velocity spectra-digital computer derivation and applications of velocity functions: Geophysics, **34**, 859–881.

Tatham R., and M. McCormick, 1991, Multicomponent seismology in petroleum exploration: SEG.

Tessmer G., and A. Behle, 1988, Common reflection point data-stacking technique for converted waves: Geophysical Prospecting, **36**, 671–688.

Tsvankin, I., and L. Thomsen, 1994, Nonhyperbolic reflection moveout in anisotropic media: Geophysics, **59**, 1290–1304.

Section 3

Reservoir Monitoring

Chapter 15

Geostatistical Reservoir Modeling Focusing on the Effect of Mudstone Clasts on Permeability for the Steam-assisted Gravity Drainage Process in the Athabasca Oil Sands

Koji Kashihara,[1] Akihisa Takahashi,[2] Takashi Tsuji,[1] Takahiro Torigoe,[2] Koji Hosokoshi,[3] and Kenji Endo[4]

Introduction

Steam-assisted gravity drainage (SAGD) method is a heavy-oil in situ recovery technique used for bitumen production of the Athabasca Oil Sands, where bitumen reserves from oil sands are estimated at 173 billion barrels (Alberta Energy and Utilities Board, 2008). The typical configuration of the SAGD includes two horizontal wells of 750 m in length and vertically separated by 5 m, in which the upper well is used for steam injection to increase the mobility of the bitumen and the lower well is for bitumen production. Feasible bitumen recovery from oil sands by SAGD is limited within the lateral perpendicular distance of approximately 50 m from the horizontal well pairs. Therefore, profitably viable bitumen production performance requires that the SAGD well pair location be at thick reservoir sands. The complexity of facies distribution in the target formation requires effort in understanding detailed distribution of the reservoir sands. A deterministically constructed geologic model was visualized to better understand 3D distribution of the reservoir sands in the study area (Takahashi et al., 2006). Because lateral continuity of the lithologic facies in the area of interest is shorter than typical interwell distance, the deterministically predicted facies distribution leaves inherent uncertainty in terms of the bitumen production forecasting.

The reservoir sand facies can contain impermeable thin mudstone layers and impermeable mudstone clasts. Previous works including Schmitt (2004) and Takahashi et al. (2006) often refer to the thin mudstone layers adversely affecting the growth of the steam chamber during the SAGD process and consequent bitumen production performance. Although the mudstone clasts have not been often discussed in regard to the impact on the bitumen production performance, our field operation has experienced unexpectedly lowered bitumen production from the reservoir containing the mudstone clasts.

The mudstone clasts in reservoir sand facies introduce subseismic-scale heterogeneity in the distribution of petrophysical properties such as permeability. Reservoir simulation models in forecasting bitumen production performances require reconstructing the reservoir property heterogeneity caused by the existence of the mudstone clasts. Multiple realizations of the reservoir model generated from the stochastic modeling approach will be tested in future tasks quantifying the range of uncertainty in production forecasting.

Environment of Deposition

The main reservoir of the Athabasca Oil Sands comprises channel sands of the Lower Cretaceous McMurray Formation deposited on an erosional surface of Devonian limestone (Flach and Mossop, 1985; Wightman and Pemberton, 1997). The sedimentation occurred in response to a gradual rise in a relative sea level, resulting in the McMurray Formation grading from fluvial at the base to marine at the top with an upward-fining character in general. The lower part of the formation is dominated by crosslaminated sand with medium to coarse grain size in common and sometimes contains mudstone clasts resulting from cut-bank caving, whereas the middle to upper part of the formation consists of beds of very-fine- to fine-grained sand separated by thin mudstone layers that represent mud

---
[1]Research Center, JAPEX, Chiba, Japan.
[2]Exploration Division, JAPEX, Tokyo, Japan.
[3]Development Division, JAPEX, Tokyo, Japan.
[4]Reservoir Engineering, JACOS, Calgary, Alberta, Canada.

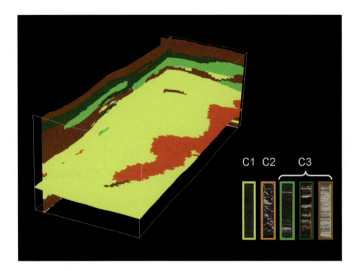

Figure 1. Deterministically constructed facies model from Takahashi et al. (2006). Facies proportion is locally varied in the area of interest.

drapes on the point-bar surface becoming more abundant upward (Flach and Mossop, 1985). Overall changes in the lithologic character in the study area are captured in the deterministically constructed geologic model (Figure 1) from the previous study (Takahashi et al., 2006).

In this study, the lithologic facies described above are classified into three groups depending on the nature of mudstones: sand facies without mudstone (C1), sand with mudstone clast facies (C2), and sand with thin mudstone layers facies (C3). Typical examples of the three facies are shown in the core photographs in Figure 2, where bitumen-saturated sand appears in black, and water-saturated mudstone clasts and mudstone layers appear in light color. The mudstone clasts (in C2) and the thin mudstone layers (in C3) can reduce the facies permeability and can adversely affect the steam chamber growth and consequent SAGD performance. The impact of mudstone clasts on permeability of the C2 facies is the interest of this study and is addressed in the next section.

Lateral facies continuity is very small, and deterministic prediction of the facies distribution is essentially difficult in the McMurray Formation. Thus, the stochastic approach is used for the facies modeling in this study. The overall trend in the facies distribution (shown in Figure 1) needs to be considered for an appropriate implementation of the geostatistical tools that ideally assume the stationary feature of the facies distribution in 3D space. The facies modeling section describes a method to obtain locally varying facies proportion from seismic data to provide to geostatistics.

Permeability of Sand with Mudstone Clast Facies

One of the most challenging issues in reservoir modeling of this study is to obtain appropriate representative

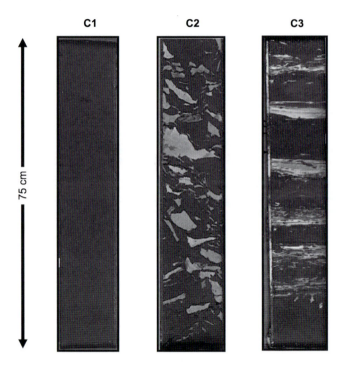

Figure 2. Three facies classified for this study: (C1) massive sand facies, (C2) sand facies with mudstone clasts, and (C3) sand facies with thin mudstone layers.

permeability values of the sand with mudstone clast facies to be assigned to model's grid block. Permeability of the grid block elements, $10 \times 10 \times 0.5$ m in size, cannot be obtained directly from field or laboratory measurements. Core plugs of 1 inch (2.54 cm) in diameter by a few centimeters in length prepared for permeability measurements had been preferentially selected from the sand matrix without mudstone fractions (Figure 3). Core gamma-ray measurement recognizes invisibly buried mudstone clasts so as to prevent them from being plugged. Thus, the core permeability measurements shown in Figure 4 are completely biased toward sand permeability. Neither any of the core measurements nor averages of the measurements provide the facies permeability to be assigned to a grid block.

Permeability of heterogeneous media consisting of sand and mudstone clasts can be calculated by single-phase steady-state flow simulation performed on permeability models (Journel et al., 1986; Desbarats, 1988; Deutsch, 1989). A discrete binary-type permeability distribution is adopted for the permeability model of the sand with mudstone clast facies (C2). First, sand/mudstone configuration is stochastically simulated by using an object modeling technique in which ellipsoidal objects mimic the mudstone clasts and are randomly distributed in background sand matrix. Constant size and shape of the mudstone clasts are assumed and the axis lengths of the ellipsoids are fixed at 3 and 3 cm in horizontal directions and 1 cm in the vertical direction after the investigation of core photographs. One realization of the generated sand/mudstone clast configuration is shown in Figure 5, in which the ellipsoidal objects (mudstone clasts) are distributed in the space of 1 × 1 × 0.5 m. This size is large enough compared with the mudstone clasts that the block size effect discussed in Deutsch (1989) and Dimitrakopoulos (1993) seems to be eliminated. Then, each grid block is assigned sand permeability or mudstone permeability. Sand permeability is assumed to be constant at an average value of the core permeabilities (7300 mD) for simplification, although variability is indicated in the frequency distribution of the core measurements in Figure 4. Mudstone permeability is much smaller than sand permeability by more than two orders of magnitude; therefore, only sand can have fluid flow paths in the time range of economically feasible SAGD operation. Zero permeability is simply assigned for mudstone.

Single-phase steady-state flow simulation is applied to the permeability model to calculate effective permeability for the sand with mudstone clast facies to be assigned to grid blocks. The procedure is repeated with different volume fractions of mudstone clasts varying up to 0.8 with a peak at 0.4 (Figure 6), providing an empirical relationship between permeability and mudstone volume fraction. Calculated permeability is standardized on original permeability (7300 mD) at zero mudstone volume and expressed in the form of permeability reduction with increasing mudstone volume fraction, as shown in Figure 7. Permeability reductions at three particular mudstone volume fractions (0.2, 0.4, and 0.6) are represented by averaged values from 20 realizations. At 0.1, 0.3, 0.5,

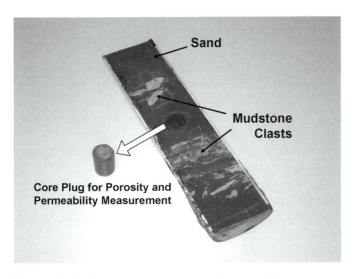

Figure 3. Scale issue in porosity and permeability measurements. Core samples are preferentially selected from sand matrix and do not represent porosity and permeability of grid model.

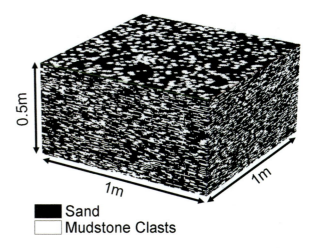

Figure 5. Minimodel for sand with mudstone clast facies (C2). Several models with different mudstone volume are generated to examine the impact of mudstone clasts on permeability.

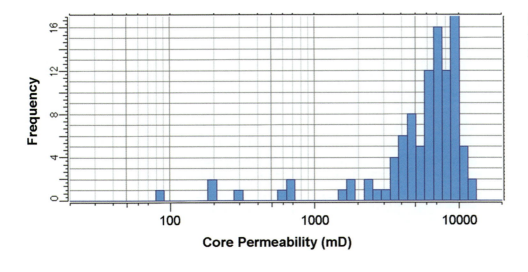

Figure 4. Frequency distribution of permeability measurements obtained from core plugs.

Figure 6. Frequency distribution of mudstone clast volume fraction.

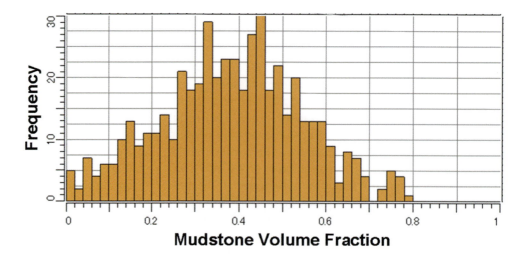

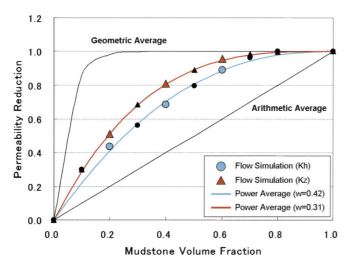

Figure 7. Mudstone volume-to-permeability reduction relationship obtained from flow simulation study using the minimodels shown in Figure 4.

0.7, and 0.8 of mudstone volume fractions, the permeability reductions are calculated on single realization and plotted for reference. Note that horizontal and vertical permeabilities reduce slower than the geometric average, also shown in Figure 7. Geometric averaging is a good estimator for a random system assuming 2D flow (Jensen et al., 2000; Desbarats and Dimitrakopoulos, 1990). The deviation from the geometric average may represent the effect of 3D flow, of which theoretical analysis is currently difficult. Percolation threshold is varied depending on the shape of mudstone clasts in terms of the horizontal-to-vertical anisotropy ratio (Deutsch, 1989). The critical mudstone volume fraction beyond which the flow rate drops drastically may be near 0.7 for horizontal and vertical flows in case of the object shape assumed in this study. Almost all observed mudstone volume fractions are less than 0.7 (Figure 6) and categorized in the low mudstone fraction regime in which fluid flow is dominated by flow-through sand (Ringrose et al., 2003).

Mudstone permeability currently assuming zero permeability has little impact on fluid flow.

The relationships between permeability and mudstone volume fraction are well approximated by power average (Journel et al., 1986; Deutsch, 1989) with a power constant $w = 0.42$ for horizontal permeability and $w = 0.31$ for vertical permeability, respectively. These curves are used in the section Reservoir Modeling Workflow to obtain the facies permeability at an arbitrary mudstone volume fraction for the sand with mudstone clast facies (C2). The facies discrimination between the C2 and C3 facies should be accomplished before estimating permeability distribution because the mudstone clasts and thin mudstone layers should have different impacts on permeability and the approximated function in Figure 7 cannot be applied to the C3 facies. The next section describes seismic attributes to be used in the facies discrimination.

Seismic Attribute Selection for Facies Discrimination

The three facies defined in the section on environment of deposition should be determined at each grid block before permeability estimation because a different function should be used for different facies to estimate permeability. The large-scale trend in the facies distribution, as is seen in Figure 1, becomes an issue in geostatistical facies modeling assuming stationarity. In this situation, facies discrimination using 3D seismic data can provide the large-scale trend that is required for an appropriate implementation of geostatistical facies modeling. This section discusses the seismic attributes selected for better facies discrimination resulting in better trend extraction from seismic data.

The three histograms in Figure 8 show frequency distributions of density log response readings for the C1, C2, and C3 facies, respectively. Separation of the

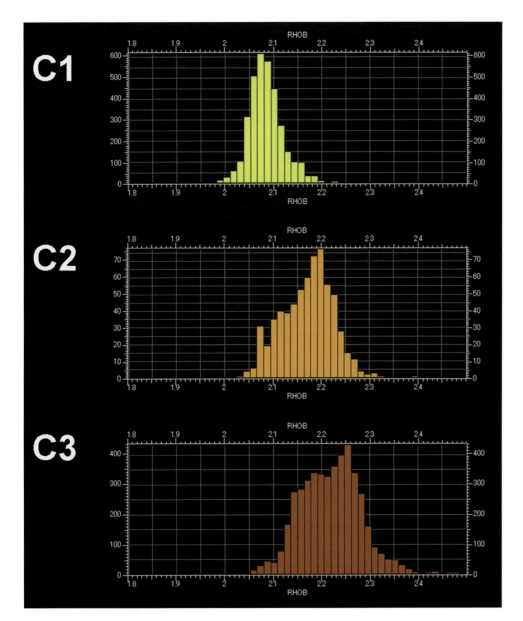

Figure 8. Histograms of density. Density can be a discriminator between C1 and the other two facies.

histograms indicates ability of density to discriminate C1 from the other two facies. This observation is consistent with the fact that, in general, mudstone has higher density than sand in the study area. On the other hand, no significant separation is seen in the histograms between the C2 and C3 facies, indicating difficulty in discriminating these two facies using density. Figure 9 shows histograms of sonic log responses for the three facies. Separation between the C2 and C3 facies indicates that P-wave velocity can be a discriminator between these two facies. The difference in P-wave velocity between the C2 and C3 facies might be explained by the well-known negative correlation between velocity and porosity. Occurrence of mudstone clasts may reflect higher flow velocity washing small-sized grains out of the C2 facies, resulting in well-sorted sand with high porosity (lower P-wave velocity).

The C3 facies deposited in relatively slower flow velocity consists of poorly sorted sands with low porosity (higher P-wave velocity).

Successful discrimination among the three facies can be expected by implementing multiple seismic attributes, including density and P-wave velocity. The prediction power can be increased by adding several other instantaneous attributes such as instantaneous frequency, instantaneous phase, amplitude-weighted cosine phase, cosine of instantaneous phase, and apparent polarity, which are selected through histogram analysis in a similar way as density and P-wave velocity.

Multiattribute analysis (Hampson et al., 2001) implementing the selected multiple seismic attributes is applied to the field data to predict the three facies. The prediction can be checked at the locations of wells and found to be

Figure 9. Histograms of P-velocity. P-velocity can be a discriminator between C2 and C3.

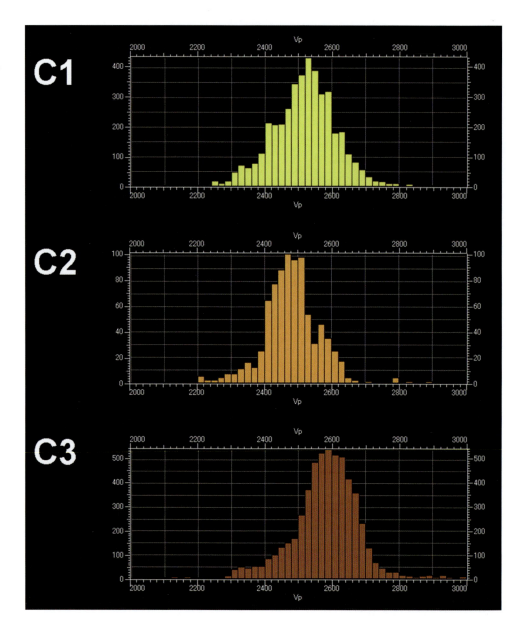

successful at 73.9%, 73.1%, and 88.9% of total samples of the C1, C2, and C3 facies, respectively. Although the prediction is not perfect, as indicated by the overlapping feature of the histograms in Figures 8 and 9, the prediction result can be used as the large-scale trend in the facies distribution required in the geostatistical facies modeling.

Reservoir Modeling Workflow

The two main challenging issues of reservoir modeling in the target area are estimating permeability in consideration of the mudstone clasts and discriminating facies between C2 and C3. The reservoir modeling workflow in this study addressing the issues is conceptually drawn in Figure 10 and consists of three main procedures: facies modeling to discriminate the facies, mudstone volume

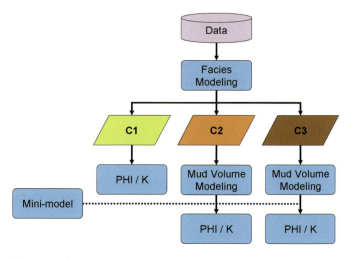

Figure 10. Conceptual reservoir modeling workflow in this study.

modeling, and porosity and permeability transformations in consideration of mudstone clasts. These procedures are described in the following subsections.

Facies modeling

The objective of facies modeling is to discriminate the three facies (C1, C2, and C3) having different mudstone impacts on permeability and to predict their spatial distributions. The large-scale trend of the facies distribution causing a "nonstationary" environment should be provided for an appropriate implementation of geostatistics assuming a "stationary" environment. As described in the section on seismic attribute selection for facies discrimination, the multiattribute analysis implementing multiple seismic attributes can be an efficient tool to provide the large-scale trend of the facies distribution. This study combines multiattribute analysis capturing the large-scale trend with geostatistics reconstructing small-scale heterogeneity of the facies distribution. The way to combine these two technologies is described below.

Probabilistic neural network (PNN) used in the multiattribute analysis calculates facies probability at every grid cell location behind estimating facies. PNN is one of the statistical pattern recognition techniques with significant classification performance. Human face recognition (Mao et al., 2000) and the satellite cloud classification (Tian et al., 2000; Tian and Azimi-Sadjadi, 2001) are successful examples of the recent applications of this technique to practical pattern recognition problems. PNN requires training samples in the training process to optimize kernel function parameters. Kernel function is a function of the distance from training samples to a test sample in the n-dimensional attribute space where n is a number of seismic attributes used in the analysis (Figure 11). A Gaussian function or uniform function is often used as a kernel function. Summation of the kernel functions provides an approximated probability density function for each category. Probability of a test sample X for a category C_i is given by

$$P(\mathbf{X}|C_i) = \frac{1}{N}\sum_{j=1}^{N} K\left(\mathbf{X} - \mathbf{X}_i^{(j)}\right), \quad (1)$$

where K is kernel function and $\mathbf{X}_i^{(j)}$ is the jth training sample in the ith category. Figure 12 shows the probability distributions estimated by PNN for the three facies categories. For the purpose of validation, the probabilities obtained by PNN are compared with the facies proportion observed at the locations of wells. Figure 13 shows facies proportion binned by the PNN probability. Each facies is actually shown with a higher proportion in which PNN calculates higher probability of the corresponding facies.

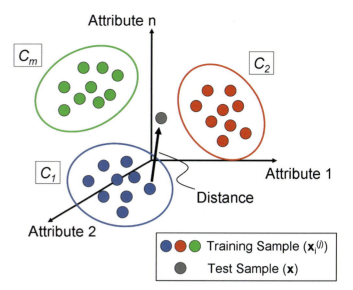

Figure 11. Distance from training sample to test sample in n-dimensional attribute space. PNN calculates the probability of the test sample being a member of each category (C_1, C_2... C_m) in accordance with the distance.

These figures indicate the validity of the PNN probabilities that is utilized in geostatistics as the locally varying facies proportion.

The facies probabilities obtained by PNN are used as the locally varying facies proportion required for the facies modeling in this study. Sequential indicator simulation with gridded indicator prior mean (Deutsch and Journel, 1998) is implemented with the facies probability obtained by PNN. Figure 14 shows realizations of the facies modeling. All of the realizations successfully reflect the locally varying facies proportion in Figure 12. These realizations also reconstruct small-scale heterogeneity required from the viewpoint of flow simulation. (Compare with the deterministically constructed geologic model in Figure 1.)

Mudstone volume modeling

The next step is modeling the mudstone volume fraction in the region of the C2 facies where mudstone clasts reduce the facies permeability. Instead of using conventional log predictions, the mudstone volume fractions are measured on core photographs obtained from the whole of the McMurray Formation interval at all the wells in the study area. Because mudstones appear on the core photographs in light color and sand appears in black color (as seen in Figure 2), the volume fraction of the mudstones can be given by the number of light-colored pixel elements (representing mudstones) divided by the total pixel number within a specified area of a process window. Sliding the window along a well axis provides

Figure 12. Facies probabilities at each grid location provided by PNN.

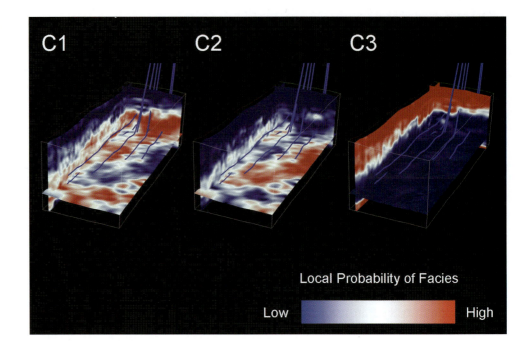

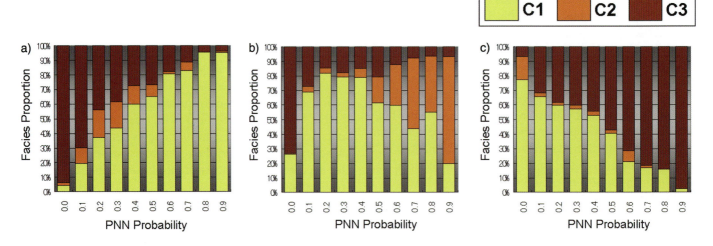

Figure 13. Facies proportion at the locations of wells binned by PNN probability for (a) C1, (b) C2, and (c) C3.

the mudstone volume fraction curve as a function of depth in the form of logging data.

Figure 15 shows a crossplot comparing the photographically predicted mudstone volume fraction on the horizontal axis with the gamma-ray log response on the vertical axis. Although a general linear correlation is observed, gamma-ray log response can vary from approximately 20 to 60 whereas mudstone volume fraction remains constant at zero (no mudstone). The variability of gamma-ray response independent of the mudstone volume is thought to be due to the interstitial clay in sands. Existing methods such as calculation from gamma ray or calculation from neutron/density crossplots are not able to separate the mudstone from the interstitial clay, resulting in poor prediction of the mudstone volume. Thus, this study uses core identification to quantify the mudstone volume separated from the interstitial shaliness of sand.

Mudstone volume modeling is implemented by using sequential Gaussian simulation with collocated cokriging (Deutsch and Journel, 1998) using the photographically predicted mudstone volume fraction and seismic-derived density volume as hard and soft conditioning data, respectively. Realizations of the stochastic simulation are shown in Figure 16. The yellow color denoting zero mudstone

Chapter 15: *Geostatistical Reservoir Modeling Focusing on the Effect of Mudstone Clasts* 211

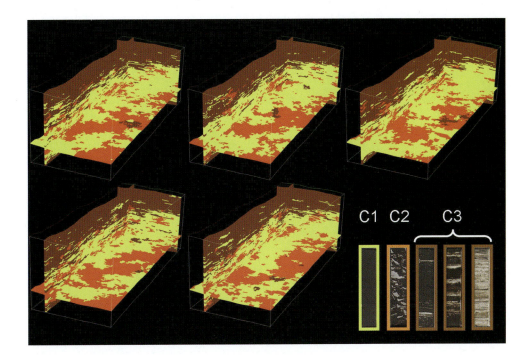

Figure 14. Realizations of facies modeling. All of the realizations successfully reflect the locally varying facies proportion and also reconstruct the small-scale heterogeneity required from the viewpoint of flow simulation.

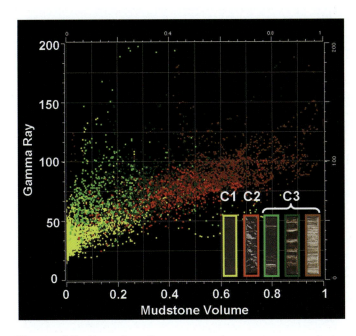

Figure 15. Crossplot of gamma-ray log responses against mudstone volume fraction derived from core photographs. The variability in gamma-ray log response at a certain mudstone volume fraction may be caused by a variable amount of clay minerals in the sand matrix.

volume corresponds to the C1 facies having no reduction in permeability due to the mudstones. The C3 facies corresponding to the brown color is regarded as nonproducible facies in this study and zero permeability is to be assigned to this region. Variation in the mudstone volume indicates variability in permeability of the C2 facies. Using a single constant permeability value as a representative permeability of the C2 facies in flow simulation may result in wrong bitumen production forecasting.

Porosity and permeability estimations

The final step of the reservoir modeling is to estimate the porosity and permeability of the C2 facies in consideration of the mudstone clasts. Porosity and permeability measurements obtained from core plugs are biased to sand because the core samples were preferentially selected from sand matrix so as to exclude mudstone fractions. The core scale properties can be transformed to the grid scale porosity and permeability to be assigned to grid blocks for the C2 facies by using the following equations:

$$\phi_{grid} = \phi_{core} \times (1 - V_{mud}), \qquad (2)$$

and

$$K_{grid} = f(K_{core}, V_{mud}), \qquad (3)$$

where ϕ_{grid} and K_{grid} are the porosity and permeability to assign to the model's grid block, ϕ_{core} and K_{core} are the porosity and permeability from core measurements, and V_{mod} is the estimated mudstone volume. The relational function f describing the mudstone volume-permeability relationship is obtained from the minimodel study in the section Permeability of Sand with Mudstone Clast Facies.

Figure 16. Realizations of mudstone volume modeling. Variation in the mudstone volume distribution is observed in the C2 category region, which can cause the wide range of variations in permeability.

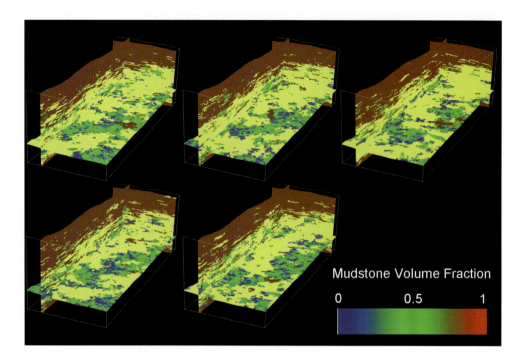

Figure 17. Realizations of permeability modeling. Small-scale heterogeneity in permeability is reconstructed in the realizations, which can cause variations in SAGD performance.

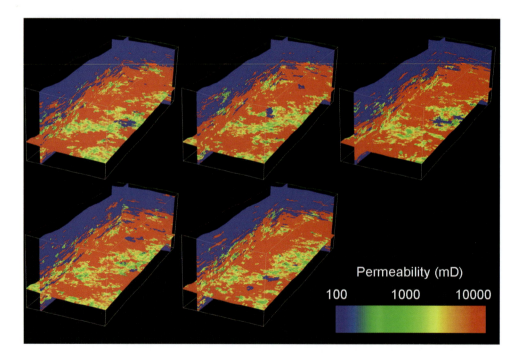

Although variation in porosity and permeability of the sand is indicated by the core permeability distribution in Figure 4, P-wave velocity distribution in Figure 9, and the crossplot in Figure 15, they are assumed to be simply constant at the averaged values of the core measurements.

Resulting permeability distributions are shown in Figure 17. Small-scale heterogeneity in permeability is reconstructed in the realizations, which can cause variations in SAGD performance. Flow simulation is to be applied to the realizations for the assessment of uncertainty in the SAGD performance. The facies C3 is regarded as nonproducible, and very small values are assigned for porosity and permeability in this study.

Discussion and Conclusions

Geostatistical reservoir modeling reconstructing the subseismic-scale heterogeneity in reservoir properties is conducted for reservoir simulation in forecasting SAGD production performance. Lithologic facies are reclassified into three facies depending on the type of mudstones having a different effect on permeability. This study focuses on the mudstone clasts and quantitatively evaluates permeability of the sand with mudstone clast facies. Facies discrimination is also discussed to separate the mudstone clasts from the thin mudstone layers. The following are the notable features of the modeling workflow in this study:

- facies modeling utilizing locally varying facies proportion calculated by PNN in multiattribute analysis implementing multiple seismic attributes
- core photographic quantification of mudstone volume fraction to separate the visible mudstones from interstitial shaliness
- flow simulation applied to minimodels to provide a permeability reduction curve as a function of increasing mudstone volume fraction

The facies modeling procedure combines geostatistics with multiattribute analysis to address the trend in the facies distribution. 3D seismic data play an important role in generating the locally varying facies proportion in 3D space. Selection of the seismic attributes used in the multiattribute analysis is based on the histogram analysis. Further study for selecting seismic attributes, perhaps with the approach from rock physics, may increase the facies discrimination performance. Although counting the pixel elements of core photographs at every sampling depth is time-consuming work that requires patience, it is adopted to identify the visible mudstone volume from interstitial shaliness that reduces the accuracy of the log-predicted mudstone volume. Studies on the interstitial shaliness of sands may improve the log-predicted mudstone volume. Thin sections and grain size analysis may provide information about sands. The minimodels in this study to obtain grid block permeability by applying flow simulation assume constant size and shape of the mudstone clasts. Changes in these parameters probably result in different permeability, and the sensitivity study can be conducted by changing the size and shape of the objects. This study may lead to the future task of addressing the thin mudstone layers that should reduce vertical permeability much more rapidly than mudstone clasts.

The three ideas introduced in the workflow described above can be easily and practically accomplished using published software and require no special leading-edge tools.

Acknowledgments

We thank Japan Petroleum Exploration for permission to publish the results derived from this study. We also thank Japan Oil, Gas, and Metals National Corporation and Japan Canada Oil Sands for permission to disclose the log and seismic data for this chapter. Finally, we express our gratitude to all of the members of the oil-sands team for the precious discussion.

References

Alberta Energy and Utilities Board, 2008, Alberta's reserves 2007 and supply/demand outlook 2008–2017: ST98-2008.

Desbarats, A., 1988, Estimation of effective permeabilities in the Lower Stevens Formation of the Paloma Field, San Joaquin Valley, California: SPE Reservoir Engineering, **3**, 1301–1307.

Desbarats, A., and R. Dimitrakopoulos, 1990, Geostatistical modeling of transmissibility for 2D reservoir studies: SPE Formation Evaluation, **5**, 437–443.

Deutsch, C., 1989, Calculating effective absolute permeability in sandstone/shale sequences: SPE Formation Evaluation, **4**, 343–348.

Deutsch, C. V., and A. G. Journel, 1998, Geostatistical software library and user's guide, 2nd ed.: Oxford University Press.

Dimitrakopoulos, R., 1993, Geostatistical modeling of gridblock permeabilities for 3D reservoir simulators: SPE Reservoir Engineering, **8**, 13–18.

Flach, P., and G. Mossop, 1985, Depositional environments of Lower Cretaceous McMurray Formation, Athabasca Oil Sands, Alberta: AAPG Bulletin, **69**, no. 8, 1195–1207.

Hampson, D. P., J. S. Schuelke, and J. A. Quirein, 2001, Use of multiattribute transforms to predict log properties from seismic data: Geophysics, **66**, 220–236.

Jensen, J., L. Lake, P. Corbett, and D. Goggin, 2000, Statistics for petroleum engineers and geoscientists: Elsevier.

Journel, A., C. Deutsch, and A. Desbarats, 1986, Power averaging for block effective permeability: SPE Paper 15128.

Mao, K. Z., K. C. Tan, and W. Ser, 2000, Probabilistic neural-network structure determination for pattern classification: IEEE Transactions Neural Networks, **11**, 1009–1016.

Ringrose, P., E. Skjetne, and A. C. Elfenbein, 2003, Permeability estimation functions based on forward modeling of sedimentary heterogeneity: SPE Paper 84275.

Schmitt, D., 2004, Oil sands and geophysics, CSEG Recorder, November.

Takahashi, A., K. Kashihara, S. Mizohata, N. Shimada, T. Nakayama, M. Kose, and T. Torigoe, 2006, Construction of three-dimensional geological models for oil sands reservoir in Athabasca, Canada: Geophysical Exploration, **59**, 233–244 (in Japanese with an English abstract).

Tian, B., M. R. Azimi-Sadjadi, T. H. Vonder Haar, and D. Reinke, 2000, Temporal updating scheme for probabilistic neural network with application to satellite cloud classification: IEEE Transactions Neural Networks, **11**, 903–920.

Tian, B., and M. R. Azimi-Sadjadi, 2001, Comparison of two different PNN training approaches for satellite cloud data classification: IEEE Transactions Neural Networks, **12**, 164–168.

Wightman, D. and S. Pemberton, 1997, The Lower Cretaceous (Aptian) McMurray Formation: an Overview of the Fort McMurray Area, Northeastern, Alberta, Canadian Society of Petroleum Geologists, Memoir 18, 312–344.

Chapter 16

Monitoring an Oil-sands Reservoir in Northeast Alberta Using Time-lapse 3D Seismic and 3D P-SV Converted-wave Data

Toru Nakayama,[1] Akihisa Takahashi,[1] Leigh Skinner,[2] and Ayato Kato[3]

Time-lapse 3D seismic monitoring was conducted in the Japan Canada Oil Sands Limited (JACOS) Hangingstone steam-assisted-gravity-drainage (SAGD) operation area in Alberta, Canada, to delineate steam-affected areas. The time-lapse surveys, acquired in February 2002 and March 2006, show distinct response changes around the SAGD well pairs. In addition, 3D P-SV converted-wave processing and analysis were applied on the second 3D data set [recorded with three-component (3-C) digital sensors] for a reservoir characterization study. Background information on the Hangingstone SAGD operation is contained in Kato et al. (2008).

Time-lapse 3D Seismic Data

Figure 1 shows a map of the two 3D seismic surveys and SAGD well locations in the field. Black solid lines represent SAGD well paths, and the red dotted line indicates a north–south line (referred as NS hereafter) that will be used throughout this study. The first 3D seismic (baseline survey = 5.4 km^2) was acquired in 2002 to construct detailed 3D geologic models for reservoir characterization. The production of the five eastern SAGD well pairs (A, B, C, D, and E in Figure 1) commenced prior to the recording of the first 3D survey. After the first 3D survey, steam injection was implemented in four stages at 10 additional western SAGD well pairs before 2006. The second 3D survey was conducted in 2006 in the northern part (4.3 km^2) of the 2002 baseline survey, where the active 15 SAGD wells exist.

The seismic data sets were recorded with nearly identical field acquisition parameters. One major difference between the surveys is the receiver type; analog geophone arrays were used in 2002, and 3-C digital sensors were used in 2006.

Both data sets underwent an identical processing flow at the same time. In addition to the basic seismic data processing, spectral decomposition for the surveys and P-SV data processing of the 3-C monitoring survey were conducted.

Core Velocity Data and Rock Physics Model

As described in Kato et al. (2008), P- and S-wave velocities of oil-sand core plugs from the field were measured under various pressure and temperature conditions to understand the relationship between elastic properties of the oil sands and changes in temperature and pore pressure.

As a result, a rock physics model was developed to predict velocity changes of the oil sands caused by any variations of pore pressure, temperature, fluid saturation, and fluid phase changes expected during SAGD operations.

We used this rock physics model in a relative manner for our interpretations of the time-lapse 3D seismic and P-SV data. This will be described in the following sections.

Seismic Calibration and Interpretation

The 2002 baseline data were used as a reference data set, and the 2006 monitoring data were calibrated to the 2002 data to remove differences between them that were not production-related response changes.

---

[1]Exploration Division, JAPEX, Tokyo, Japan
[2]Geology and Geophysics, JACOS, Calgary, Alberta, Canada
[3]University of Houston, Department of Earth and Atmospheric Sciences, Houston, Texas
This paper appeared in the September 2008 issue of THE LEADING EDGE and has been edited for inclusion in this volume.

Figure 2 shows the final migration of the NS line from the 2002 and 2006 surveys. Calibration has not yet been applied. The southern half of this line intersects with the SAGD well pairs H, I, K, L, M, and N; the northern part of the line is away from the steam-injected areas. Although Figure 2 shows remarkable changes in seismic response (around and below the red oval), there were also some differences to be compensated for between the two sections with regard to frequency contents and static shifts.

Figure 3 compares the calibrated 2006 data with the 2002 data. Our calibration consisted of band-pass filter, trace scaling, phase and amplitude correction, and static time correction.

Figure 4, our interpretation of the line, clearly shows time delays of seismic events at the reservoir bottom (Top Devonian) and below the reservoir around the active SAGD well pairs that passed through the left half of each section.

Figure 5 shows a time-difference map between the Top Devonian horizons of the 2002 and the 2006 data. Positive traveltime differences in Figure 5 represent decreases in V_P within the reservoir because reservoir thickness did not change with the injection of steam. Significant V_P decreases

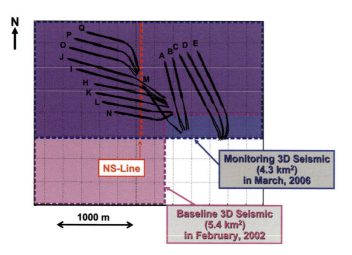

Figure 1. Map of the time-lapse 3D seismic survey and the SAGD well locations. Black solid lines represent SAGD well paths. The red dotted line shows the location of a representative NS seismic section described in Figures 2, 3, 4, and 6.

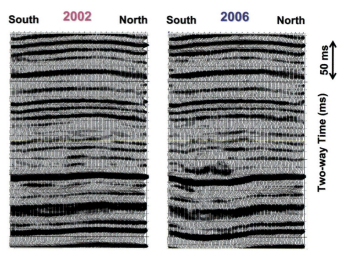

Figure 3. Sections from the calibrated 2006 data and the 2002 data.

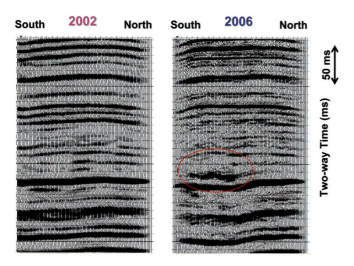

Figure 2. Migrated sections (NS line) for the 2002 and 2006 surveys to which calibration has not been applied. The southern half of each section intersects with the SAGD well pairs H, I, K, L, M, and N. The northern part is away from the steam-injected areas.

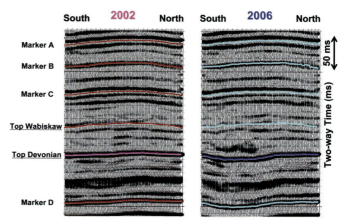

Figure 4. Interpreted NS line. Top Devonian is regarded as the reservoir bottom (Base McMurray) and Top Wabiskaw is approximately 5 m shallower than the reservoir top (Top McMurray).

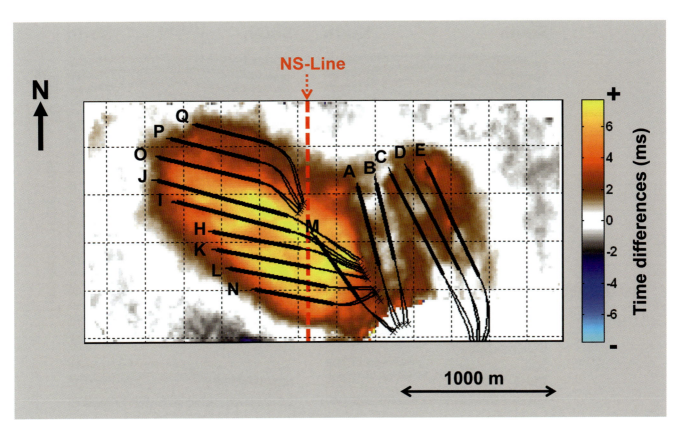

Figure 5. Time differences (ms) between the 2002 and 2006 Top Devonian horizons. The positive time areas show that the 2006 Top Devonian horizon is deeper than the 2002 Top Devonian horizon in the two-way time domain.

in the oil-sands layer were observed around the SAGD well pairs in the western part of the survey area. As previously stated, the production of the five eastern well pairs commenced prior to the recording of the baseline survey. Therefore, the reservoir property changes between the two surveys were relatively small in the eastern areas.

Figure 4 also shows large differences in seismic character within the reservoir around the active SAGD well pairs. To analyze seismic response changes (such as amplitude) within the reservoir, we applied a horizon-based time-shift correction to the 2006 data. This correction means the times of the 2006 interpreted horizons were adjusted to their corresponding 2002 horizons in a "stretch and squeeze" manner to remove production-related time delays (Figure 6).

We calculated trace shape similarity and amplitude difference within the reservoir between the two data volumes on a trace-by-trace basis. Figures 7 and 8 can be considered trace shape similarity maps that show maximum crosscorrelation values and root-mean-square (rms) values of amplitude differences within the reservoir on each trace location, respectively. Around the 10 western SAGD well pairs, Figure 7 shows low crosscorrelation areas (green to magenta), and Figure 8 shows dominant high-amplitude difference areas (red to dark red). These low-similarity areas indicate production-related trace shape changes in the field.

Detailed velocity analysis at every common midpoint (CMP) location using two 3D seismic prestack gather data sets was also conducted to detect small P-wave velocity changes within the oil-sands reservoir caused by the steam injection. Figure 9 is a time slice intersecting the reservoir showing interval velocity differences between the 2002 and 2006 smoothed interval velocity cubes.

As Figures 5 and 7–9 are quite consistent with each other, these four maps were integrated into one map using self-organizing maps (SOM) (Kohonen, 2001) and K-means methods (Matos et al., 2007) to analyze regional seismic characteristics. These are effective methods for nonsupervised clustering analysis using seismic attributes.

Our basic procedure for integrating the four maps is as follows. First, 4D vector space consisting of (a) time difference, (b) maximum crosscorrelation, (c) rms value of amplitude difference, and (d) interval velocity difference was assumed. The number of vectors is therefore equal to the number of CMP bins. Next, prototype vectors (representative vectors calculated from the data) were selected by the SOM technique. Assuming the number of the prototype vectors to be 100, the 100 prototype vectors were defined, each input 4D vector was grouped into the

218 Heavy Oils: Reservoir Characterization and Production Monitoring

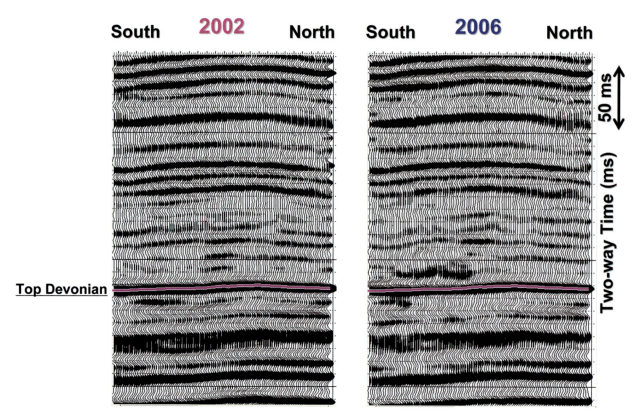

Figure 6. Example of the horizon-based, time-shift correction results on the NS line.

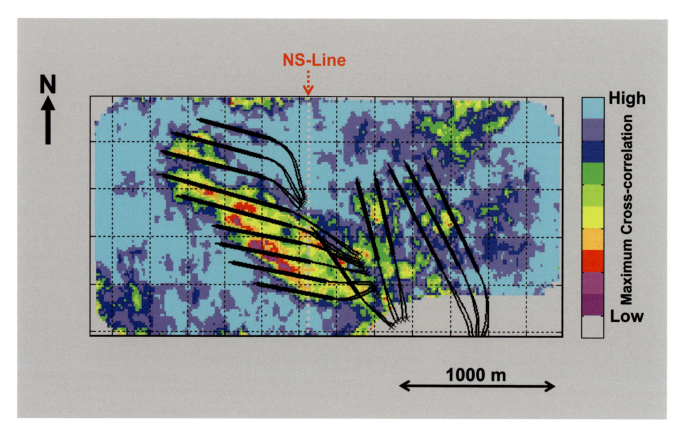

Figure 7. Maximum crosscorrelation values, within the reservoir on each trace location. Warm colors represent lower maximum crosscorrelation values, and cold colors represent higher values.

Chapter 16: *Monitoring an Oil-sands Reservoir in Northeast Alberta* 219

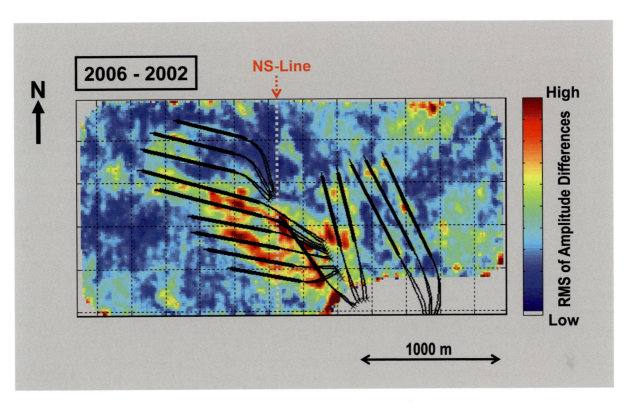

Figure 8. The rms values of trace amplitude differences within the reservoir on each trace location. Warm colors represent higher rms values of amplitude differences, and cold colors represent lower values.

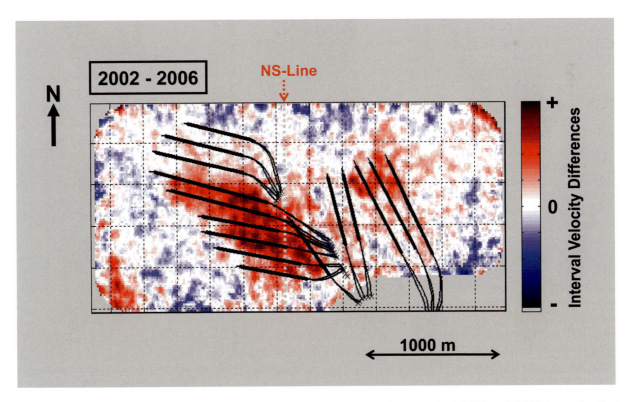

Figure 9. Interval velocity differences between time slices extracted from the smoothed 2002 and 2006 interval velocity cubes. The red areas (positive values) show that the interval velocities in 2006 are lower than those in 2002. This time slice intersects with the reservoir interval. The velocity data picked at every CMP location were smoothed to eliminate the erroneous values and then converted into the interval velocity cubes.

nearest prototype vector, and then all of the input vectors were classified into 100 classes. These 100 classes can be assumed to represent 100 patterns of reservoir change. Finally, the 100 SOM prototype vectors were classified into six classes by the K-means method to reduce the class number, and then the original input 4D vectors were grouped into their nearest SOM + K-means prototype vector.

Figure 10 shows crossplots between the four map attributes with their SOM 100 prototype vectors. Figure 11 shows an SOM classification map.

Figure 12 shows crossplots between the four map attributes with their classification by the SOM and SOM + K-means methods. Figure 13 is a classification map using the SOM + K-means combination method showing the clustering of the data of the four-attribute vectors into the six groups.

Figures 11 and 13 can classify the extent of the rock property changes caused by the steam injection into 100 classes and six classes, respectively. In Figure 11, steam chambers were considered to have grown sufficiently in the red and orange areas but did not develop in the green and blue areas. We also consider that yellow areas in Figure 11 indicate transition zones. In Figure 13, our interpretation was that steam chambers grew sufficiently in the orange and cyan areas (class ID = 2 and 4) but did not develop in the yellow and blue areas (class ID = 3 and 6). We also consider that brown areas (class ID = 1) indicate the transition zones.

Since SAGD operation of the five eastern well pairs (A, B, C, D, and E pairs) started before the baseline seismic survey, and the reservoir property changes around them were relatively small, it is difficult to interpret the rock property changes on those areas. However, the

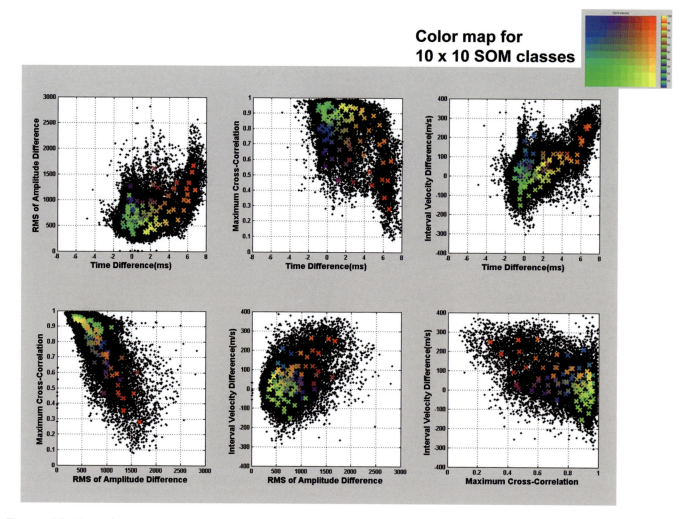

Figure 10. Crossplots between the four map attributes with their SOM prototype vectors. Black dots are raw attribute data and colored crosses show the SOM prototype vectors (100 classes). Close prototype vectors in the SOM domain have similar colors.

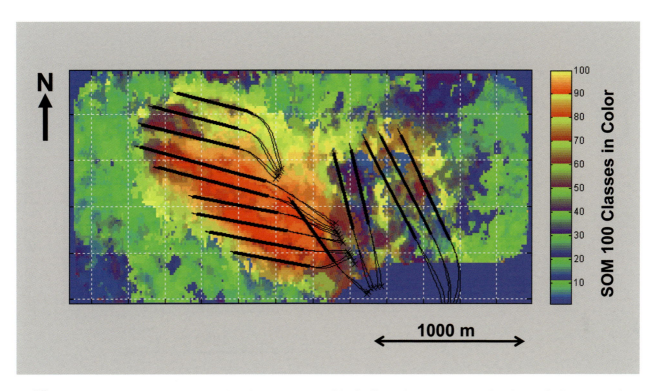

Figure 11. SOM classification map with 100 classes. Areas with similar colors are assumed to have similar reservoir conditions.

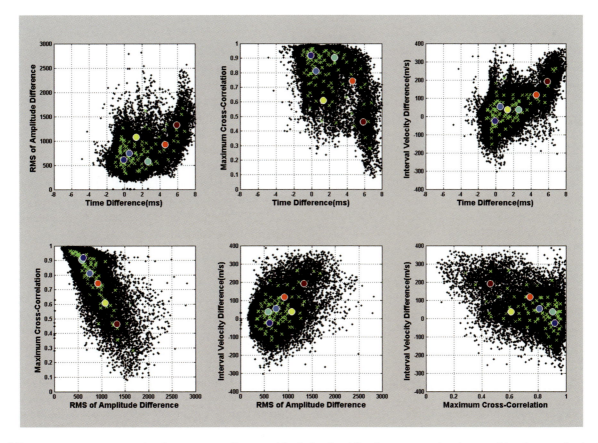

Figure 12. Crossplots between the four map attributes with their classification vectors by the combination method of SOM and K-means. Black dots are raw data and green crosses are the SOM prototype vectors (the same as in Figure 10). Colored circles are the six vectors selected by SOM + K-means.

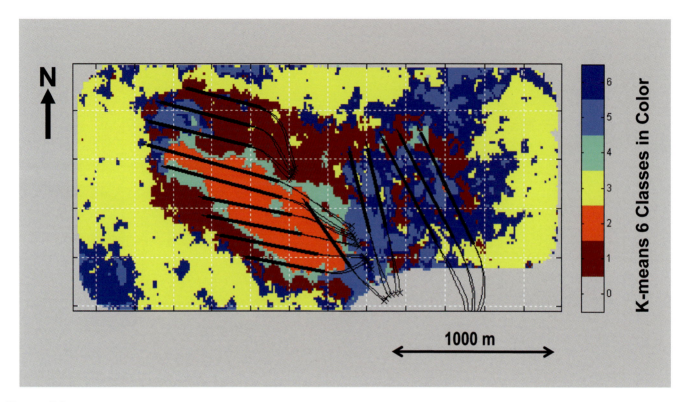

Figure 13. Classification map (six classes) using the SOM + K-means combination. Areas with similar colors are assumed to have similar reservoir conditions on the basis of the four seismic map attributes. Class ID numbers themselves are not meaningful.

integrated maps like Figures 11 and 13 help qualitatively delineate the steam-affected zones.

Spectral Decomposition

Spectral decomposition, based on a wavelet-transform method, was conducted to detect detailed changes of the spectrum components of the seismic data due to the steam injection.

Figure 14a and b shows time slices of 10-Hz isofrequency amplitude spectrum cubes for the 2002 and 2006 data, respectively. The 10-Hz isofrequency amplitude cube is composed of the 10-Hz component value of the amplitude spectrum calculated at each time sample. The maps in Figure 14 show noticeable seismic anomalies around the active SAGD wells at each survey time and enable us to qualitatively delineate the steam-affected zones.

P-SV Analysis and Interpretation

P-SV seismic data processing using the 3-C 2006 seismic data was conducted to characterize the oil-sands reservoir simultaneously using V_P and V_S information.

First, we carried out modeling of the P-SV and P-P seismic responses with the well log data. Figure 15 shows the P-SV and P-P synthetic seismograms (angle gathers) at a well located outside of the steam-affected areas. Next, the synthetic P-P and P-SV reflection events were correlated with the well markers, and then the P-P and P-SV synthetic data were tied to the P-P and P-SV seismic data. Figure 16 shows an example of the time sections. Three horizons (including Top Devonian and Top Wabiskaw) are shown on the P-SV and P-P sections.

Figure 17 shows the V_P/V_S ratio map between Top Wabiskaw and Top Devonian horizons based on the P-P and P-SV data. The V_P/V_S ratios were calculated using the following formula:

$$\frac{V_P}{V_S} = \frac{2TWT_{PSV} - TWT_{PP}}{TWT_{PP}}, \quad (1)$$

where TWT_{PSV} and TWT_{PP} are the P-SV isochron and P-P isochron, respectively, between the two horizons.

The V_P/V_S ratio of the interval between the Top Wabiskaw and Top Devonian approximately represents the V_P/V_S ratio of the reservoir zone as most of that interval corresponds to the reservoir zone in the steam-injected area. Figure 17 also shows comments regarding the V_P/V_S ratio. On this map, we classified the areas with visually high or low V_P/V_S ratios into three groups. Note that our qualitative interpretation of Figure 17 was done in combination with the results of Kato et al. (2008).

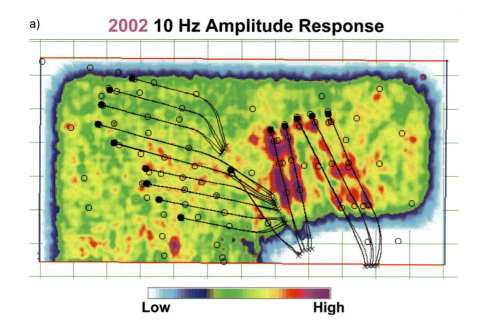

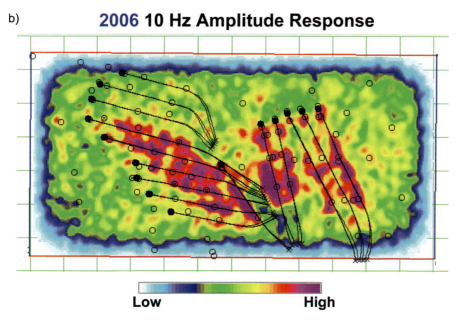

Figure 14. Time slices from spectral amplitude cubes at 10 Hz of (a) the 2002 3D seismic data, and (b) the 2006 3D seismic data. These time slices are within the reservoir interval. Warm colors represent higher amplitude values, and cold colors represent lower values.

1) Areas with low V_P/V_S ratios (enclosed by black dashed polygons around A, B, C, D, E, H, I, J, and K well pairs) are ones in which the steam chamber grew sufficiently. These low V_P/V_S ratio areas were probably under the conditions around step 21 as described by Kato et al. (2008).
2) Areas of high V_P/V_S ratios surrounded by white dashed polygons (in front of the toe of well pair H and between well pairs J and O) are probably steam affected, but the steam chambers did not develop sufficiently. Although the steam chambers partly developed, the reservoir conditions of these high V_P/V_S ratios were, on the whole, somewhere in the high V_P/V_S ratio steps — steps 6–20 in Kato et al. (2008).
3) Areas of high V_P/V_S ratios enclosed by brown dashed polygons (in front of the toes of the I, J, and N well pairs) were little affected by steam because these areas are known from well data to have poor reservoir development.

The interval V_P/V_S ratio information is useful for estimation of the rock property changes of the oil-sands reservoir. However, we have to consider that the zone used for the V_P/V_S ratio calculation was thicker than the real oil-sands reservoir interval, and the V_P/V_S ratio map did not show the accurate V_P/V_S values of the reservoir itself. Thin shale layers within the reservoir were also neglected, and the reservoir interval was treated as one homogeneous oil-sands

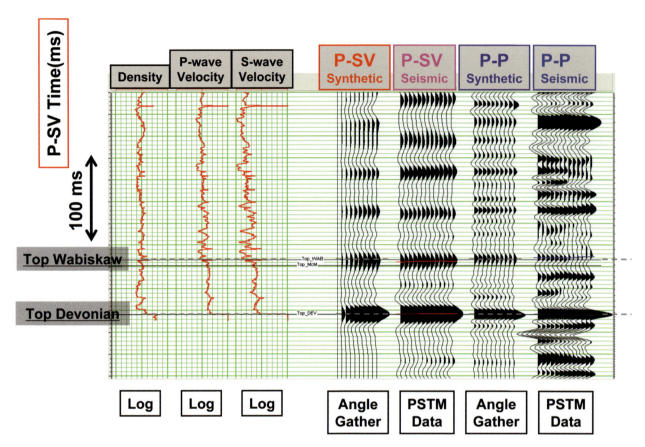

Figure 15. Example of the P-SV and P-P synthetic seismograms (angle gathers) with density log, V_P log, V_S log, and the 2006 PSTM P-SV and the 2006 PSTM P-P seismic sections around the well. Time scale is P-SV time domain.

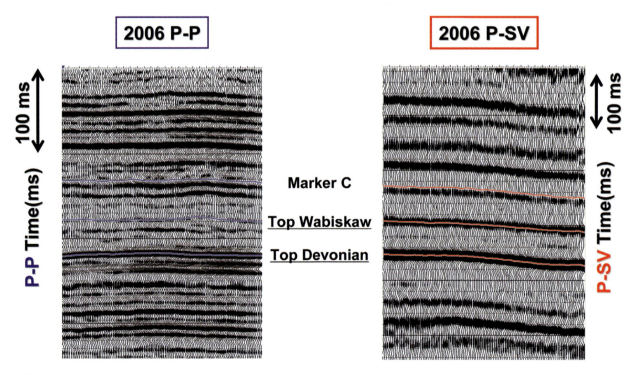

Figure 16. Example of the P-SV and P-P time sections along a sample line with three horizons, including Top Devonian, Top Wabiskaw, and another seismic event. Time scales of the two sections are different.

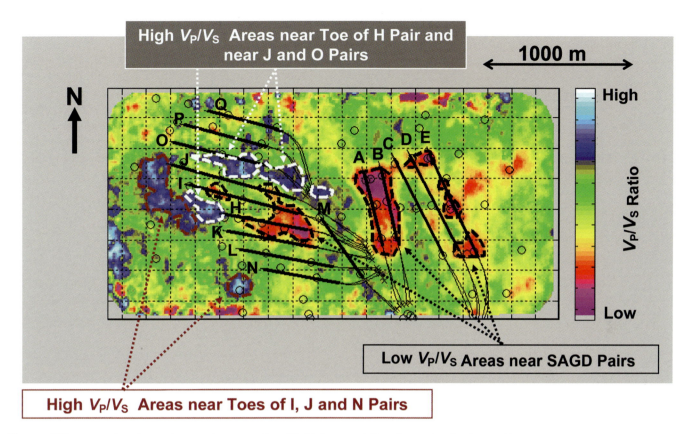

Figure 17. Interval V_P/V_S ratio map between Top Wabiskaw and Top Devonian horizons. Warm colors represent lower V_P/V_S ratios, and cold colors represent higher V_P/V_S ratios. Areas with high or low V_P/V_S ratios were selected and classified into three groups. Each group was annotated with an individual color, and the area of each group was outlined with a colored dashed line.

layer. This is mainly because the temporal resolution of the P-SV data was much lower than the P-P seismic data, and it was too difficult to interpret the horizons within the reservoir in detail. However, V_P/V_S ratio maps such as Figure 17 are qualitatively valuable for estimating reservoir properties.

Discussion

As stated before, Figures 5 and 7–9 are qualitatively consistent with each other, especially around the 10 western SAGD well pairs. However, when these maps are compared in detail, some inconsistencies between the edges and the shapes of the anomaly areas on these maps were noted. The following are our interpretations of the inconsistencies using our rock physics model.

Large velocity decreases should occur in the high-temperature zone in the steam chamber, and smaller velocity decreases are predicted in the high pore-pressure zone on the basis of our rock physics model.

Thermal conductivity of the oil sands is so low that the heat front is relatively close to the steam chamber front. Conversely, the pressure fronts are expected to spread wider than the heated zones. As a result, a larger area would be influenced by pressure than by temperature.

Large time delays of the Top Devonian horizon and large trace shape changes in the reservoir are expected in the extent of the steam chamber zones, and small time delays of the Top Devonian horizon and small trace shape changes in the reservoir are expected in the extent of the pore-pressure increase zones. Currently, we consider that Figure 5 basically shows anomalous areas caused by the combination of high temperature and high pore pressure, and Figures 7–9 mainly represent areas heated during the time between the two seismic surveys.

Conclusions

The time-lapse seismic survey and analysis were conducted in the JACOS Hangingstone SAGD operations area to monitor steam chamber development. The two seismic P-P volumes acquired in 2002 and 2006 show large differences in seismic character within the reservoir around the SAGD well pairs, and the V_P/V_S ratio map from the P-SV and P-P volumes of the 2006 3-C data clearly shows areas of variation.

The time-lapse seismic monitoring and the P-SV seismic data are useful for investigating the rock property changes of the interwell reservoir sands in the field. The results of the geophysical study are now being integrated with geologic and production information for efficient reservoir management.

Acknowledgments

This study was conducted jointly between JACOS and Japan Oil, Gas and Metals National Corporation (JOGMEC). The authors thank JOGMEC, JACOS, and JAPEX for permission to publish these data.

References

Kato, A., S. Onozuka, and T. Nakayama, 2008, Elastic property changes of bitumen reservoir during steam injection: The Leading Edge, **27**, 1124–1131.

Kohonen, T., 2001, Self-organizing maps: Springer-Verlag.

Matos, M. C., P. Leo Manassi Osorio, and P. R. Schroeder Johann, 2007, Unsupervised seismic facies analysis using wavelet transform and self-organizing maps: Geophysics, **72**, no. 1, P9–P21.

Chapter 17

Oil-sands Reservoir Characterization for Optimization of Field Development

Akihisa Takahashi[1]

Introduction

The proved remaining reserves of bitumen from oil sands in Canada are estimated as approximately 170 billion barrels (Alberta Energy and Utilities Board, 2007). Including conventional crude oil, the total number of proved remaining reserves in Canada takes second place in the world after Saudi Arabia. From the production perspective, more than 40% of the crude oil production in Canada is from oil sands. Figure 1 shows the index map of the area.

Japan Canada Oil Sands Ltd. (JACOS), a subsidiary of JAPEX, has been operating the development and production of bitumen resources using an in situ steam-assisted gravity drainage (SAGD) method since 1997 in an area approximately 50 km southwest of Fort McMurray. The oil-sands reservoirs here exist at a depth of approximately 300 m.

The SAGD method is a type of enhanced oil recovery using steam injection (Figure 2). Horizontal well pairs (vertically parallel) are drilled, and steam is injected from the upper well (injector). Oil sands are heated by thermal conduction, and bitumen becomes less viscous and more mobile. In the next stage, a steam chamber is generated by replacing bitumen with steam. Then, along the rim of steam chamber, movable bitumen and oversaturated steam runs down because of gravity and is recovered through the lower well (producer).

For economically successful SAGD recovery, detailed description of local subsurface geology is essential because the bitumen is recovered from only close proximity to the SAGD well pairs because of the high viscosity of bitumen under in situ reservoir temperatures. Recognition of the distribution of shale is also important because the low-permeability shale inhibits the development of the steam chamber in the SAGD process.

Oil-sands reservoir characterization was thus conducted to delineate detailed oil-sands reservoir distribution. For this purpose, considerable amounts of core and well logs have been acquired and a complex depositional system that consists of several incised valley fill sandstone units has been developed. In 2002, a 3D seismic survey was conducted (Figure 3) and geological models were constructed by integrating core, log, and seismic data.

Geologic Background

Sedimentary environments of the McMurray formation (Lower Cretaceous), which contains the main deposit of oil sands (bitumen) here, are considered to be fluvial-to-upper-estuarine channel fill deposits around the mouth of a river where tidal influence is dominant. Figure 4 shows the sequence-stratigraphic framework of the McMurray formation in the JACOS oil-sands area. Incised valleys cut during an early lowstand period are filled with massive sandstones with an upward-fining stacking pattern, and a shale-dominant facies with an upward-coarsening pattern was deposited during the transgressive to highstand period. In the next lowstand period, another set of incised valley fill is created.

The McMurray formation is composed of several sedimentary facies such as the massive sandstone (channel sand-bar facies: channel sands), sandstone with muddy rip-up clasts (channel sand-bar with rip-up clasts facies: clasts), interbedded sandstone/mudstone (lower and upper point-bar facies: LPB and UPB), and mudstone-abandoned channel fill (tidal flat facies: TF). Figure 5 shows an example of cores representing each sedimentary facies. Channel sands, clasts, and LPB are considered as the reservoir facies for the SAGD method, whereas UPB and TF are not. The McMurray formation

[1]Exploration Division, JAPEX, Tokyo, Japan

is overlain by the Wabiskaw shale and underlain by the massive Devonian limestone as shown in the well log example (Figure 6).

Geophysical Background

Figure 7 shows an example of seismic data (prestack time migration) in this area. The predominant frequency of the seismic data in the McMurray interval is approximately 100 Hz, which suggests vertical resolution to be approximately 6.5 m according to the one-quarter wavelength rule ("separability"). On the other hand, "visibility" [resolution limit for amplitude analysis defined by Brown (2004)] should be around 1/30th of the wavelength in such a high signal-to-noise ratio (S/N) case, which corresponds to 1.0 m (Takahashi et al., 2006a).

As can be seen in Figure 7, strong reflections occur at the base of the McMurray formation as the Devonian carbonate beneath it has quite higher impedance. Sidelobes of this reflection cover almost all of the interval of the oil-sands reservoir.

For the seismic amplitude analysis, it is important to consider the range of reflection intensity from each boundary of lithofacies. Average velocity and density of each lithology under the reservoir temperature and pressure are as follows.

Lithology	Velocity (m/s)	Density (g/c^3)
Sand (McMurray formation)	2600	2.09
Shale (McMurray formation)	2660	2.26
Limestone (Devonian)	4700	2.70

Reflection strengths in terms of reflection coefficient at each boundary are as below.

Boundary	Reflection coefficient
Sand—limestone	0.4
Sand—shale	0.05

The reflection coefficient at the sand and shale boundaries in the McMurray formation is much less than that at the boundary of the McMurray formation and Devonian (basement). Within the McMurray formation, density

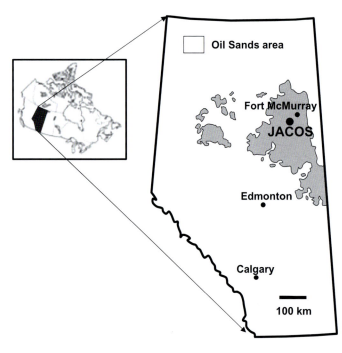

Figure 1. Index map of the oil-sands area in Canada.

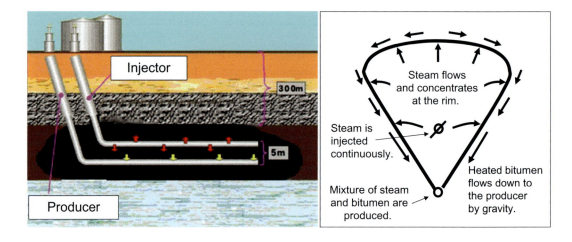

Figure 2. Principle of the SAGD method. Courtesy of JACOS.

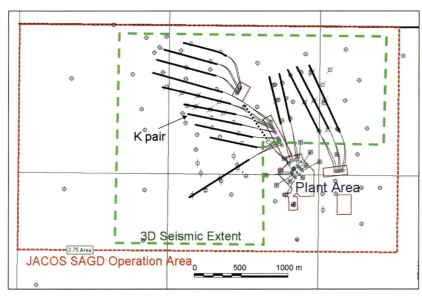

Figure 3. Index map of the JACOS Hangingstone SAGD operation area. Appraisal well locations (circles) and 3D seismic survey extent (dotted line) are shown. Total number of 16 horizontal well pairs (solid lines) for bitumen production were drilled by 2005.

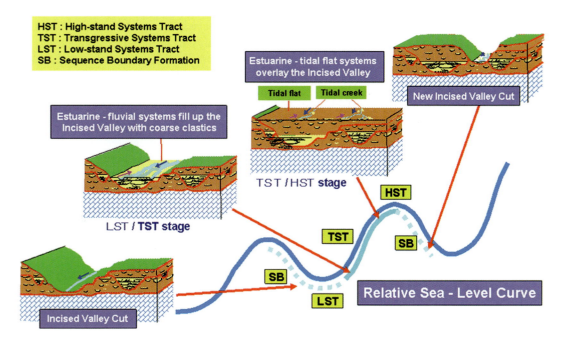

Figure 4. Sequence stratigraphic framework of the McMurray formation in the survey area.

contrast (8%) is the main cause of reflectivity because velocity contrast is merely 2%.

Workflow

A flow diagram of the analysis methodology is shown in Figure 8. Initially the sedimentary facies of every 10 cm was determined from existing core data, and the relationships between core facies and well log responses are examined. Next, sedimentary facies of the wells without core data were estimated from the well log data.

Sequence stratigraphic correlation was then conducted by using well log data, and a framework of the stacked incised valley system was established (Torigoe et al., 2005; Takahashi et al., 2006b). Further detailed sequence structural models were constructed using seismic data at a later stage. An example of the structural model showing one of

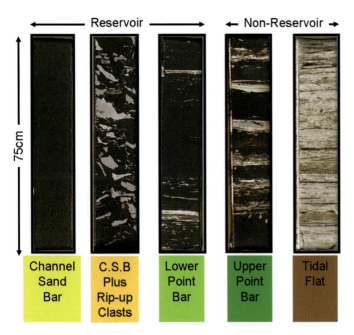

Figure 5. Core photos showing each sedimentary facies. Classification of reservoir/nonreservoir facies is also shown.

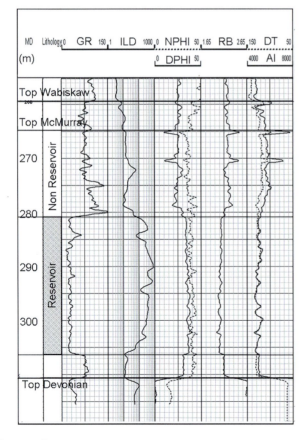

Figure 6. Example of well logs in the McMurray formation, showing both reservoir and nonreservoir facies.

the sequence boundaries (bottom of the incised valley) is shown in Figure 9.

Model-based acoustic impedance (AI) inversion was carried out to obtain better resolution and suppress the effect of sidelobes. In general, model-based inversion results are highly dependent on the initial model. In the McMurray formation, as the sand and shale distribution is not simply stratified, it is felt that the initial model constructed by the interpolation of well-logs would not be adequate. It was then decided to adopt a simple three-layer model with constant velocities and leave all the higher frequency determination to the seismic trace itself.

Multiattribute analysis integrating well and seismic data was carried out using the EMERGE software package. As the purpose of this analysis was to distinguish between reservoir/nonreservoir facies, sedimentary facies index logs were adopted as the target logs as shown in Figure 10. In this figure, indices of 1-5 correspond to channel sands, clasts, LPB, UPB, and TF. As the seismic resolution in this area is considered to be around 1.0 m from the point of visibility, sedimentary facies logs were upscaled from 10 cm to 50 cm. These sedimentary facies logs were further simplified to two categories, reservoir and nonreservoir facies, to stabilize the result of multiattribute analysis. In order to obtain an equivalent time interval in the seismic data, the output sampling interval for the multiattribute analysis was downscaled from 2.0 to 0.5 ms. When selecting seismic attributes, two attributes such as AI inversion and instantaneous frequency were a priori selected because of good correlation, and the others, such as integrated seismic trace, second derivative of instantaneous amplitude, and amplitude weighted cosine of instantaneous phase, were selected based on the stepwise regression as described in Hampson et al. (2001). Full automatic stepwise regression did not give satisfactory results in this case. The probabilistic neural network was adopted as the method of multiattribute transformation using 23 wells.

Figures 11 and 12 show examples of AI inversion and multiattribute analysis sections along the same line as Figure 7. In Figure 11, density logs from three control wells are indicated. Actually, these wells were drilled after the analysis. In Figure 12, probability values as the reservoir facies are shown in the range between 0.0 and 1.0. Sedimentary facies index logs from control wells are also shown.

On the basis of the core and log observation at the wells and 3D seismic data, physical property models showing sedimentary facies were constructed as shown in Figure 13.

Evaluation of Facies Prediction

Multiattribute analysis enables basal shale (base of reservoir) imaging and the identification to a certain degree of

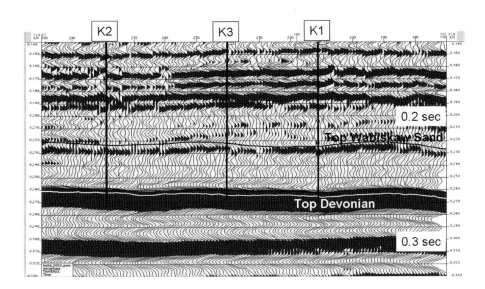

Figure 7. Prestack time-migrated section along K-pair (location indicated in Figure 3) with control well locations. Interpreted horizons of Top Devonian (Bottom McMurray) and Top Wabiscaw Sand (Above Top McMurray) are also shown. Lateral change of lithology cannot be seen clearly at this stage.

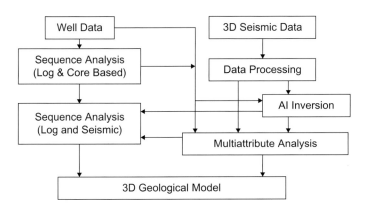

Figure 8. Work flow of the construction of the 3D geological models by integrating well and seismic data.

thin shales within the reservoir. Top and bottom depths of the reservoir are estimated in the range of 2.0 m even in such a complex channel sands environment.

As the target log is simplified to reservoir facies, whose index is one, and nonreservoir facies, whose index is zero, probabilistic expression of multiattribute transformation varies between 0.0 and 1.0. These values can be considered as confidence levels of being reservoir based on seismic data. Prediction would have more confidence for reservoir facies if values approached 1.0, and for nonreservoir facies if values approached 0.0. At the existing well locations, predicted target log values for each actual sedimentary facies were compiled, and frequency of predicted log values of every 5% were put together in Figure 14. Considering target log values above 0.5 to be reservoir and those below 0.5 to be nonreservoir, prediction success ratios for each sedimentary facies of existing wells are summarized in Table 1. For example, in channel sands, clasts, and LPB, prediction, success ratio as reservoir should be 100% ideally. Channel sands facies are predicted as reservoir with the probability of 89%, and TF facies are predicted as nonreservoir with the probability of 75%. These two end members are well predicted as correct facies. In the case of UPB and LPB facies, prediction remains 50%, which means it is very difficult to distinguish between these two. One possible reason is that the classification of UPB and LPB is somewhat arbitrary because the percentage of intercalated shale layers in point bars changes continuously. Regarding clasts, they are predicted as reservoir facies with the probability of 91% although percentage of shale contents varies within a wide range. A possible reason for this high prediction success is that clasts are distinguished from LPB, UPB, and TFs as they are a seismically homogeneous media rather than a stratified media (Takahashi et al., 2006a). Because the distribution of LPB and UPB in the study area is rather limited, reservoir facies distribution can be conducted with high precision.

Horizontal Well Pair Planning

The analyses described above contribute to reserves estimation and the optimization of development plans. This chapter describes an example of the actual procedure for horizontal well pair deployment.

During the horizontal well planning of the K-pairs, locations of control wells were considered. Multiattribute analysis results indicated that the thickness of the reservoir was reduced toward the west, and an abrupt change of character was observed as indicated by an arrow in Figure 12. Basal shale was considered to exist along the entire planned K-pair trajectory with quite consistent thickness. The AI inversion section along the K-pair (Figure 11) shows that AI values are increasing toward the west near the base of the reservoir interval. The existence of shale was inferred from this phenomenon, but it was not certain whether it was shale clasts (reservoir facies)

232   Heavy Oils: Reservoir Characterization and Production Monitoring

Figure 9. Structure model showing the base of one of the major sequences. Shape of the incised valley created during the lowstand period can be seen.

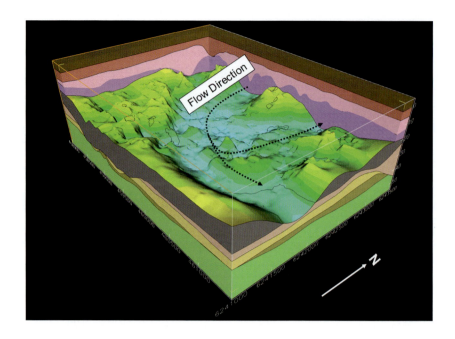

Figure 10. Target log of the multiattribute analysis. Original sedimentary facies classification of every 10 cm was upscaled to every 50 cm, and simplified to two categories (reservoir and nonreservoir).

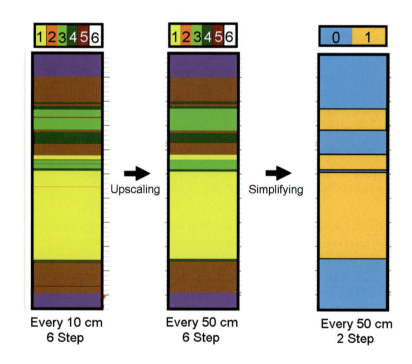

Figure 11. Cross section of the acoustic impedance inversion along the K pair. Density logs along the three wells are also indicated. There are thick channel sands in the lower part of K1 and K3 expressed by lower acoustic impedance value. Acoustic impedance value increases toward K2 due to the existence of mud clasts.

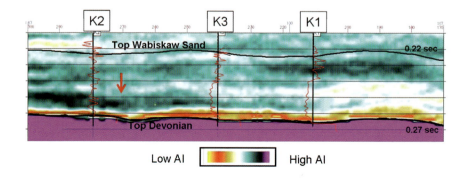

Chapter 17: *Oil-sands Reservoir Characterization for Optimization of Field Development* 233

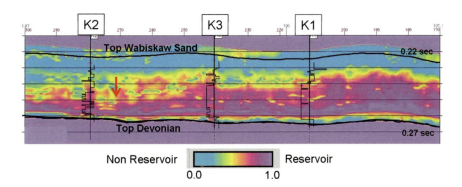

Figure 12. Cross section of the multiattribute analysis along the the K pair. Sedimentary facies index logs from appraisal wells are also shown. Reservoir distribution was laterally and vertically well predicted. Distribution of basal shale was also well described.

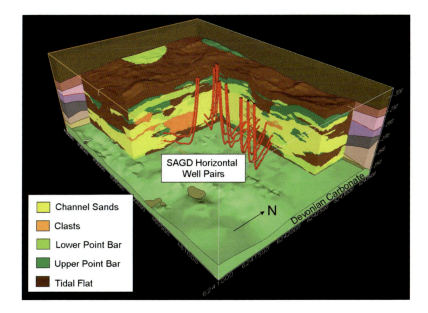

Figure 13. Physical property model showing the distribution of the sedimentary facies. The structure model of the basement and locations of the horizontal well pairs are also shown.

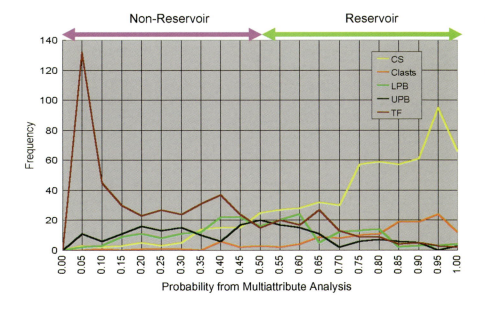

Figure 14. Frequency of predicted values for each sedimentary facies from multiattribute analysis at the well locations.

or layered shale. In the case of layered shale, this zone would become nonproductive as the shale fraction increased. On the basis of these observations, three control wells were planned along the K-pair.

First, K1 was drilled, sufficiently thick channel sands were encountered, and the prediction of reservoir bottom was confirmed. The K2 well was then drilled to investigate the high AI and anomalous multiattribute zone, and

Table 1. Prediction success ratio for reservoir/nonreservoir at well locations by multiattribute analysis.

Lithology	Predicted as Reservoir	Predicted as Nonreservoir
Channel Sand Bar	89%	11%
Channel Sand Bar with Rip-up Clasts	91%	9%
Lower Point Bar	54%	46%
Upper Point Bar	47%	53%
Tide Flat Shale	25%	75%

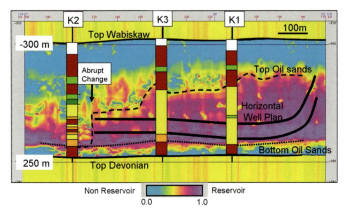

Figure 15. Depth section of the multiattribute analysis along the K pair with sedimentary facies logs. Relatively thick reservoir observed at K1 is thinning toward K3, and drastically reduces its thickness near K2.

the zone was shown to be alternating tidal flat shale and clasts. Finally, the K3 well was drilled to confirm the extent of the channel sands observed in the K1 well. The K3 well found consistent thick channel sands and a relatively thicker basement shale.

Figure 15 shows the depth converted section of the multiattribute analysis along the K-pair (the same line with Figure 12). Depth conversion was conducted using the geostatistical method of kriging with external drifts. Sedimentary facies indices of control wells are overlain on the section.

The detailed SAGD well plan for the K-pair is shown in the multiattribute analysis section in Figure 15. Interpretation of Top Devonian, Bottom Reservoir, and Top Reservoir is indicated. Between K3 and K2, an abrupt change in sedimentary facies is observed approximately 70 m west of K2. The reservoir is thinning toward the west and temporal thickening of the reservoir is observed around K3. The reservoir to the east of K1 is thick enough for development for at least 200 m.

On the basis of the observations above, the K-pair was planned to start 100 m west of K1 and to end 75 m east of K2. Because high-quality reservoir extends to the east of K1, another north–south well pair was planned later on. Referring to the multiattribute analysis section, the toe of the well was planned to stop before the observed abrupt facies change to avoid early breakthrough of steam because of thin reservoir.

Conclusions

Reservoir characterization procedures described in this chapter significantly improved the prediction accuracy of oil-sands reservoir distribution. Among them, multiattribute analysis plays an important role because of its high resolution and predictability of basal shale. These procedures should be widely applicable to these kinds of complex reservoirs, not just oil sands.

Acknowledgments

I thank Japan Petroleum Exploration Co., Ltd.; Canada Oil Sands Ltd.; and Japan Oil, Gas, and Metals National Corporation for permission to use the data and publish this chapter.

References

Alberta Energy and Utilities Board, 2007, Alberta's energy reserves 2006 and supply/demand outlook 2007–2016: ST98-2007.

Brown, A. R., 2004, Introduction, resolution, and interpretation of three-dimensional seismic data, 6th ed.: AAPG Memoir 42 and SEG Investigation in Geophysics 9, 3–5.

Hampson, D. P., J. S. Schuelke, J. A. Quirein, 2001, Use of multiattribute transforms to predict log properties from seismic data: Geophysics, **66**, 220–236.

Takahashi, A., K. Kashihara, S. Mizohata, N. Shimada, T. Nakayama, M. Kose, and T. Torigoe, 2006a, Construction of three-dimensional geological models for oil sands reservoirs in Athabasca, Canada, Butsuri-Tansa: Geophysical Exploration, **59**, 233–244 (in Japanese with an English abstract).

Takahashi, A., T. Torigoe, T. Tsuji, K. Kashihara, T. Nakayama, M. Kose, L. Skinner, and R. Nasen, 2006b, Geological modeling of the oil sands reservoir by integrating the borehole and seismic data in Athabasca, Canada: Journal of the Japanese Association for Petroleum Technology, **71**, 54–63 (in Japanese with an English abstract).

Torigoe, T., T. Tsuji, A. Takahashi, 2005, Sequence stratigraphic framework of the incised valley system in the Athabasca Oil Sands Area, Canada: Proceedings of the 11th Formation Evaluation Symposium of Japan, Japan Formation Evaluation Society, 1–5.

Section 4

Production

Chapter 18

The Effects of Cold Production on Seismic Response

Fereidoon Vasheghani,[1] Joan Embleton,[1] and Larry Lines[1]

Introduction

Cold production is a nonthermal recovery mechanism in which a progressive cavity pump simultaneously produces oil, water, gas, and sand. This extraction decreases the reservoir pressure to values less than bubble point; therefore, gas comes out of solution and forms a foam-like material called foamy oil. On the other hand, because of sand production, high-porosity and high-permeability channels known as wormholes are created with diameters ranging from 10 cm to as much as 1 m (Tremblay et al., 1999).

It is very important to avoid drilling into the wormholes; therefore, petroleum engineers need to know the location of wormholes and the extent of depleted zones. Fortunately, the reservoir undergoes significant changes during cold production that we can monitor using seismic information. In this modeling study, we evaluate the influence of changes in porosity and foamy-oil effects caused by cold production on seismic data.

Model and Methodology

We used a simple three-layer homogeneous reservoir model with wormholes in the x and y directions and a vertical production well. This creates the L-shaped wormhole model shown in Figure 2. The top of the reservoir is at a depth of 200 m, and the reservoir is surrounded by overburden and underburden of constant properties. Modeling foamy-oil flow is based on some empirical adjustments to the solution-gas drive models (Maini, 2001). Some practical modifications are critical gas saturation, oil/gas relative permeability, fluid and/or rock compressibility, pressure-dependent oil viscosity, absolute permeability, and bubble point pressure. In this modeling study, we increased the critical gas saturation to 20% and modified the relative permeability curve for gas. Figure 1 shows the fluid properties used in the model.

There are two practical approaches for modeling wormhole behavior. One method uses coupled geomechanical modeling. Fluid-flow equations are solved simultaneously with sand-flow relations. In each time step, the pressure field in the reservoir is calculated using the fluid-flow equations. The new pressure values are used in sand-flow equations to calculate the new porosity and therefore the permeability of each grid block. Then, in the next time step, the updated porosities and permeabilities are used to calculate the saturation and pressure, and this loop continues until the desired results are achieved. This method is accurate but time-consuming and computationally expensive.

The other approach uses "static" wormholes in the reservoir; that is, horizontal or other directional wells are used as wormholes. One major difference between these models is that static wormholes cannot be updated at each time step, so the reservoir specifications remain constant during the experiment. In this study, we used two horizontal wells to represent the static wormholes in our model.

The black oil model is used in this study. Black oil simulation is a standard model that assumes that three components (oil, gas, and water) exist in the reservoir in three phases (liquid, free gas, and dissolved gas) and no change occurs in the composition of components during the life of the reservoir. We modeled production for 24 months at the rate of 500 barrels per day and the minimum bottom-hole pressure of 500 psi. Table 1 summarizes the simulation parameters. Figure 2 shows the plan view of the pressure in the reservoir after the simulation and a cross section at $x = 720$ m. As expected, the pressure drop is higher in the vicinity of the wormholes, which explains the higher gas saturations around them. Figure 3 shows the plan view and a cross section of gas saturation in the reservoir at $x = 720$ m.

---

[1]University of Calgary, CHORUS, Department of Geoscience, Calgary, Alberta, Canada
This paper appeared in the September 2008 issue of THE LEADING EDGE and has been edited for inclusion in this volume.

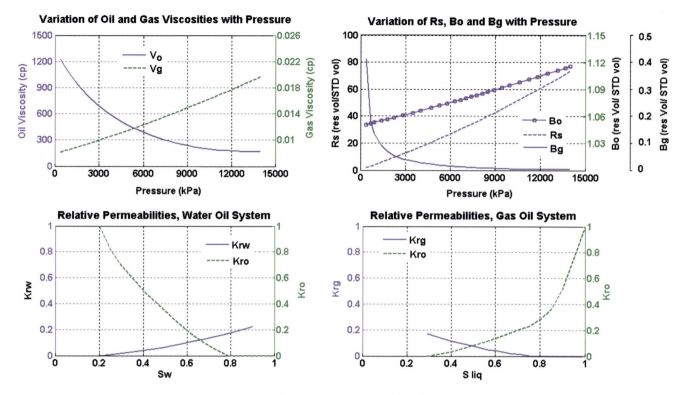

Figure 1. Fluid properties: (a) oil and gas viscosities versus pressure; (b) variation of dissolved gas, oil, and gas-formation volume factors with pressure; (c) oil and water relative permeabilities in a water-oil system versus water saturation; and (d) gas and oil relative permeabilities versus oil saturation in a gas-oil system.

Seismic Modeling and Imaging

The first step in our seismic modeling is calculating P-wave velocities for the reservoir using

$$V_P = \sqrt{\frac{K + \frac{4}{3}\mu}{\rho}} \qquad (1)$$

where K denotes the saturated bulk modulus, μ is saturated shear modulus, and ρ is density. Density is given by the weighted average of fluid and matrix densities:

$$\rho = \varphi(S_o \rho_o + S_w \rho_w + S_g \rho_g) + (1 - \varphi)\rho_m \qquad (2)$$

where S denotes saturation and subscripts o, w, g, and m represent oil, water, gas, and matrix, respectively. The saturated shear modulus is constant and equal to the dry shear modulus; this means that fluids do not affect the shear properties of the reservoir.

$$\mu = \mu_{\text{dry}} \qquad (3)$$

where μ_{dry} is the dry frame shear modulus of the rock. To calculate the saturated bulk modulus, we used Gassmann's equation:

$$K = K_{\text{dry}} + \frac{\left(1 - \frac{K_{\text{dry}}}{K_m}\right)^2}{\frac{\varphi}{K_f} + \frac{1-\varphi}{K_m} - \frac{K_{\text{dry}}}{K_m^2}} \qquad (4)$$

where K_{dry}, K_m, and K_f are dry, matrix, and fluid bulk modulus, respectively, and φ is porosity. One important parameter is dry bulk modulus. Toksöz et al. (1976) defined the dry bulk modulus as

$$K_{\text{dry}} = K_m \left(\frac{1 - \varphi}{1 + \frac{3K_m \varphi}{4\mu_m}}\right). \qquad (5)$$

This means that if the porosity is constant, the dry bulk modulus will remain constant.

K_f, on the other hand, varies with pressure, temperature, and fluid saturation. Depending on the fluid distribution in the porous media, we can use the harmonic or arithmetic average to calculate the overall bulk modulus of the reservoir fluids. Because we are dealing with foamy oil, the gas is distributed uniformly in the reservoir, and we should use the harmonic average (Kirstetter et al., 2006):

$$\frac{1}{K_f} = \frac{S_o}{K_o} + \frac{S_w}{K_w} + \frac{S_g}{K_g} \qquad (6)$$

Chapter 18: The Effects of Cold Production on Seismic Response

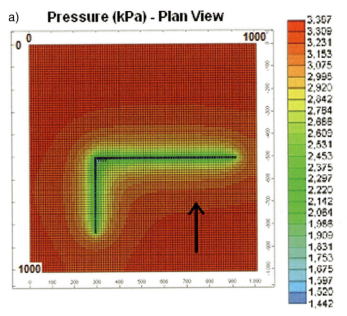

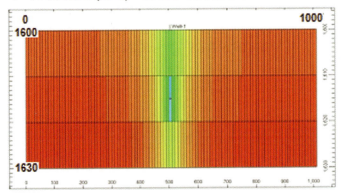

Figure 2. Pressure in the reservoir: (a) plan view of the middle layer, and (b) cross section at $x = 720$ m. The shape on the plan view is in the form of the letter L because two wormholes are in the x and y directions. The arrow shows the extent of the foamy-oil zone (low-pressure zone).

Table 1. Reservoir simulation parameters.

Reservoir size (m)	$1000 \times 1000 \times 30$
Simulation grid size (m)	$10 \times 10 \times 10$
Vertical well perforation (m)	30
Vertical well radius (m)	0.0762
Porosity (fraction)	0.3
Horizontal permeability (md)	2000
Vertical permeability (md)	200
Initial pressure (kPa)	3200
Temperature (°C)	35
Production time (days)	720
Min BHP (kPa)	500

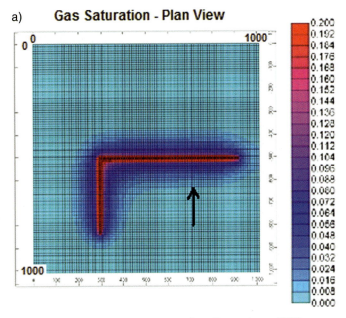

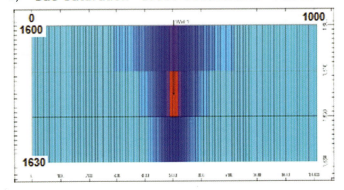

Figure 3. Gas saturation in the reservoir: (a) plan view of the middle layer, and (b) cross section at $x = 720$ m. The arrow shows the extent of the foamy-oil zone (high gas-saturation zone).

Figure 4 shows the velocity maps calculated for the reservoir. Comparison of Figures 3 and 4 shows that P-wave velocity is very sensitive to changes in gas saturation, and even in zones with minimal changes in gas saturation we observe a significant drop in velocity.

The P-wave seismic response was calculated by solving the 3D acoustic wave equation using a finite-difference scheme with second-order accuracy in time and fourth-order accuracy in space. The selection of grid spacing, time steps, and dominant frequency should be such that the solution is stable and grid dispersion is minimal. To avoid dispersion, we used at least five grids per wavelength (Alford et al., 1974). To achieve a stable solution, we used the following relation (Lines et al., 1999):

$$\frac{v \Delta t}{h} \leq \frac{1}{2} \qquad (7)$$

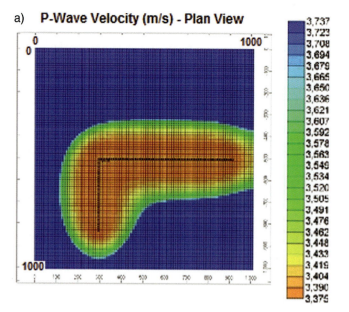

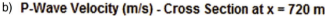

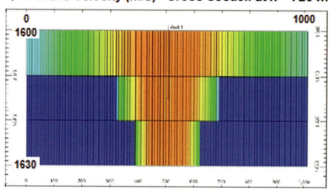

Figure 4. Velocity maps of the reservoir calculated using Gassmann's equation: (a) plan view, and (b) cross section at $x = 720$ m.

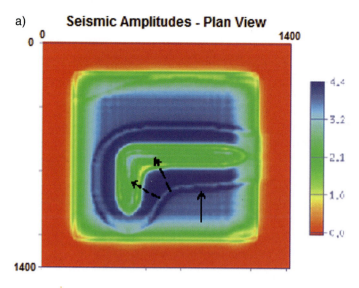

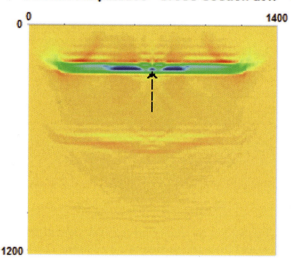

Figure 5. Seismic response of the reservoir, depth-migrated: (a) plan view, and (b) cross section at $x = 720$ m. Solid arrow shows the extent of the foamy-oil zone. Dashed arrows indicate wormholes.

where v is the seismic wave velocity, h is the spatial grid spacing, and Δt is the time increment. A damping zone at the boundaries prevented reflections off the nonphysical boundaries of the grid. Reverse-time depth migration is used for the imaging. Figure 5 shows the seismic results after the production.

Comparison of images generated by the modeling and reservoir simulation (Figures 2, 3, and 5) reveals that the extent of the foamy oil is indicated on the seismic response (solid arrows on the figures). However, the sizes of the zones on these maps are not equal. This is because even a very small change in the gas saturation decreases compressional-wave velocity dramatically (Domenico, 1976). On the other hand, the change in porosity caused by sand production does not have as much effect on the seismic response as expected. This is because the wormhole size is orders of magnitude smaller than the resolving power of conventional seismic surveys. However, the combined effects of gas saturation, pressure, and porosity at grid points around the wormholes can help us find an approximate location of the wormholes (dashed arrows on Figure 5).

Conclusions

Cold production of heavy oils creates foamy oil and wormholes in the reservoir, and for optimum recovery it is very important to define these zones. Although the exact location of individual wormholes cannot be determined, we can find the approximate vicinities on seismic maps. This will help the engineers find the best locations for infill wells.

Acknowledgments

We thank the Consortium of Heavy Oil Research by University Scientists (CHORUS) and its sponsors for their technical and financial support. We appreciate the contributions from Alberta Ingenuity Centre for In Situ Energy (AICISE).

References

Alford, R. M., K. R. Kelly, and D. M. Boore, 1974, Accuracy of the finite-difference modeling of the acoustic wave equation: Geophysics, **39**, 834–842.

Domenico, S. N., 1976, Effect of brine-gas mixture on velocity in an unconsolidated sand reservoir: Geophysics, **41**, 882–894.

Kirstetter, O., P. Corbett, J. Somerville, and C. MacBeth, 2006, Elasticity/saturation relationships using flow simulation from an outcrop analogue for 4D seismic modeling: Petroleum Geoscience, **12**, 205–219.

Lines, L. R., R. Slawinski, and R. P. Bording, 1999, A recipe for stability of finite-difference wave-equation computations: Geophysics, **64**, 967–969.

Maini, B. B. 2001, Foamy oil flow: Journal of Petroleum Technology, **53**, 54–64.

Toksöz, M. N., C. H. Cheng, and A. Timur, 1976, Velocities of seismic waves in porous rocks: Geophysics, **41**, 621–645.

Tremblay, B., 1999, A review of cold production in heavy oil reservoirs: Presented at the 10th European Symposium on Improved Oil Recovery, EAGE.

Chapter 19

Effects of Heavy-oil Cold Production on V_P/V_S Ratio

Duojun (Albert) Zhang,[1] Larry Lines,[1] and Joan Embleton[1]

Introduction

Heavy-oil reservoirs are an abundant resource, particularly in Canada, Venezuela, and Alaska. By some estimates, heavy oils represent as much as 6.3 trillion barrels of oil in place. This matches available quantities of conventional oil. More than 50% of Canada's oil production is now from heavy oil (Batzle et al., 2006). Much of the heavy-oil recovery in Western Canada involves steam injection, called "hot production." An alternative to thermal heavy-oil production in the field is known as "cold production," which is a primary nonthermal process in which reservoir temperature is not affected. The cold production process has been economically successful in several unconsolidated heavy-oil fields in Alberta and Saskatchewan, Canada (Sawatzky et al., 2002). During the cold production process, sand and oil are produced simultaneously by progressive cavity pumps, generating high-porosity channels termed "wormholes." The development of wormholes causes reservoir pressure to fall below the bubble point, resulting in dissolved gas coming out of solution to form foamy oil. Foamy oil and wormholes are believed to be two key factors in the enhancement of oil recovery (Metwally et al., 1995; Maini, 2004).

The development of wormholes and the formation of foamy oil will disturb fluid properties in the reservoir during heavy-oil cold production. Batzle et al. (2006) showed that the bulk modulus of heavy oil drops to near zero very quickly from approximately 2.6 GPa after pressure is lower than the bubble point line at approximately 2 MPa. This disturbance will probably be detectable for seismic survey.

To detect what kind of roles seismology can play to map the disturbance of the initial reservoir state after heavy-oil cold production, Lines et al. (2003) revealed the possibility of detecting wormhole presence instead of imaging individual wormholes by normal seismic method.

Chen et al. (2004) calculated elastic parameters of a heavy-oil reservoir before and after cold production on the basis of Gassmann's equation and discussed the use of time-lapse reflection seismology for theoretically detecting the presence of foamy oil and wormholes. Zou et al. (2004) analyzed a repeated 3D seismic survey over a cold production field in eastern Alberta and showed an interesting correlation between time-lapse seismic changes and heavy oil cold production. Lines and Daley (2007) showed that 3D depth migration can delineate cold production zones to within the Fresnel resolution limits. All of the above research is encouraging and confirms that time-lapse seismology can play an important role in mapping the disturbance of the initial reservoir state due to heavy-oil cold production.

Among many seismic properties that can be analyzed from seismic survey, we researched how cold heavy-oil production affects the V_P/V_S ratio to reveal the feasibility of using V_P/V_S ratios to monitor the recovery process of cold heavy-oil production.

Fluid Substitution: Gassmann's Equation

Gassmann's (1951) equation has been used for calculating the effect of fluid substitution on seismic properties using the matrix properties. It predicts the bulk modulus of a fluid-saturated porous medium using the known bulk moduli of the solid matrix, the frame and the pore fluid in the following manner:

$$K^* = K_d + \frac{(1 - K_d/K_m)^2}{\frac{\phi}{K_f} + \frac{1-\phi}{K_m} - \frac{K_d}{K_m^2}}, \quad (1)$$

where, K^*, K_d, K_m, K_f, and Φ are the saturated porous rock bulk modulus, the frame rock bulk modulus, the matrix

[1]University of Calgary, Department of Geoscience, Calgary, Alberta, Canada

bulk modulus, the fluid bulk modulus, and the porosity, respectively. It is assumed that the shear modulus μ^* of the saturated rock is not affected by fluid saturation, so that

$$\mu^* = \mu_d, \qquad (2)$$

where μ_d is the frame shear modulus.

P-wave and S-wave velocities, V_P and V_S, respectively, for an isotropic, homogeneous, elastic material are given by

$$V_P = \sqrt{\frac{K^* + 4\mu^*/3}{\rho^*}}, \qquad (3)$$

and

$$V_S = \sqrt{\frac{\mu^*}{\rho^*}}, \qquad (4)$$

where ρ^* is the saturated rock bulk density and can be calculated as

$$\rho^* = \rho_m(1 - \phi) + \rho_f \phi, \qquad (5)$$

where ρ_m and ρ_f are the densities of solid grains and the fluid mixture, respectively, at reservoir conditions.

Equations 1–5 establish the relationships between the rock moduli and the seismic velocities. There are several assumptions for the accuracy of the Gassmann's equation to calculate the seismic velocities; one of them is that the pores are filled with a frictionless fluid (liquid, gas, or mixture). This assumption implies that the viscosity of the saturating fluid is zero. This may be the most questionable assumption for heavy oil, especially at cold temperatures (approximately 20–40°C).

Fortunately, Batzle et al. (2006) found that although viscosity is influenced by pressure and gas content, it is primarily a function of oil gravity and temperature. Increasing the temperature will decrease the sample's viscosity, bulk and shear moduli decrease approximately linearly with increasing temperature, and the shear modulus approaches zero at approximately 80°C. Moreover, the frequency also plays an important role for traveling waves in heavy oil. At high frequencies (e.g., with laboratory ultrasonics), a heavy-oil sample is still effectively a solid at low temperature (0°C), but not for extremely heavy oil, at seismic frequencies, by +20°C, the shear modulus of heavy oil is negligible and heavy oil acts still like a liquid, especially after cold production when foamy oil is created. In this case, Gassmann's equation can still help us understand the response of heavy-oil reservoir to seismic survey for pre- and postcold production.

Heavy-oil cold production is being carried out at the Plover Lake oil field, as described by Lines et al. (2005).

The in situ reservoir parameters from a Plover Lake oil well are listed in Table 1, the reservoir temperature is 27°C, and the American Petroleum Institute (API) gravity of heavy oil is 12.1. From Batzle et al. (2006), we know that the heavy-oil sample with a gravity of API = −5 can go through shear relaxation and acts like a liquid with shear modulus of zero at seismic frequencies by +20°C. So, for the in situ heavy oil in Plover Lake with an API gravity of 12.1, it should be acceptable to assume that the heavy oil acts like a liquid at seismic frequencies by 27°C. To test the feasibility of Gassmann's equation, one in situ well with dipole sonic log data and density log data is selected from Plover Lake oil field to do the calculation. To simplify the calculation, average values of P-wave velocity, S-wave velocity, and density for preproduction conditions are estimated for the production zone (Table 2).

From the reservoir parameters in Table 1, we can calculate fluid properties on the basis of Batzle–Wang formulas (Batzle and Wang, 1992). The physical properties of a solid matrix mineral can be examined based on mineral composition, distribution, and in situ conditions (Han and Batzle, 2004). From these solid matrix mineral properties and porosity from well log data based on equation 5, unknown parameters K_d and μ_d can be given in the following equations (Mavko and Mukerji, 1998a, b):

$$K_d = K_m(1 - \phi/\phi_c), \quad \text{and} \qquad (6)$$

$$\mu_d = \mu_m(1 - \phi/\phi_c), \qquad (7)$$

Table 1. Reservoir parameters for the in situ well.

Heavy-oil API	12.1
Specific gravity of methane	0.574
Solution gas-oil ratio (m^3/m^3)	16.64
Reservoir temperature (°C)	27
Reservoir pressure (MPa)	6.4
Water saturation (%)	25
Oil saturation (%)	75
Gas saturation (%)	0
Water salinity (ppm)	19,280

Table 2. Estimated average values of production zone for V_P, V_S, and ρ^*.

P-wave velocity V_P (km/s)	S-wave velocity V_S (km/s)	Density ρ^* (g/cc)
3.02	1.55	2.13

Table 3. Calculated saturated moduli from well log data and Gassmann's equation.

Parameters	Well log	Gassmann's equation
Saturated bulk modulus K^* (GPa)	12.60	11.61
Saturated shear modulus μ^* (GPa)	5.12	4.97

where, ϕ_c is the critical porosity, which separates mechanical and acoustic behavior of rocks into two distinct domains: load bearing and suspension. For sandstone, $\phi_c \approx 38\%$. At this time, the saturated moduli can be calculated from equations 1 and 2, and the results are listed in Table 3, together with the calculated saturated moduli from well log data based on equations 3 and 4.

In the real world, we can think that the calculated saturated moduli from well log data are reliable if the quality of well log data is good. From Table 3, we can see that Gassmann's equation gives very good estimations of saturated bulk modulus and shear modulus, especially saturated shear modulus. As stated previously, for oil that is not extremely heavy, the shear modulus of heavy oil is negligible and Gassmann's equation is still applicable at seismic frequencies for temperatures of $+20°C$.

Difference of Heavy-oil Physical Properties between Pre- and Postproduction

As described previously, heavy-oil reservoirs experience a dramatic change as a result of cold production: porosity increases because of sand extraction, pore pressure decreases because of porosity increase, and there is a phase transition of heavy oil to foamy oil because of pore pressure decrement. Table 4 lists a typical comparison of reservoir parameters between pre- and postcold production in Plover Lake oil field. These changes of reservoir parameters, especially the decrement of reservoir pressure from 6.4 MPa for preproduction to 0.6 MPa for postproduction, will absolutely change the physical properties of heavy oil in the reservoir. Table 5 shows calculated physical properties of reservoir fluids before and after cold production on the basis of the Batzle–Wang formulas (Batzle et al., 1992) and reservoir parameters are from Table 4.

Compared with the bulk modulus of heavy oil for preproduction (2.2166 GPa), the bulk modulus of foamy oil for postproduction is just approximately 0.0636 GPa, which is a dramatic decrease. Such a decrease will cause

Table 4. A typical comparison of reservoir parameters between pre- and postcold production in Plover Lake oil field.

Parameters	Preproduction	Postproduction
Heavy-oil API	12.1	12.1
Specific gravity of methane	0.574	0.574
Solution gas-oil ratio (m^3/m^3)	16.64	0.9
Reservoir temperature (°C)	27	27
Reservoir pressure (MPa)	6.4	0.6
Water saturation (%)	25	19
Oil saturation (%)	75	62
Gas saturation (%)	0	19
Water salinity (ppm)	19,280	19,280

the reduction of P-wave velocity and will absolutely affect the response of seismic survey. However, regional and lithologic variations in P-wave velocity may be even greater than these anomalies. Hence, observations of P-wave velocity alone may not be sufficient to identify zones of interest. Theoretically and experimentally, the S-wave velocity of a porous rock has been shown to be less sensitive to fluid saturants than P-wave velocity. It can be used as a normalizing quantity with which to compare P-wave velocity, and observations of the ratio of the seismic velocities for P- and S-wave velocity that traverse a changing or laterally varying zone could produce an observable anomaly that is independent of the regional variation in P-wave velocity (Tatham et al., 1976). Moreover, the V_P/V_S ratio is especially sensitive to the pore fluid found in sedimentary rocks. In particular, the V_P/V_S value is much lower (10%–20%) for gas saturation than for liquid saturation, and there is a characteristic decrease in the V_P/V_S ratio for gas-saturated sandstones (Tatham, 1982).

Effects of Heavy-oil Cold Production on V_P/V_S Ratio

As discussed previously, for heavy oil with an API greater than 10, the shear modulus of heavy oil is negligible for seismic frequencies at $+20°C$, and heavy oil still acts like a liquid, especially after cold production when foamy oil is created because of the dissolved gas from heavy oil, and the mobility of reservoir fluids is much improved. In this case, Gassmann's equation can still help us understand the response of a heavy-oil reservoir to seismic survey for pre- and postcold production.

Table 5. Calculated physical properties of reservoir fluids for pre- and postcold production.

Parameters	Preproduction			Postproduction		
	Heavy oil	Gas	Water	Heavy oil	Gas	Water
Bulk modulus (GPa)	2.2166	0.01	2.37	0.0636	0.0008	2.34
Density (g/c^3)	0.97	0.048	1.01	0.97	0.004	1.0088

Using the patchy model, where $K^* = K_P + K_d$, Murphy et al. (1993) introduced another expression of Gassmann's equation 1 as

$$\rho V_P^2 = K_P + K_d + \frac{4}{3}\mu^*, \quad (8)$$

where K_P is the pore space modulus and other parameters are same as those described previously. If we recall Gassmann's equation 1, K_P can be expressed as (Murphy et al., 1993)

$$K_P = \frac{\alpha^2}{\frac{\alpha - \phi}{K_m} + \frac{\phi}{K_f}}, \quad (9)$$

where α is the compliance of the frame relative to that of the solid grains and is defined as (Murphy et al., 1993)

$$\alpha = 1 - K_d/K_m. \quad (10)$$

From equation 6, equation 10 can be written as

$$\alpha = 1 - (1 - \phi/\phi_c) = \phi/\phi_c. \quad (11)$$

To explicitly reveal the dependence of K_P on porosity, equation 9 can be simplified as

$$K_P \approx \frac{\phi}{\phi_c^2} K_f. \quad (12)$$

This simplification uses the result shown by Zhang (2007) that the second term in the denominator of equation 9 is much larger than the first term.

Equation 12 explicitly reveals the proportional dependence of K_P on porosity and K_f. This relationship reveals the fact that the contribution of K_P to V_P is quite significant at high porosities compared with that at low porosities. The contribution of K_f to V_P is the same fact.

By dividing equation 2 into equation 8, the velocity ratio may be naturally expressed in the terms of the moduli that are introduced above.

$$R^2 = \left(\frac{V_P}{V_S}\right)^2 = \frac{K_P}{\mu^*} + \frac{K_d}{\mu^*} + \frac{4}{3}. \quad (13)$$

From equations 2, 6, and 7, we obtain

$$\frac{K_d}{\mu^*} = \frac{K_m}{\mu_m}. \quad (14)$$

So, the ratio of the frame moduli K_d/μ^* is independent of the pore fluid. Finally, from the above discussion, K_P/μ^* represents the pore fluid contribution, which is an important factor at high porosity and is insignificant at low porosity. This is the source that we can use in time-lapse technology to monitor the recovery process of an unconsolidated reservoir.

From equations 2, 7, 12, 13, and 14, we can further reveal the contributions of K_f and porosity to the V_P/V_S ratio.

$$R^2 = \left(\frac{V_P}{V_S}\right)^2 \approx \frac{K_P}{\mu^*} + \frac{K_m}{\mu_m} + \frac{4}{3} = \frac{K_f \phi}{\mu_m \phi_c (\phi_c - \phi)} + \frac{K_m}{\mu_m} + \frac{4}{3}. \quad (15)$$

Equation 15 explicitly reveals the dependence of the V_P/V_S ratio on porosity and fluids saturation. For a completely gas saturated reservoir, $K_f \approx 0$, $K_P/\mu^* \approx 0$, and equation 15 reduces to

$$R^2 = \left(\frac{V_P}{V_S}\right)^2 \approx \frac{K_m}{\mu_m} + \frac{4}{3}. \quad (16)$$

The V_P/V_S ratio is constant and the smallest compared with other fluid saturations. For partial fluids saturation, K_f and ϕ have opposite effects on the V_P/V_S ratio after heavy-oil cold production: The smaller value of K_f will decrease the V_P/V_S ratio, and by contrast, larger porosity values will increase the V_P/V_S ratio. Let us examine these two points further. For the Plover Lake oil sands, typical values are $\phi = 0.31$, $\phi_c = 0.38$, $K_m = 39$ GPa, $\mu_m = 27$ GPa, and equation 15 becomes

$$R^2 = 0.432 K_f + 2.778. \quad (17)$$

Figure 1 displays the effect of K_f on the V_P/V_S ratio in this case, and the V_P/V_S ratio will decrease with the reduction of K_f.

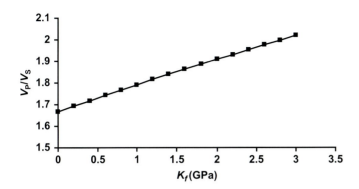

Figure 1. The effect of K_f on the V_P/V_S ratio.

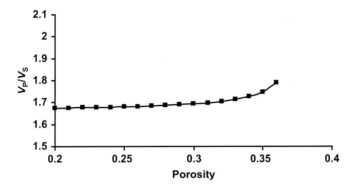

Figure 2. The effect of porosity on the V_P/V_S ratio.

As to the effect of porosity on the V_P/V_S ratio, it is a little bit complicated. Murphy et al. (1993) pointed out that the V_P/V_S ratio will increase at different rates for different fluid partial saturation. The V_P/V_S ratio keeps constant for gas-saturated sands and will increase more for water-saturated sands with the increment of porosity. From Tables 4 and 5, we can obtain $K_f \approx 0.244$ GPa on the basis of the V-R-H model (Hill, 1952) after heavy-oil cold production, and equation 15 becomes

$$R^2 = \frac{0.244\phi}{10.26(0.38-\phi)} + 2.778. \qquad (18)$$

Figure 2 shows the result from equation 18. For $\phi < 0.30$, the V_P/V_S ratio almost keeps constant and has very little increment with the improvement of porosity, but for $\phi > 0.30$, the V_P/V_S ratio will increase relatively quickly.

For the in situ case, let us see how the V_P/V_S ratio changes after heavy-oil cold production. From Tables 4 and 5 and the V-R-H model (Hill, 1952), we can obtain $K_f \approx 2.254$ GPa for preproduction, then from equation 17, $V_P/V_S \approx 1.937$. For postproduction, from previous context, $K_f \approx 0.244$ GPa, which decreases dramatically because of the creation of foamy oil. Usually, the reservoir porosity will have much less improvement after

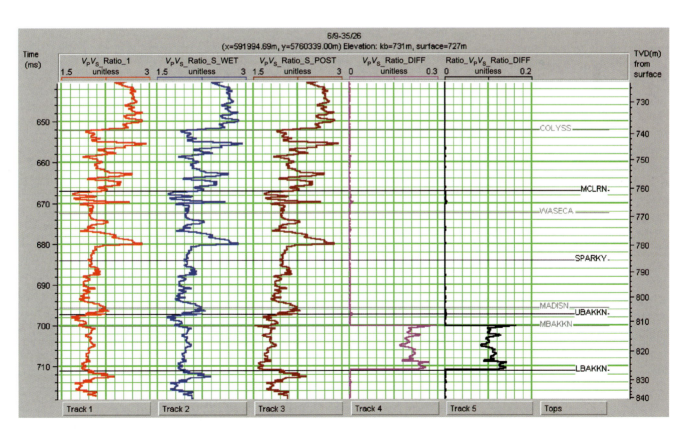

Figure 3. All three V_P/V_S ratios for preproduction, wet, and postproduction conditions after fluid substitution and the difference in V_P/V_S ratios between post- and preproduction conditions.

heavy-oil cold production; for example, the reservoir porosity is improved to 0.32 from 0.31, then from equation 18, $V_P/V_S \approx 1.704$, and the reduction of the V_P/V_S ratio is approximately 0.233 because of cold production. This value is for the assumption that fluids are mixed together between patchy and uniform. If fluids are mixed together uniformly, the bulk modulus K_f will be decreased to 0.004 GPa from 2.254 GPa because of cold production and the creation of foamy oil, the V_P/V_S ratio will be reduced to 1.667 from 1.937, and the reduction value is 0.270. So generally, although porosity has an opposite effect on the V_P/V_S ratio, the reduction of fluid bulk modulus will have a more significant effect on the V_P/V_S ratio, and the V_P/V_S ratio will decrease after heavy-oil cold production.

Figure 3 shows all three V_P/V_S ratios for preproduction, wet, and postproduction conditions after fluid substitution. The curve in track 4 is the difference of the V_P/V_S ratios between post- and preproduction conditions (tracks 3 and 1). There is a reduction in the V_P/V_S ratio of approximately 0.2 after heavy-oil cold production and an approximate 10% reduction shown in track 5. This figure provides a similar result with that calculated previously.

Conclusions

The generation of wormholes and the formation of foamy oil from simultaneous extraction of oil and sand during heavy-oil cold production will disturb fluid properties in the reservoir. This disturbance will be detectable by seismic surveys. For heavy oil in the 10–20 API range at an ambient temperature of 20°C, the shear modulus is negligible and heavy oil still acts like a liquid at seismic frequencies, especially after cold production. Gassmann's equation can still help us understand the seismic response of heavy-oil reservoirs for pre- and postcold production. The V_P/V_S ratio is a function of fluid bulk modulus and porosity. For unconsolidated sands with high porosity, pore fluids have a significant influence on the final V_P/V_S ratio. Because of the dramatic reduction of fluid's bulk modulus after heavy-oil cold production, the V_P/V_S ratio will have a detectable reduction, although the increasing porosity from wormholes slightly increases the V_P/V_S ratio. This significant result should greatly help us to interpret time-lapse multicomponent seismic surveys in cold production fields.

Acknowledgments

We thank the sponsors of this research, including the Consortium of Heavy Oil Research by University Scientists (CHORUS), the Consortium for Research in Elastic Wave Exploration Seismology (CREWES), and the Natural Science and Engineering Research Council of Canada (NSERC). We especially thank Nexen, a CHORUS sponsor, for providing data from the Plover Lake heavy-oil field.

References

Batzle, M., R. Hofmann, and D. H. Han, 2006, Heavy oils—seismic properties: The Leading Edge, **25**, 750–756.

Batzle, M., and Z. Wang, Z, 1992, Seismic properties of pore fluid: Geophysics, **57**, 1396–1408.

Chen, S., 2004, Time-lapse seismology to determine foamy oil and wormhole footprints in heavy oil cold production reservoir: M.S. thesis, University of Calgary.

Chen, S., L. Lines, and P. Daley, 2004, Foamy oil and wormhole footprints in heavy oil cold production reservoirs: CSEG Recorder, **29**, 49–51.

Gassmann, F., 1951, Uber die elastizitat poroser medien: Vierteljahrsschrift der Naturforschenden Gesellschaft in Zurich, **96**, 1–23.

Han, D. H., and M. L. Batzle, 2004, Gassmann's equation and fluid-saturation effects on seismic velocities: Geophysics, **69**, 398–405.

Hill, R. W., 1952, The elastic behavior of crystalline aggregate: Proceedings of the Physical Society of London, **A65**, 349–354.

Lines, L. R., S. Chen, P. F. Daley, and J. Embleton, 2003, Seismic pursuit of wormholes: The Leading Edge, **22**, 459–461.

Lines, L. R., and P. F. Daley, 2007, Seismic detection of cold production footprints in heavy oil extraction: Journal of Seismic Exploration, **15**, 333–344.

Lines, L. R., Y. Zou, D. A. Zhang, K. Hall, J. Embleton, B. Palmiere, C. Reine, P. Bessette, P. Cary, and D. Secord, 2005, V_P/V_S characterization of heavy-oil reservoir: The Leading Edge, **24**, 1134–1136.

Maini, B. B., 2004, Foamy oil flow in cold production of heavy oil: Distinguished lecture, Petroleum Society of Canadian Institute of Mining.

Mavko, G., and T. Mukerji, 1998a, Bounds on low-frequency seismic velocities in partially saturated rocks: Geophysics, **63**, 918–924.

Mavko, G., and T. Mukerji, 1998b, Comparison of the Kiref and critical porosity models for prediction of porosity and V_P/V_S: Geophysics, **63**, 925–927.

Metwally, M., and S. C. Solanki, 1995, Heavy oil reservoir mechanisms, Linbergh and Frog Lake Fields, Alberta, Part 1: Field observations and reservoir simulation: 46th Annual Technical Meeting of the Petroleum Society of Canadian Society of Mining.

Murphy, W., A. Reischer, and K. Hsu, 1993, Modulus decomposition of compressional and shear velocities in sand bodies: Geophysics, **58**, 227–239.

Sawatzky, R. P., D. A. Lillico, M. J. London, B. R. Tremblay, and R. M. Coates, 2002, Tracking cold production footprints: Presented at the 2002 Canadian International Petroleum Conference.

Tatham, R. H., 1982, V_P/V_S and lithology: Geophysics, **47**, 336–344.

Tatham, R. H., and P. L. Stoffa, 1976, V_P/V_S–a potential hydrocarbon indicator: Geophysics, **41**, 837–849.

Zhang, D. A., 2007, Applications of V_P/V_S and AVO modeling for monitoring heavy oil cold production: M.S. thesis, University of Calgary.

Zou, Y., L. R. Lines, K. Hall, and J. Embleton, 2004, Time-lapse seismic analysis of a heavy oil cold production field, Lloydminster, Western Canada: 74th Annual International Meeting, SEG, Expanded Abstracts, 1555–1558.

Chapter 20

Collaborative Methods in Enhanced Cold Heavy-oil Production

Larry Lines,[1] Hossein Agharbarati,[1] P. F. Daley,[1] Joan Embleton,[1] Mathew Fay,[1] Tony Settari,[1] Fereidoon Vasheghani,[1] Tingge Wang,[1] Albert Zhang,[1] Xun Qi,[2] and Douglas Schmitt[2]

Introduction

Heavy-oil reservoirs are an abundant hydrocarbon resource, which will in all probability comprise a significant portion of long-term world oil production. The world's heavy-oil reserves have been estimated to be approximately 6 trillion barrels — roughly equivalent to conventional reserves. The largest heavy-oil reserves are in Canada, Venezuela, the United States, Norway, Indonesia, China, Russia, and Kuwait.

Cold production is a low-energy production method that has been widely used in Western Canada. Although the primary recovery rates are relatively modest, cold production of heavy oil requires much less energy than hot production methods such as cyclic steam injection (CSS) or steam-assisted gravity drainage (SAGD), and as a consequence it results in much less hydrocarbon use in the recovery stage.

Wormholes, Foamy Oil, and Cold Production

During the cold production process, sand, oil, water, and gas are produced simultaneously using progressive cavity pumps that generate high-porosity channels termed "wormholes" (Figure 1). The characteristics of wormholes were described by Tremblay et al. (1999), Sawatzky et al. (2002), and Lines et al. (2003). Figure 2 shows a wormhole model from a western Canadian oil field. Wormholes have a fractal-like pattern similar to tree branches (Yuan et al., 1999). Typical dimensions of wormholes can be 100–200 m in length with circumference believed to be approximately 10–20 cm after several years of uninterrupted production.

Wormhole evolution causes reservoir pressure to decrease below the bubble point, resulting in gas coming out of solution to form foamy oil. The bubbles are trapped in oil of extremely high viscosity. The phenomenon of foamy-oil development is similar to creation of bubbles in shaving cream, except that it is viscous oil that traps the bubbles rather than soap. Figure 3 shows a sample of foamy oil created at the Imperial Oil Research Laboratory in Calgary. Foamy oil and wormholes are believed to be major driving factors in the cold production of heavy-oil recovery.

Cold production is somewhat "miraculous" in that there is intentional sand production, along with oil, water, and gas by progressive cavity pumps. Initially the sand cut is very high, with a marginal amount of oil recovery. However, after a few weeks of pumping sand and fluids, an unexpectedly large amount of oil is produced and the sand cut diminishes exponentially. It is believed that this high oil recovery is the result of microbubbles in the solution gas drive. In effect, a "horizontal well" has been created without actually drilling one.

In cold production of heavy oil, it is fundamental to delineate the depletion zones or footprints to optimize drilling strategies. Figure 4 illustrates this production strategy. Once the depletion footprints have been created by drilling a set of wells, we do not want to drill another well into these depletion zones. This would be a "wasted well" because the zone is depressurized and circulation would be lost from existing wells. We hope to delineate the production zones so that new infill wells will be productive and eventually allow maximum recovery with a minimal number of wells. Our objective is to delineate these cold production footprints by seismic methods.

Seismic Resolution of Cold Production Zones

Lines et al. (2003) revealed the possibilities of detecting wormhole distribution rather than attempting to image

[1]University of Calgary, Calgary, Alberta, Canada
[2]University of Alberta, Institute for Geophysical Research, Department of Physics, Edmonton, Alberta, Canada
This paper appeared in the September 2008 issue of THE LEADING EDGE and has been edited for inclusion in this volume.

individual wormholes by normal seismic method. Chen (2004) calculated elastic parameters of a heavy-oil reservoir before and after cold production using Gassmann's equation and discussed the possible use of time-lapse reflection seismology (theoretically for the detection of the presence of foamy oil and wormholes). Zou et al. (2004) analyzed a repeated 3D seismic survey over a cold production field in eastern Alberta, which showed an interesting correlation between time-lapse seismic changes and heavy-oil cold production. Lines and Daley (2007) showed that 3D depth migration can delineate cold production zones to within Fresnel resolution limits.

Figure 5 shows time-lapse surveys illustrating the effects of cold production. In this figure, a 3D survey near Provost, Alberta, acquired in 1989 is compared with a survey in 2000. The red arrows denote the changes in seismic amplitude and traveltime for reflectors at the Mannville and Rex sandstone levels. The decrease in the P-wave velocity during cold production causes changes in seismic amplitudes and delays in traveltimes.

Figure 6 compares a velocity model for a wormhole network (top) with the depth slice from a 3D depth-migrated seismic section. Because of the band-limited nature of the wavelength, the seismic image is a somewhat blurred image of the actual model. However, note that the edges of the model are reasonably well defined. Although the production effects are seen in the amplitudes and traveltimes, they can also be seen in inversion estimates of the seismic impedance. (These are to be shown as part of Wang's M.S. thesis.)

All the above research is encouraging and confirms that time-lapse seismology can play an important role in mapping the variation from the initial state because of an interval of cold heavy-oil production.

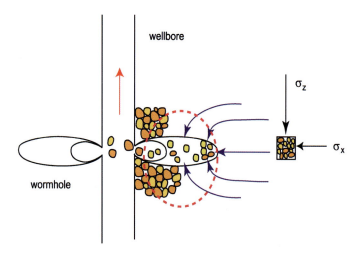

Figure 1. In the cold production of heavy oil, progressive cavity pumps extract sand, oil, water, and gas from the borehole. This figure shows the geomechanics of the cold production process. Figure courtesy of Jen Wang and Tony Settari of Taurus Corporation.

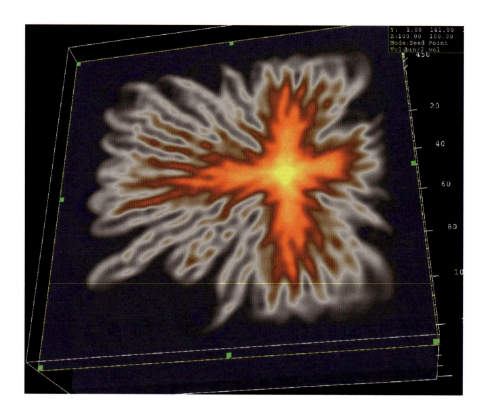

Figure 2. Depth slice showing cold production wormholes from a study at Plover Lake, Saskatchewan. The well is centered at the bright yellow zone. The yellow-orange zones are high-porosity, high-permeability regions of low seismic velocity within the layer of undisturbed oil sands (dark blue).

Figure 3. A sample of foamy oil from Chen et al. (2004). The figure is originally from the Imperial Oil Research Laboratory and David Greenridge.

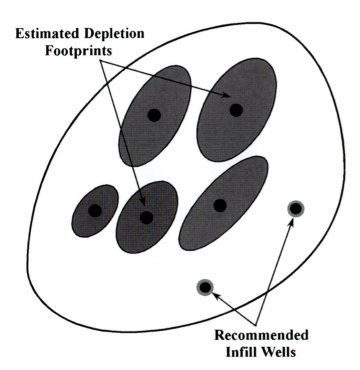

Figure 4. The challenge of cold production. As cold production wells begin producing, cold production footprints consisting of foamy oil and wormholes are created. For infill drilling, we seek to drill new wells outside of these depletion footprints to avoid loss of circulation and to optimize the production of the entire field. From Lines and Daley, 2007.

A particular emphasis of our research concerned how cold heavy-oil production affects the V_P/V_S ratio and the feasibility of using it to monitor recovery. Recent studies, focused on rock physics, showed that pressure reduction in the reservoir should cause a 10%–15% reduction in the V_P/V_S ratio.

Rock Physics of Cold Production of Heavy Oil

Gassmann's equation predicts the bulk modulus of a fluid-saturated porous medium using the known bulk moduli of the solid matrix, the frame, and the pore fluid in the following manner:

$$K^* = K_d + \frac{(1 - K_d/K_m)^2}{\frac{\phi}{K_f} + \frac{1-\phi}{K_m} - \frac{K_d}{K_m^2}}, \quad (1)$$

where K^*, K_d, K_m, K_f, and ϕ are the saturated porous rock bulk modulus, the frame rock bulk modulus, the matrix bulk modulus, the fluid bulk modulus, and the porosity, respectively. It is assumed that the shear modulus μ^* of the saturated rock is not affected by fluid saturation, so that

$$\mu^* = \mu_d, \quad (2)$$

with μ_d being the dry frame shear modulus.

The P- and S-wave velocities, V_P and V_S, respectively, for an isotropic, homogeneous, elastic material are

$$V_P = \sqrt{\frac{K^* + 4\mu^*/3}{\rho^*}}, \quad (3)$$

and

$$V_S = \sqrt{\frac{\mu^*}{\rho^*}}, \quad (4)$$

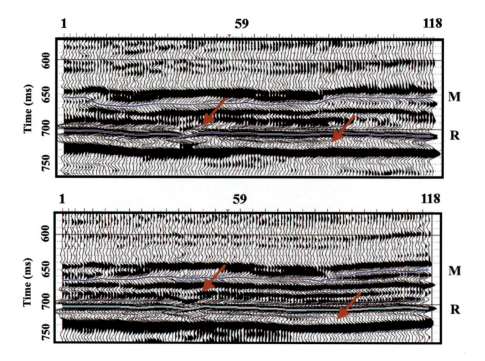

Figure 5. These time-lapse seismic results show that cold production will affect seismic amplitudes and traveltimes. The upper line is from a 3D survey near Provost, Alberta, in 1989, and the lower figure shows the line from a 2000 survey over the same location. Red arrows denote that amplitudes are altered and traveltimes are delayed by lower seismic velocity in the reservoir zones. From Zou et al., 2004.

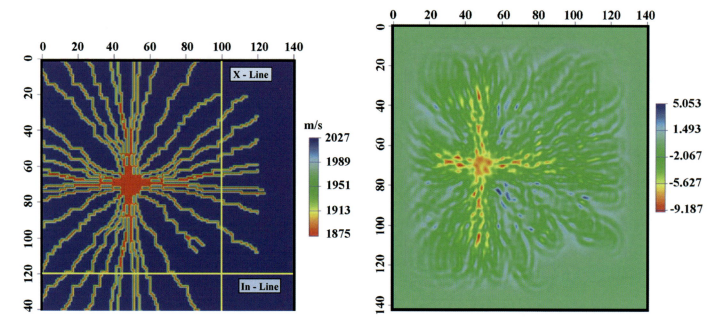

Figure 6. A comparison of (left) an idealized wormhole model with (right) its seismic image at typical seismic frequencies. This figure illustrates that seismic resolution is limited by the Fresnel zones. Although the individual wormhole zones may be blurred, the edge of the cold production zones can be delineated. From Lines and Daley, 2007.

where ρ^* is the saturated rock bulk density and can be calculated as

$$\rho^* = \rho_m(1 - \phi) + \rho_f \phi, \tag{5}$$

where ρ_m and ρ_f are the densities of solid grains and the fluid mixture at reservoir conditions, respectively.

Equations 1–5 establish the relationships between rock moduli and seismic velocities. However, there are several assumptions that could impact the accuracy of Gassmann's equation for calculating the seismic velocities in a reservoir (Han and Batzle, 2004). One is that the pores are filled with a frictionless fluid (liquid, gas, or mixture). This implies that the viscosity of the saturating fluid is zero, which may

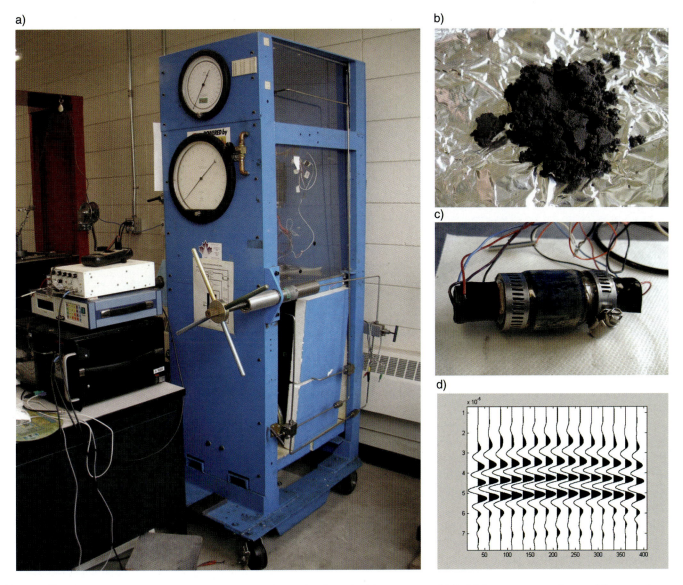

Figure 7. (a) Rock physics apparatus for measuring seismic velocities on core (b) held between transducers (c) to produce seismic response (d).

be questionable for heavy oil, especially at cold temperatures (approximately 20–40°C).

Despite these simplifying assumptions and somewhat surprisingly, Gassmann's equation provides a reasonable description of the elastic moduli for heavy-oil sands even at the low temperatures of cold production. Zhang (2007) analyzed heavy-oil sands from cold production fields at Plover Lake, Saskatchewan, comparing the bulk modulus and shear modulus from equations 1 and 2 with the elastic moduli computed using equations 3 and 4 on dipole sonic velocities. He found that the estimates of the Gassmann estimates for bulk moduli and shear moduli differed by less than 10% and 3%, respectively, for the values obtained using dipole sonic logs.

Rock physics measurements help to provide an essential link between seismic velocities and the physical properties used in petroleum reservoir simulation. The measurements of seismic properties (such as amplitudes and traveltimes) as a function of varying reservoir properties (such as pressures, temperatures, fluid properties, and lithologies) provide the crucial relationships that relate seismic models to reservoir models. Figure 7a shows the rock physics apparatus in Schmitt's laboratory at the University of Alberta where seismic measurements on rocks have been made for more than 20 years. Rock cores (Figure 7b) are placed between transducers (Figure 7c) to produce the seismograms (Figure 7d).

Reservoir Simulation of Heavy Oil Cold Production

Relationships between seismic velocity and reservoir model parameters will allow us to enhance reservoir

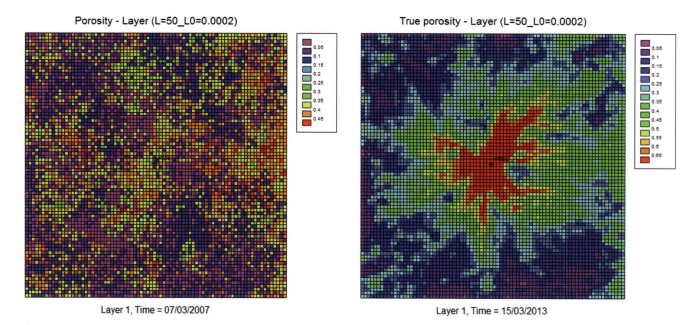

Figure 8. A comparison of porosity models for cold production (left) before reservoir production and (right) after 10 years.

modeling. Reservoir models for cold production of heavy oil are shown in Figure 8. The reservoir model before cold production (left) shows random distribution of porosities with most cells having 20%–40% porosity. After 10 years of cold production for a well at the center of the model, we see extremely high porosities exceeding 60% emanating out from the well (right). The map of porosities is highest at the well location but does show preferential branching directions that can be determined by time-lapse seismology.

Our experience has shown that time-lapse seismology and rock physics can aid in the reservoir characterization of porosities. Our next immediate goal will be to use seismic and rock physics data to characterize reservoir fluids, in particular the heavy-oil viscosity.

Conclusions

The generation of wormholes and the formation of foamy oil from simultaneous extraction of oil and sand during heavy-oil cold production will change fluid properties in the reservoir. This change will be detectable by seismic surveys. For heavy oil in the 10–20 American Petroleum Institute (API) range at ambient temperature of 20°C, the shear modulus is negligible and heavy oil still acts like a liquid at seismic frequencies, especially after cold production. Gassmann's equation can still help us understand the seismic response of heavy-oil reservoirs before and after cold production. The V_P/V_S ratio is a function of fluid bulk modulus and porosity. For unconsolidated sands with high porosity, pore fluids have a significant influence on the final value of the ratio. Because of the dramatic reduction of fluid's bulk modulus after heavy-oil cold production, the ratio will have a detectable reduction. This should help interpret time-lapse multicomponent seismic surveys in cold production fields.

Acknowledgments

We thank the sponsors of this research including the Consortium of Heavy Oil Research by University Scientists (CHORUS), the Consortium for Research in Elastic Wave Exploration Seismology (CREWES), and the Natural Science and Engineering Research Council of Canada (NSERC). We especially thank Nexen, a CHORUS sponsor, for providing data from the Plover Lake heavy-oil field.

References

Chen, S., L. Lines, and P. Daley, 2004, Foamy oil and wormhole footprints in heavy oil cold production reservoirs: CSEG Recorder, October, 49–52.

Chen, S., 2004, Time-lapse seismology to determine foamy oil and wormhole footprints in heavy oil cold production reservoir: M.S. thesis, University of Calgary.

Han, D. H., and M. L. Batzle, 2004, Gassmann's equation and fluid-saturation effects on seismic velocities: Geophysics, **69**, 398–405.

Lines, L., S. Chen, P. F. Daley, and J. Embleton, 2003, Seismic pursuit of wormholes: The Leading Edge, **22**, 459–461.

Lines, L. R., and P. F. Daley, 2007, Seismic detection of cold production footprints in heavy oil extraction: Journal of Seismic Exploration, **15**, 333–344.

Sawatzky, R. P., and D. A. Lillico, 2002, Tracking cold production footprints: Presented at the 2002 Canadian International Petroleum Conference.

Tremblay, B., G. Sedgwick, and D. Vu, 1999, A review of cold production in heavy oil reservoirs: EAGE 10th European Symposium of Improved Oil Recovery.

Yuan, J. Y., B. Tremblay, and A. Babchin, 1999, A wormhole network model of cold production in heavy oil: SPE Paper 54097.

Zhang, D., 2007, Application of V_P/V_S and AVO modeling for monitoring heavy oil cold production: M.S. thesis, University of Calgary.

Zhou, Y., L. R. Lines, K. Hall, and J. Embleton, 2004, Time-lapse seismic analysis of a heavy oil cold production field, Lloydminster, Western Canada: 74th Annual International Meeting, SEG, Expanded Abstracts, 1555–1558.

Chapter 21

Crosswell Seismic Imaging — A Critical Tool for Thermal Projects

Mark McCullum[1]

Introduction

The goal of any operator involved in heavy oil whether flowing or not is to maximize production and minimize costs. With the rapid growth of projects in oil sands, the SAGD process has become commonplace. Although widespread, SAGD is still a relatively new process and operators are discovering that developing an adequate steam chamber requires careful planning. An integral part of this planning process is detailed characterization of the reservoir. With the current operating experience, it is apparent that relatively small features in the reservoir, once thought to be of little or no concern, have been a major impediment or permanent barrier to steam chamber development. The lack of steam chamber growth or inconsistent growth along the horizontal wells is a major contributor to underperforming wells. Therefore, upfront reservoir characterization and the identification of baffles and barriers to steam growth are critical to the process. This chapter details how operators are utilizing crosswell seismic imaging to increase reservoir knowledge and plan SAGD well pair placement and assess overall performance of the steam injection process.

Background

Heavy oil is characterized by an American Petroleum Institute (API) gravity of less than 22.3°. Bitumen at 8–10° API will not flow at normal reservoir conditions. The API gravity of the oil in the reservoir predicates the production methods. Various production methods are now being used in heavy-oil reservoirs. They range from cold heavy-oil production with sand (CHOPS), cyclic steam injection (CSS), steam-assisted gravity drainage (SAGD), steam-assisted gas push (SAGP), vapor extraction (VAPEX), fire flood, and Toe-to-Heel Air Injection (THAI™). With the exception of CHOPS and VAPEX, the other methods rely on heat to mobilize the oil. In the case of fire flood and THAI, heat is generated internal to the reservoir by active combustion, and the other methods rely on steam injected from the surface. It is these processes that will be focused on here.

Steam is generated by burning natural gas in boilers to produce high-quality steam. This steam is then piped to the reservoir and injected into the producing zone. Obviously a key criterion is to inject as little steam as necessary to produce a given amount of oil. This is referred to as steam-oil ratio (SOR), and the usual goal in SAGD is a 2:1 ratio, meaning that for every two volumes of steam, one volume of oil is produced. The challenge then is to achieve or go lower than the 2:1 ratio. This can only be achieved with proper well placement and careful monitoring of the process over time. To achieve this goal, a thorough knowledge of the reservoir and the process over time must be gained. Crosswell imaging is becoming the recognized tool to provide the detailed reservoir knowledge necessary for success.

Crosswell imaging provides a very high-resolution image between two wellbores. From the surface, crosswell imaging looks like a conventional wireline logging operation. A typical crosswell acquisition will have a logging unit at the source well and one or more logging units at nearby receiver wells (Figure 1).

Seismic energy is focused on the zone of interest and does not travel through the highly attenuating unconsolidated layers of earth near the surface. The result is much higher frequencies at the receiver array. This higher frequency content produces the higher resolution achieved with this technique.

In a crosswell acquisition, two types of information are acquired and processed on a routine basis: first, the direct arrivals between source and receivers, and second, reflection information from horizons above and below the source and receiver positions. In addition, there are many other modes and attributes of the crosswell seismic data field that can be captured and processed to yield specific information about the subsurface, such as guided waves

[1] Schlumberger Deeplook-CS, Houston, Texas, U.S.A.

Figure 1. Source (top) and receiver units (bottom).

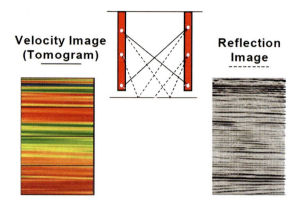

Figure 2. Raypaths for direct and reflected arrivals.

and converted waves. In Figure 2, the direct paths shown as solid lines and the dotted reflection paths allow direct measurement of seismic velocity.

An inversion procedure produces a 2D or 3D map of seismic velocity between the intervals logged in the two wellbores, often called a tomogram. The reflection paths are shown in Figure 2 as the dotted lines. The reflection data are imaged to produce what is typically shown as a wiggle trace section, a high-resolution 2D or 3D seismic reflectivity section between two wellbores. A typical flow for crosswell seismic data processing (Washbourne and Bube, 1998; Washbourne et al., 2002) is shown in Figure 3. The resultant information is contained in the combination of the velocity image produced using traveltime inversion and the reflection image. One fundamental advantage of crosswell versus surface seismic — the ability to directly estimate a velocity field for reflection imaging — is exploited in the general flow, in which the traveltime inversion is used to produce an initial velocity model for reflection imaging. The survey geometry, formation velocity, attenuation, structure, and imaging objectives control the exact sequence of steps followed in the imaging flow. Some steps in the processing flow are fully elective; they are used for some data sets and not for others. In some steps, various methods (e.g., filters) may be selected to achieve the objective of the step. Steps such as wavefield separation may be performed selectively to remove arrivals one at a time, and efficient processing of crosswell data requires access to a suite of software tools that are optimized for crosswell data set geometries, bandwidths, and general data characteristics.

The ultimate resolution of the data is driven largely by the signal-to-noise ratio (S/N) and the frequency content of the raw data. Great strides have been made in source technology to increase the amplitude or power, allowing for increased S/N at the receiver well. In highly attenuating shallow heavy-oil reservoirs, interwell distances of up to 500 m have been successfully imaged. In more competent formations, well separations of greater than 1 km are possible.

In typical clastic bitumen reservoirs, frequencies from 600 to 800 Hz are achieved with crosswell as compared with the 80–110 Hz for surface seismic. The result is that small features in the reservoir such as mudstone up to 1.5 m thick are capable of being imaged.

Imaging Reservoir Features

As thermal projects (primarily SAGD) continue to expand and operating experience is developed, it is becoming more and more apparent that small reservoir features can have a large impact on steam chamber development. The goal now is to identify the complexity of the reservoir

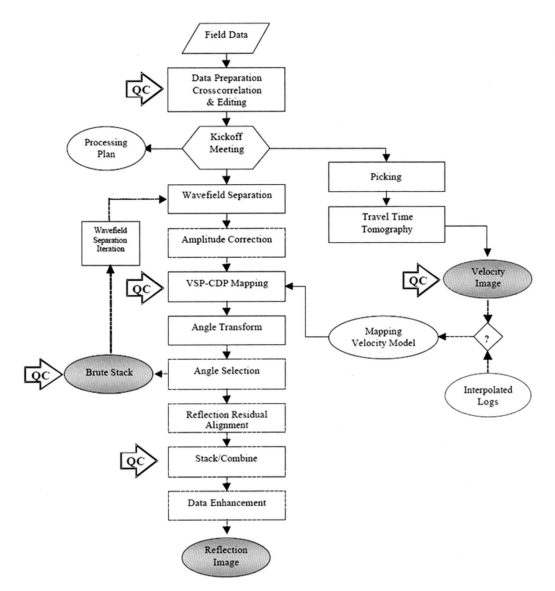

Figure 3. Data processing flow.

before drilling costly well pairs and avoid features that might impinge on steam chamber growth. Applying technologies that can successfully image reservoir features such as shale, mudstone, inclined heterolithic stratification (IHS) beds, and scour features is key to understanding such complexity. The example in Figure 4 displays the types of features that can be resolved from crosswell imaging.

From this example (Zhang et al., 2002, 2005), it is clear that the reservoir has significantly more heterogeneity than could be characterized from the surface seismic image. As one moves along the horizontal well pair, there is an evident mudstone close to the injector. Such mudstone was thought at one time to be only a minor factor in the performance of a well pair because over time the mudstone would break down or be bypassed. Operational experience has shown that neither occurs and mudstone can cause significant reductions in the overall production rate of a well pair. Clearly geology affects the results obtained from the individual well pairs. When using crosswell, it is typical to utilize existing observation wells before drilling the horizontal wells. By obtaining multiple profiles using common wellbores, a fence diagram of the reservoir (Figure 5) can be derived. A 3D volume can be easily interpolated from this.

Crosswell images are obtained in depth, eliminating the time-to-depth ambiguity. Crosswell images also have the benefit of tie points at the wellbores with well logs. This helps the interpreter determine the seismic signatures for specific lithologies, such as shale, mudstone, or clean sand.

Another benefit of crosswell is that because the image is derived in depth, it is not affected by surface features such as muskeg or younger depositional features such as the Quaternary Channel encountered in the oil sands of Canada. Many operators have been faced with the challenge of developing oil-sands plays in areas cut

Figure 4. Comparison of crosswell image (left) versus surface seismic (right).

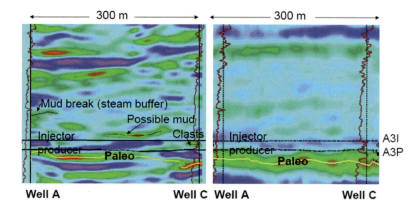

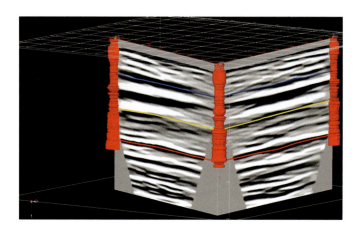

Figure 5. Multiple crosswell images in 3D plane.

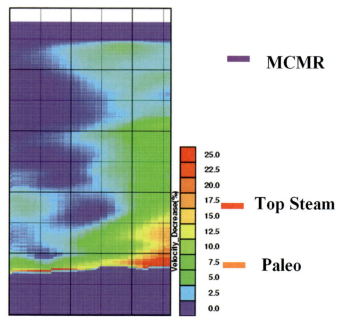

Figure 6. Cross section of a SAGD steam chamber.

by the Quaternary Channel. This younger fluvial system has deposited a thick layer of unconsolidated sediment ranging from coarse sand and gravel to rubble. This layer has proven to absorb most seismic energy generated at surface, resulting in poor or no image below the channel itself. Crosswell imaging provides one of the only techniques to image the producing interval below this channel feature.

Once the reservoir has been characterized, the wells drilled, and steaming commenced, there is a need to evaluate the effectiveness of the process. Although steam injection provides some information versus production rates, technologies such as crosswell are able to provide a detailed early image into the workings of the SAGD process.

In Figure 6, a cross section of a steam chamber is displayed. The image is a tomographic velocity difference comparing the velocity profiles pre- and poststeam. Values are presented as percent reduction in velocity. On the basis of corroborative thermocouple data, we know that the top of the steam correlates to approximately an 18% velocity reduction. The steam chamber is clearly outlined in the warmer red and orange colors. Also of significant interest is the additional velocity change in the reservoir not associated with temperature and hence no steam propagation. This area of the reservoir has clearly been impacted by steam injection and may prove to be an early indicator of how the steam chamber is likely to grow in the future because it is logical to assume that the steam will follow to areas of the reservoir already affected by the injection process. This then provides the operator with a powerful tool to address issues regarding the shape of the steam chamber and potential impact on production much earlier in the process than with other imaging technologies. In addition to the velocity difference image, the image can be combined with the presteam structural or reflection image to provide knowledge into how structure affects steam growth.

Even without a baseline or presteam image, crosswell imaging has proven to be a powerful diagnostic tool. Because the reduction in velocity and density contrast in the reservoir are so dramatic, clear images of steam are possible without the necessity of a baseline image.

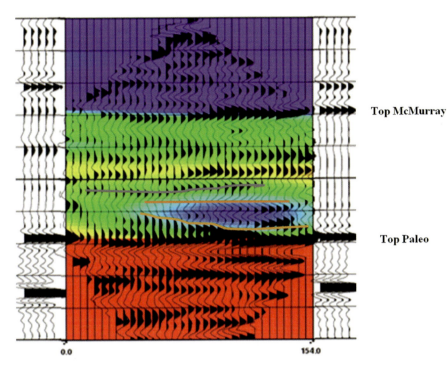

Figure 7. Combined velocity and reflection image.

The example above (Figure 7) demonstrates how powerful a tool crosswell imaging can be in diagnosing what is occurring in the reservoir. In this example, lower than expected production and steam chamber development were causing concerns. As can clearly be seen in the crosswell image, steam is being forced to propagate laterally versus vertically as was designed. The top of the steam chamber is not only indicated by the blue velocity profile but also by the strong trough in the reflections. The base of steam is also indicated by the velocity deep blue color and the strong reflection near the Paleo formation at the base of the reservoir that provides a strong impedance contrast. This type of combining of the amplitudes at the base of steam is common when it occurs close to the top of the Paleo. Further processing can be performed to further separate the reflection characteristics of the base of steam and top of Paleo. We also observe a feature in the reservoir immediately above the steam chamber, indicated by the gray line. This feature is likely a shale stringer or continuous mudstone that is forming an impervious barrier to vertical steam growth. Having this powerful diagnostic tool available allows operators to assess what is happening in their current wells in addition to providing detailed knowledge of the reservoir with which to make intelligent exploitation plans for the future.

Conclusions

Crosswell seismic imaging has moved from the realm of scientific research to mainstream reservoir and process imaging. As operators in thermal projects come to understand the large impact of relatively small features within the reservoir, the need for higher resolution images and model refinement only becomes greater. Crosswell seismic is one technology that is filling the information void with near outcrop-scale resolution of the interwell space.

References

Washbourne, J. K., and K. P. Bube, 1998, 3D High-resolution imaging from crosswell seismic data: SPE Paper 49176.

Washbourne, J. K., F. Miranda, M. Antonelli, and K. P. Bube, 2002, Crosswell reflection tomography: Presented at the 64th EAGE Annual Conference.

Zhang, W., G. Li, and J. Meyer, 2002, Understanding reservoir architectures at Christina Lake, Alberta with crosswell seismic imaging; CSEG Recorder, 33–35.

Zhang, W., S. Youn, and Q. Doan, 2005, Understanding reservoir architecture and steam chamber growth at Christina Lake, Alberta by using 4D seismic and crosswell seismic imaging: SPE/PS-CIM/CHOA 97808.

Chapter 22

The Impact of Oil Viscosity Heterogeneity on Production from Heavy Oil and Bitumen Reservoirs: Geotailoring Recovery Processes to Compositionally Graded Reservoirs

Ian D. Gates,[1] Jennifer J. Adams,[2] and Steve R. Larter[3]

Most of the world's petroleum resources are contained in heavy oil and oil-sands reservoirs. Average recoveries from heavy oil and oil-sands reservoirs are typically low ranging from 5% to 15% for cold heavy-oil production and from 30% to 85% for steam-based in situ processes. Two reasons account for this: (a) geologic heterogeneity in the form of variable rock and rock-fluid properties, and (b) fluid heterogeneities in the form of variable fluid composition. Geologic heterogeneities refer to spatial variations of porosity, permeability, relative-permeability curves, shale, and mud layers, etc. Fluid heterogeneities refer to spatial variations of the fluid composition and properties such as viscosity and density. We will show that the controlling variable on recovery of these resources is fluid compositional variations.

Figure 1 displays the three axes that define a recovery process for oil sands: tolerance to geologic and fluid heterogeneity (geotolerance), environmental impact (gas emissions and water use), and energy efficiency. A recovery process such as mining is geotolerant; that is, heterogeneity does not matter because all of the oil sand is processed and the oil recovery factor is high — typically higher than 90%. However, mining is only suitable for shallow resources, is very costly, and has high carbon dioxide emission and other environmental penalties. In situ processes to produce viscous and poor-quality oils rely on high-pressure primary production, as in cold heavy-oil production, or thermal and/or solvent-based methods to mobilize the oil by reducing its viscosity. The key problem of these processes is that they are not very geotolerant; that is, their performance is adversely affected by the reservoir geology and fluid heterogeneity. Also, profit margins are small because of high capital and operational costs. Furthermore, thermal processes produce large amounts of carbon dioxide emissions and use huge volumes of water. We believe this geo-intolerance, the excessive emissions, the environmental impact, and the consequent energy losses can be reduced and are due to insufficient tailoring of recovery processes to geologic and fluid property variability commonly seen in heavy-oil and bitumen reservoirs.

Heavy-oil and bitumen reservoirs exhibit significant variation of oil composition and thus fluid properties, such as oil viscosity and density vertically and laterally throughout the reservoirs (Larter et al., 2008). Fluid properties commonly vary by one or more orders of magnitude across the thickness of a reservoir or laterally over the distance of a single horizontal production well. The flow rate of bitumen in a reservoir is given by Darcy's law.

$$\mathbf{u}_{bit} = -\underbrace{\frac{k_{bit}}{\mu_{bit}}}_{\text{Oil Mobility}} \underbrace{\nabla(P - \rho g)}_{\text{Driving Force}} \qquad (1)$$

There are two components to this equation: (a) the oil phase mobility (the quotient of the oil phase effective permeability and its viscosity), and (b) the driving force. This means that increasing the oil mobility or raising the driving force (i.e., the pressure gradient) enhances the flow rate of oil. If in gravity drainage mode, as is the case in steam-assisted gravity drainage (SAGD), then the only control on flow rate is the oil mobility because the acceleration due to gravity is constant. Thus, the key control on a gravity drainage recovery process is the oil mobility. In a reservoir where the permeability varies 2–3 times over the vertical range of the reservoir, if the oil viscosity varies by orders of magnitude, then the control on the variability of

[1]University of Calgary, Department of Chemical and Petroleum Engineering, Schulich School of Engineering, Calgary, Alberta, Canada
[2]ConocoPhillips, Houston, Texas, U.S.A.
[3]University of Calgary, Petroleum Research Group, Department of Geosciences, Calgary, Alberta, Canada

Figure 1. The three performance axes for thermal solvent oil-sands recovery processes. Mining is tolerant to geological and fluid heterogeneities but has high emissions and water use and relatively low energy efficiency. SAGD and CSS have slightly better energy efficiency and emissions and water use but are not as tolerant to geological and fluid heterogeneities. gSAGD is SAGD optimized for well placement based on the mobility distribution (Larter et al. 2008). JAGD is described in the text (Larter et al. and 2008; Gates et al. 2007). The ideal process is one which is tolerant of heterogeneities, i.e., it is robust and has low environmental impact and high energy efficiency.

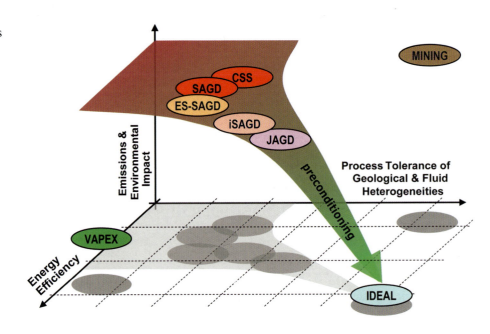

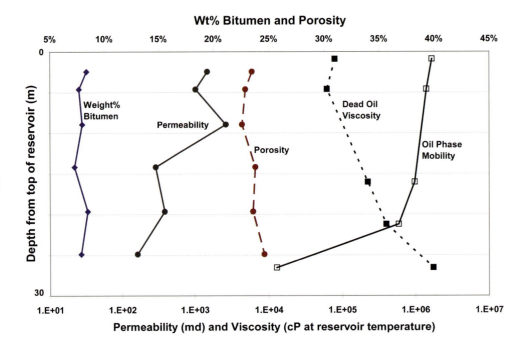

Figure 2. Reservoir (permeability, oil saturation), dead oil viscosity along a vertical profile in a typical Bluesky Formation reservoir of the Peace River oil sands. Oil viscosity increases significantly with depth (ostensibly toward the oil-water contact) while permeability decreases toward the base of the reservoir. The oil phase mobility varies by over two orders of magnitude with depth if relative permeability, absolute permeability, and viscosity at reservoir conditions are considered. N.B. mobility values here are multiplied by 10^8 for visualization purposes only.

the oil mobility is the viscosity and thus the viscosity is the most important factor controlling flow in heavy-oil and bitumen gravity drainage processes.

It has been our experience that the lower viscosity oil phase is at the top of the reservoir. For example, a Peace River viscosity profile is displayed in Figure 2. At the top of the reservoir, the oil viscosity is approximately 80,000 cP, whereas at the bottom it is approximately 1.75 million cP at 20°C. This is a 22-fold increase in the viscosity of the oil phase going from the top of the reservoir to its bottom. From Figure 2, permeability rises to the bottom of the reservoir by approximately seven times. Permeability coupled with oil viscosity results in oil phase mobility ranging from approximately 0.017 mD/cP at the top of the reservoir to approximately 0.000030 mD/cP at the bottom. That is, oil at the top is approximately 500 times more mobile than oil at the bottom. This of course has direct consequences for production of heavy oil from this reservoir. In such a vertically graded reservoir, the best place to position the production well is near the top of the reservoir in the lower viscosity material.

Figure 3 shows the variation of viscosity along four different horizontal wells. In each well, the viscosity varies significantly over distances of up to 1300 m. In a reservoir with large areal variations of oil viscosity, the production

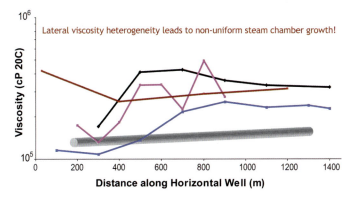

Figure 3. Dead oil phase viscosity variations along four different horizontal wells estimated from geochemical proxies measured from cuttings along the well.

well should maximally intersect the lower viscosity portions of the reservoir. For thermal process design, the better quality, lower viscosity oil is typically at the top of the reservoir. Therefore, to accelerate production, well placement should be designed to recover this oil first.

Variability, not uniformity, of oil properties seems to be the rule! Here, we summarize types of variations in fluid properties seen in heavy-oil and bitumen reservoirs, briefly describe how they are formed, and then illustrate how geotailored recovery processes can be designed to take advantage of the viscosity variation. Incorporation of oil viscosity variations into reservoir simulations is a necessary step to design optimal recovery technologies for heavy-oil reservoirs.

Fluid Heterogeneity Is Caused by Biodegradation over Geologic Timescales

Heavy oils can form via biodegradation of conventional crude oils [i.e., microbial metabolism or conversion of hydrocarbons to biogenic gas (e.g., CO_2 and methane)] over geologic timescales (Head et al., 2003; Larter et al., 2006; Larter et al., 2008). Because the microbes need to live in free water and have access to essential waterborne nutrients, biodegradation occurs at the oil-water contact (OWC) under anaerobic conditions (Aitken et al., 2004) in high water saturation zones usually at the base of pay. The degree of and rate of biodegradation is mainly controlled by reservoir temperature, feed components and nutrients, extent of OWC area and oil volume, and reservoir water salinity (Larter et al., 2006; Adams, 2008). The Western Canada heavy-oil province in general follows worldwide trends but is more complex because of long residence times of oils in the reservoir and the stopping of primary oil charge post-Laramide orogeny and basin inversion (Adams, 2008).

Compositional gradients are the result of buoyant lower-density fresh oils preferentially charging the top of the reservoir while oils are degraded in a water-derived nutrient-limited process at the OWC. Biodegradation preferentially removes the most available lower density, lower molecular weight saturated and aromatic hydrocarbons and thus causes heavy polar nonhydrocarbon and asphaltenes to accumulate toward the OWC with advective and diffusive mixing broadening concentration profiles. The result is a compositional profile through the reservoir (Adams, 2008; Larter et al., 2008). The distribution of the oil phase composition is the result of competition between the rate of charge oil and mixing and biodegradation rate as controlled by reservoir fluid dynamics over a few to a few tens of millions of years (Barnard and Bastow, 1991; Butler, 1997; Wenger et al., 2001; Larter et al., 2003; Larter et al., 2006; Adams et al., 2006; Adams, 2008). Burial history of the reservoir and source rocks in addition to porosity, permeability, and pore size distributions along migration pathways and in the reservoirs are also factors that determine the final spatial distribution of composition (Wilhelms et al., 2001; Adams et al., 2006; Adams, 2008).

Oil composition sets physical properties such as American Petroleum Institute (API) gravity and viscosity, which control production behavior. Vertical viscosity variations in oil-sands reservoirs found in Alberta have several distinct shapes. Figure 4 shows commonly found viscosity profiles. In case (a), the reservoir quickly fills to an underseal and degradation stops because of lack of nutrient transport to the OWC or charge stops and lighter oil column slowly mixes. In case (b), at the top of the oil column, we find slightly more viscous and more degraded oil, which may be related to increased degradation because of top water or late stage degradation at the top of the reservoir by influx of top water because of leakage of top gas. Below this top zone is a curved profile that develops by slow diffusive mixing of oil column after main charge stops. Case (c) indicates extensive biodegradation, which continues after charge stops with a basal burnout zone. Case (d) is representative of stacked reservoirs (only two shown here) separated by mudstones. Generally, the upper reservoir completely fills first and degradation stops, whereas degradation proceeds for a longer time in the lower zone.

As shown schematically in Figure 5, viscosity variations exhibit areal patterns; for example, lower viscosity "fingers" are often embedded between higher viscosity "islands" although variation is typically smooth and wavelike unless faulting is involved. Lateral oil viscosity variations occur smoothly with changes of between two and 10 times on lengthscales of 500–1000 m. This is a result of the interplay over geologic timescales of oil charge rate, diffusion and mixing, and biodegradation

Figure 4. Vertical viscosity variations in oil-sand reservoirs: (a) reservoir quickly fills to an underseal and degradation stops due to lack of nutrient transport to the OWC or charge stops and lighter oil column slowly mixes; (b) slow diffusive mixing of oil column after main charge stops dominates main curved profile coupled with a top burnout zone related to influx of top water due to leakage of top gas; (c) extensive biodegradation continues after charge stops and a burnout zone forms; (d) two stacked reservoirs separated by mudstones; however, many stacked reservoirs have been observed: the top zone commonly fills completely and degradation stops; whereas degradation proceeds for a long time in the bottom zone (after Adams, 2008).

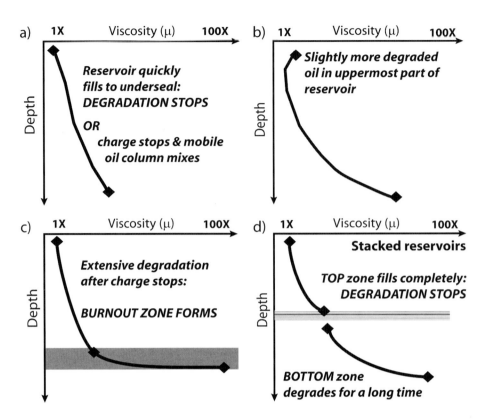

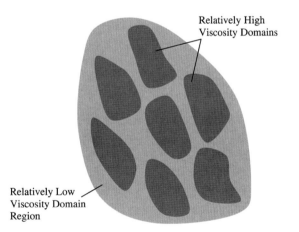

Figure 5. Schematic plan view of simplified lateral oil viscosity variations in oil sand reservoirs.

rate, which is controlled by proximity to water legs, among other factors. This process is continuous and so rather than distinct oil viscosity domains, a graded transition between relatively high and relatively low viscosity regions exist. Intersecting viscous fluid regions and complex sedimentologically controlled permeability domains result in a reservoir with complex oil mobility variations. Thus, for gravity drainage recovery processes in which oil mobility is the chief control of oil movement, the placement on wells becomes critical. Delineation of these reservoirs is necessary to create high-resolution fluid property maps. By considering a map as shown in Figure 5, for recovery processes that use long horizontal wells, the optimal well trajectory would not be a straight one. Geotailoring cold or thermal production wells for this reservoir would involve routine optimization of complex well trajectories in three dimensions, well spacing, and well length. The optimal well configuration may be quite different from a standard SAGD one!

Impact of Fluid Property Variations on Recovery

The oil mobility gradient has direct impact on how a horizontal well would inject fluid into and produce oil from a reservoir. For example, irregular steam chamber growth may well be the result of lateral viscosity gradients. The mobility heterogeneity at start-up may promote nonuniform steam conformance, which adversely impacts production. Recovery processes designed to perform in uniform fluid mobility reservoirs will have difficulties in a typical Western Canadian oil-sands reservoir.

Gravity drainage recovery processes rely on steam temperature fluid properties. Although vertical viscosity variations are common at reservoir temperature, viscosity gradients are often assumed to be negligible at steam temperature, yet varying oil composition and physical properties including temperature effects on viscosity would be expected. Figure 6 displays a plot revealing how vertical oil viscosity varies by a factor of up to four in two Peace River oil-sands reservoirs at steam temperature (215°C) and the temperature of the mobilized oil at

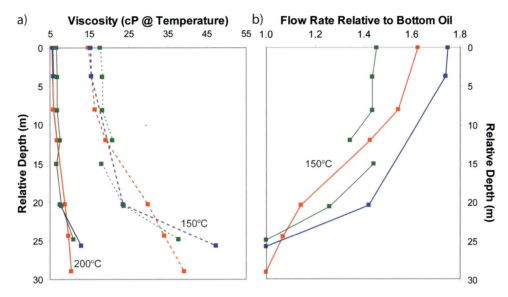

Figure 6. (a) Vertical live viscosity varies by a factor of up to four in two Peace River oil sand reservoirs at steam temperature (200°C: solid lines) and mobile oil at the edge of the steam chamber temperature (150°C: dashed lines) (data extrapolated from high temperature measurements by using the ASTM D341 equation). We assumed 6 mol% methane in the oil. (b) The vertical flow rate relative to the bottom oil flow rate along a vertical well profile at 150°C. Such viscosity variations translate into relative flow rates from top to bottom of the reservoir varying of up to 75%, assuming flow is varying by the reciprocal of the square root of the viscosity[12]. After Adams (2008).

the edge of the steam chamber (200°C). Viscosity data were extrapolated from a suite of high-temperature dead-oil viscosity measurements (up to 185°C) by using the American Society of Testing and Materials (ASTM) D341 equation. Thus, even at steam temperatures, viscosity variation in a reservoir affects production. Figure 6 also displays the oil flow rate relative to the bottom oil under gravity drainage. This means that even for a steam process, the ideal place to start a steam chamber is actually near the top of most reservoirs, not the bottom.

Geotailoring Recovery Processes for Reservoirs with Viscosity Variations

There are two requirements for any heavy-oil or bitumen recovery process: (a) the oil must be mobilized, and (b) the mobilized oil must be delivered by some driving force to the production wellbore. In SAGD, steam releases its heat to the bitumen and raises its mobility. Under gravity drainage, the mobile bitumen flows through the steam chamber and along its sides to the production wellbore. This is an elegant process that is very suitable for highly permeable, homogeneous reservoirs containing constant viscosity oil.

If a unit of energy is injected into the top part of the reservoir, a lower viscosity, more mobile oil would result at the top of the reservoir than would be the case if that unit of energy had been injected into the more viscous bottom oil. Thus, for optimal energy efficiency, the amount of mobilized oil will be greater if the steam was initially injected into the top oil rather than into the bottom oil, and to accelerate oil production and maximize recovery the top oil should be produced first. This means that steam should be injected into the upper parts of the reservoir to accelerate production and produce the highest quality oil. This can be achieved by using a horizontal steam injector located at the top of the reservoir. If a SAGD well configuration was used, the production well would be positioned approximately 5 m below the injection well. In a 30-m reservoir, the lower 23 m of the reservoir below the production well would not be recovered. There is a balance between placing the production well high enough so that it accesses relatively low viscosity oil to facilitate steam chamber growth yet not too high in the reservoir so that ultimate recovery suffers. Ultimately, the SAGD configuration is not optimal for a viscosity-graded reservoir because the oil rate or recovery will suffer at the cost of the other. The best well configuration would be one in which gravity drainage supplies reservoir fluids to the producer, the wells are close enough to efficiently establish thermal communication, initial oil production is obtained from the lower viscosity bitumen in the upper portions of the reservoir, and the wells are situated to produce the oil from the lower parts of the reservoir to obtain reasonable recovery factors.

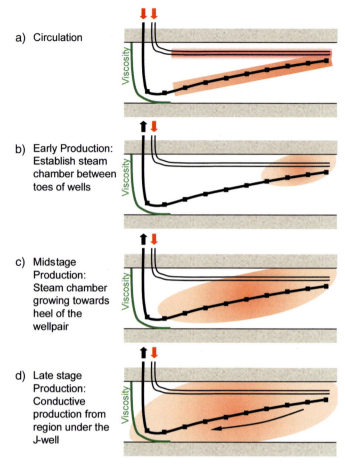

Figure 7. The JAGD (Gates et al., 2007) process: (a) steam circulation to establish thermal communication between the two wells at the toe, (b) switch to steam injection and production from J-well to establish a vapor chamber between the wells, (c) as production continues, the steam chamber extends toward the heel of the wellpair, and (d) steam chamber reaches the heel of the wellpair and conductive heating to the region under the J-well continues oil production.

J-well and Gravity Drainage Recovery Process: A Geotailored Process

A geotailored well configuration we have developed is the J-well and gravity drainage recovery process [(JAGD; formerly known as J-well and gravity-assisted steam stimulation (JAGASS)] (Gates et al., 2007) as shown in Figure 7. In this well configuration, the steam injection well is located in the top few meters of the reservoir and is substantially horizontal along its trajectory. The production well consists of an inclined J-well, which from plan view is in line with the top injection well. The toe of the production well is located several meters below the injection well whereas its heel is located near the bottom of the reservoir. Thus, the J-well intersects most of the reservoir cutting through geologic heterogeneities throughout the reservoir.

The JAGD process operates in a similar manner to that of SAGD (Gates et al., 2007). As shown in Figure 7a, before production, thermal communication is established between the toes of injection and production wells by steam circulation. Because this is only done over a relatively small length of the wells compared with SAGD, in which communication occurs along the entire length of the wells, the circulation period for JAGD is less energy intensive and of shorter duration than that of SAGD. After thermal communication is achieved, steam is then injected into the injector and production is started out of the lower well, as shown in Figure 7b. The heated mobile bitumen at the toes of the wells is flushed into the production well and a vapor depletion chamber is established there. At the edges of this small chamber, mobilized bitumen flows under gravity to the production well below and is removed from the reservoir. At the base of the chamber, steam condensate and mobilized bitumen accumulates to form a liquid pool above and within the production well. This acts to prevent live steam production from the chamber as is found in steam trap control in SAGD. The liquid bath in JAGD is shorter than that in SAGD, where it exists along the entire length of the well pair. In JAGD, this shortened interval implies more control of the steam trap above and within the production well than would be found in SAGD. As shown in Figure 7c, with continued steam injection, mobilized bitumen drains not only in the crosswell direction as is the case in SAGD but also in the up-well direction (from toe to heel) as the steam chamber evolves toward the heel of the J-well. Also, as this occurs, the steam chamber penetrates into deeper portions of the reservoir, accessing bitumen in the lower parts of the reservoir. With conductive heating from the steam chamber to the portions of the reservoir below the J-well, this oil eventually drains in the up-well direction toward the heel of the J-well (Figure 7d). Thus, ultimate recovery from the reservoir is similar to SAGD.

The key benefits of the JAGD well configuration are (Gates et al., 2007)

1) *JAGD targets more valuable, lower viscosity top oil.* JAGD first produces bitumen from the upper regions of the reservoir and thus first produces the more valuable, lower viscosity, higher API gravity oil.
2) *Reduced steam volume.* JAGD starts with a relatively small steam chamber at the well-pair toe so steam usage is lower in the earlier stages of the process than would be in the case of SAGD. Steam conformance along the injection well will be better achieved than in SAGD because the chamber grows from one end of the well to the other rather than along the entirety of the well at the same time.

3) *J-well accesses most of the reservoir.* The J-well spans nearly the entire reservoir thickness. Fluid flow that would have been impaired by vertical barriers in SAGD would not suffer to the same extent in JAGD.
4) *Local steam trap control.* The liquid bath only exists at the base of the chamber.
5) *Multiple gravity drainage directions.* Mobilized bitumen drains along the edges of the steam chamber and parallel to the wells.

Several other modifications to JAGD can be applied to improve the process further (Gates et al., 2007). For example, a movable packer or interval control valves could be used in the wells to better control the evolving steam chamber along the well pair. Also, before drilling the J-well, if the oil viscosity is low enough and solution-gas drive large enough in the upper parts of the reservoir, the top well initially could be operated as a cold production well. After the cold production has become uneconomic, it would be then converted to steam service and used as a steam injector as described above. The cold production stage of the process could provide cash that could be used to develop the thermal stage of the process.

A reservoir simulation model was used to assess how well placement can be altered to improve the recovery factor and thermal efficiency of the process. Figure 8 shows sample cross sections of the porosity, permeability, and oil saturation distributions. The reservoir thickness is 32 m. Initial reservoir conditions are 10°C and 2600 kPa at the top of the reservoir. Figure 9 shows the initial live oil viscosity profile. Viscosity variations are in the vertical direction and range from approximately 8000 cP at the top to approximately 250,000 cP at the bottom of the oil leg (Figure 9).

The JAGD process has been simulated by using the Computer Modeling Group (2005) STARS thermal simulator and compared to an equivalent SAGD process in the same reservoir. In the SAGD model, a single well pair is placed in the center of the reservoir model. The length of each well is 700 m. The location of the production well is 2 m above the bottom of the reservoir and oil leg with 5-m interwell spacing. Before SAGD, three months of steam circulation occurs to establish thermal communication. After steam circulation is done, SAGD mode starts with steam injection at 1000 kPa at 0.95 quality and fluids produced from the reservoir. In the JAGD model, the 700-m injection well is located 3 m below the top of the reservoir, and at the toe the interwell spacing is 5 m. The production well is inclined so that the heel of the well is 2 m above the base of the reservoir. The heel of the production well is directly below the heel of the injection well. Similar to the SAGD model, steam circulation occurs for three months. After the wells are placed on production, the top well injects steam at 1000 kPa.

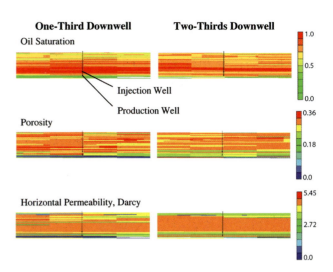

Figure 8. Sample cross sections (at locations one- and two-thirds of the length of the well) of the porosity, permeability, and oil saturation distributions, and injection and production well positions. The model contains 36 gridblocks in the crosswell direction, 35 gridblocks in the vertical direction, and 12 gridblocks in the down-well direction. Average porosity, horizontal and vertical intrinsic permeability of the oil-bearing zone are 0.32, 4100 and 1400 mD, respectively. The average oil saturation is 0.878.

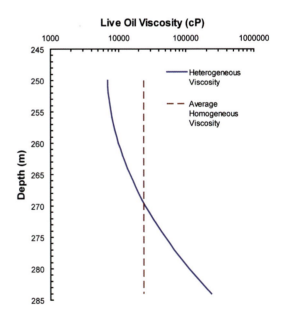

Figure 9. Live oil phase initial heterogeneous viscosity distribution versus depth used in the reservoir simulation models. The vertical homogeneous viscosity is the depth averaged viscosity of the heterogeneous distribution.

Figure 10 compares cumulative oil phase volumes in SAGD and JAGD operations in a reservoir with the viscosity profile displayed in Figure 9. The results show that JAGD produces more oil than SAGD over a seven-year period. The initial rates are higher in JAGD because the

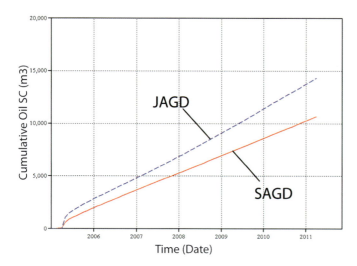

Figure 10. Comparison of oil cumulative production from SAGD and JAGD in reservoir with heterogeneous viscosity profiles.

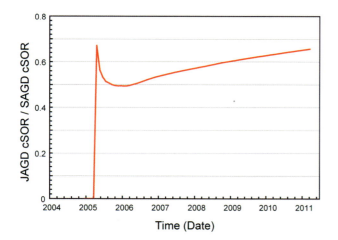

Figure 11. Comparison of energy efficiency of JAGD versus SAGD in reservoir with heterogeneous viscosity profiles.

steam chamber is accessing the lowest viscosity oil in the reservoir. Figure 11 plots the ratio of cumulative steam (expressed as cold water equivalent volume) to oil ratio (cSOR) of JAGD and SAGD. The results show that steam usage per unit bitumen volume produced is between 50% and 65% that of SAGD, and the thermal efficiency of JAGD is significantly higher than that of SAGD. Figure 12 shows cross sections, in the downwell direction, of the oil saturation distribution as JAGD evolves. The steam chamber rapidly grows in the up-well direction as bitumen is produced from the reservoir. This is especially the case in the initial period of the process because the chamber is growing in the low viscosity oil region. As the chamber grows downward by production from the J-well and starts to encounter higher viscosity oil, its growth slows down relative to that at the start of the process. As the chamber grows in the up-well direction, it also does so in the crosswell direction. The largest amount of steam chamber growth occurs at the toe of the wells where the steam chamber first starts. As the steam chamber extends in the up-well direction, the slope of the steam chamber is established along the J-well. Heat conducts from the hot chamber into the oil sand located directly underneath the J-well and mobilizes the oil there, which consequently flows down toward the heel of the J-well. This can be seen by the reducing oil saturations in the region below the J-well as the process evolves.

In conclusion, the defining characteristic of oil sands is large vertical and lateral spatial variation in fluid properties such as oil viscosity. These variations are related to the competitive effects of active oil charge, which increases oil quality and biodegradation at OWCs, which lowers oil quality and produces large vertical and longer wavelength lateral gradients in composition, API gravity, and oil viscosity. Typically the bottom of the reservoir contains the most biodegraded, most dense, and most viscous oils. Lateral viscosity variations appear to occur smoothly on viscosity scales of 2–10 times on lateral

Figure 12. Visualization of the oil saturation in the JAGD process. This cross section is taken along the JAGD wellpair. The bitumen under the J-well drains under gravity toward the heel of the wells and is eventually produced.

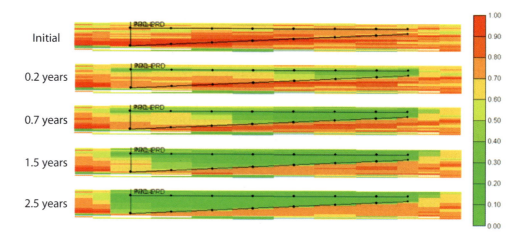

length scales from 500 m to 2 km unless faults are crossed, in which case variations can be more dramatic. The impact of oil viscosity variations on recovery process performance promotes the notion that heavy-oil production processes must take advantage of the viscosity variations to more efficiently produce these reservoirs. Recovery processes have to be geotailored to the target reservoir and its mobility ratio domain structure. This work also suggests that fixed guidelines such as locating the production well of a SAGD process a few meters above the base of pay may not always be the best design basis for heavy-oil recovery processes. The impacts of vertical gradients in oil mobility have been demonstrated in the field, where it is widely accepted that for cold production of heavy oil, the industry standard is to locate the cold production well in the top few meters of the reservoir. We have proposed the JAGD process, which takes into account the oil phase viscosity variations to improve the economic and environmental performance of the production process.

In general, the complexity of oil mobility domains laterally and vertically in oil-sands reservoirs suggests that nonlinear well configurations are likely to be more efficient than the current linear well combinations utilized in most processes. For design of heavy-oil recovery processes, the coupling of fluid property heterogeneities into compositional reservoir simulation is, we feel, a necessary step to design efficient heavy-oil production processes. Truly integrated reservoir geoscience and engineering is the key, and all of the necessary tools are available to achieve this!

References

Adams, J. J., C. L. Riediger, M. G. Fowler, and S. R. Larter, 2006, Thermal controls on biodegradation around the Peace River tar sands: paleo-pasteurization to the west: Journal of Geochemical Exploration, **89**, 1–4.

Adams, J. J., 2008, The impact of geological and microbiological processes on oil composition and fluid property variations in heavy oil and bitumen reservoirs: Ph.D. thesis, University of Calgary.

Aitken, C. M, D. M. Jones, and S. R. Larter, 2004, Isolation and identification of biomarkers indicative of anaerobic biodegradation in petroleum reservoirs: Nature, **431**, 291–294.

Barnard, P. C., and M. A. Bastow, 1991, Hydrocarbon generation, migration alteration, entrapment and mixing in the central and northern North Sea, in W. A. England and A. J. Fleet, eds., Geological Society Special Publication, No. 59, Petroleum Migration, 167–190.

Butler, R. M., 1997, Thermal recovery of oil and bitumen: GravDrain, Inc.

Computer Modeling Group, 2005, Computer Modeling Group STARS^{TM} user manual, version 2005.10.

Gates, I. D., S. R. Larter, and J. J. Adams, 2007, In situ heavy oil and bitumen recovery process. Canada Patent Application 2593585.

Head, I. M., D. M. Jones, and S. R. Larter, 2003, Biological activity in the deep subsurface and the origin of heavy oil: Nature, **426**, 344–352.

Larter, S. R., A. Wilhelms, I. Head, M. Koopmans, A. Aplin, C. Zwach, P. R. Di, M. Erdmann, and N. Telnaes, 2003, The controls on the composition of biodegraded oils in the deep subsurface. Part I: biodegradation rates in petroleum reservoirs: Organic Geochemistry, **34**, 601–613.

Larter, S. R., H. Huang, J. J. Adams, B. Bennett, F. Jokanola, T. Oldenburg, M. Jones, I. Head, C. Riediger, and M. Fowler, 2006, The controls on the composition of biodegraded oils in the deep subsurface. Part II: geological controls on subsurface biodegradation fluxes and constraints on reservoir fluid property prediction: AAPG Bulletin, **90**, 921–938.

Larter, S. R, J. J. Adams, I. D. Gates, B. Bennett, and H. Huang, 2008, The origin, prediction and impact of oil viscosity heterogeneity on the production characteristics of tar sand and heavy oil reservoirs: Journal of Canadian Petroleum Technology, **47**, 52–61.

Wenger, L. M., C. L. Davis, and G. H. Isaksen, 2001, Multiple controls on petroleum biodegradation and impact in oil quality: SPE Paper, 71450.

Wilhelms, A., S. R. Larter, I. Head, P. Farrimond, R. Di Primio, and C. Zwach, 2001, Biodegradation of oil in uplifted basins prevented by deep-burial sterilization: Nature, **411**, 1034–1037.

Chapter 23

Using Time-lapse Seismic to Monitor the Toe-to-Heel-Air-Injection (THAI™) Heavy-oil Production Process

Rob Kendall[1]

Introduction

The Whitesands project is located in the Athabasca Oil Sands near Conklin, Alberta (Figure 1). Toe-to-Heel-Air-Injection (THAI™) is an in situ controlled combustion process that cracks, upgrades, and mobilizes heavy oil. The Whitesands pilot project is designed around three well pairs producing from a central facility. Air injection commenced on the first well pair in July 2006. Air injection on the second well pair was initiated in January 2007 and on the third pair in June 2007. The project's first THAI and CAPRI™ well was drilled in the second quarter of 2008, which also incorporated a revised completion design to improve downhole sand control.

The position and progress of the combustion front is being monitored using tilt meters, passive seismic, and active 4D seismic. Each of the three methods identifies the combustion zone, but the active 4D seismic method provides the most reliable and detailed information.

In 2003, Petrobank acquired a coarsely sampled (15 × 20 m bins) 3D seismic survey for general imaging purposes. Since that time, the reservoir has undergone a SAGD style process, the initiation of the THAI process, and gas production from the overlying Clearwater formation. These changes to the reservoir and the overlying production will affect the 4D results. In 2008, Petrobank recorded a high-resolution (5 × 5 m bins) multicomponent survey that would serve as the baseline for the 4D. We were not convinced that we would be able to see a valid time-lapse signal given the differences between the 2003 and 2008 recording parameters. The 2003, data were interpolated and regularized from the 15 × 20 m bins down to 5 × 5 m bins. The data were then used for 4D analysis. In 2009, the survey was repeated with the identical geometry to the 2008 baseline. Only a few shots were missed because of a new pipeline, and the new survey confirmed the cooling of the reservoir near the newly drilled P3B well, small increases in temperature near the A1 air injector, and no change in temperature near the A2 air injector.

Reservoir description at Whitesands (Leismer)

At the Whitesands project site, the bitumen is contained within the McMurray formation; which is the basal unit of the Lower Cretaceous Manville Group. The McMurray formation is the primary reservoir for the Athabasca Oil Sands deposit of northeastern Alberta. In general, the McMurray formation is a complex estuarine depositional system divisible into a lower fluvial-dominated estuarine system and an upper sequence of bay fill deposits. A regionally mappable shale separates these two sequences and provides stratigraphic control.

The McMurray formation at the Whitesands project site is stratigraphically divided into three units: the McMurray A, B, and C, units (Figure 2). The main reservoir targeted for the Whitesands project is the McMurray B unit. The McMurray B sand is capped by the McMurray A shale marker, which is the regionally mappable shale that provides the stratigraphic control.

The thickest bitumen, with thicknesses greater than 10 m, occurs within the estuarine valley fill sands of the McMurray B unit. Preliminary stratigraphic analysis suggests that the McMurray B reservoir was deposited as a channel point bar in a northwest–southeast trending incised valley.

The McMurray B at Leismer can be further divided into three basic depositional units: a thick basal channel sand averaging 10 m thick, a middle layer of inclined heterolithic bedding (IHS) up to 12 m thick and an upper mixed sand shale unit. The McMurray B is bitumen

[1]Petrobank Energy and Resources Ltd., Calgary, Alberta, Canada

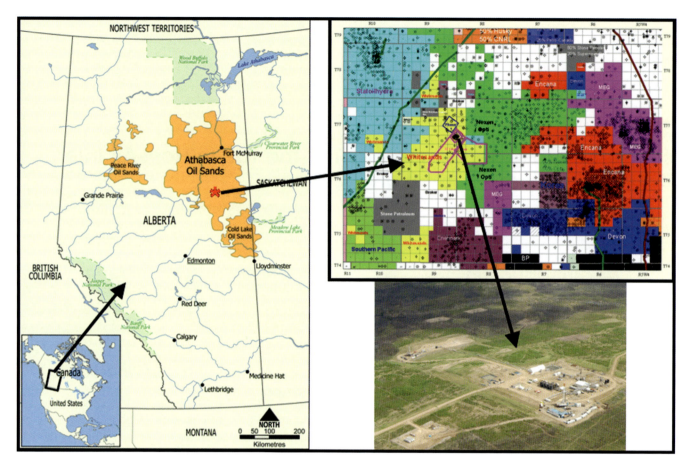

Figure 1. The Whitesands project is located in the Athabasca Oil Sands region of northeastern Alberta, Canada.

saturated (average 80% bitumen saturation) at the project site except for the presence of a thin 1 to 3 m water zone that occurs at the base of the McMurray B basal sand in some of the wells. Average measured porosities for the basal sand are 36%–38%, whereas the McMurray IHS porosities range between 31% and 35%.

Permeabilities range from 8 to 12 darcies for the basal sand and from 3 to 5 darcies for the IHS.

THAI and CAPRI

THAI is an in situ controlled-combustion process that cracks, upgrades, and mobilizes heavy oil (Figure 3). Compressed air is injected into the reservoir after a pre-injection heating cycle brings the reservoir up to an appropriate temperature for combustion to occur. The bitumen is then cracked, leaving a coke zone that becomes the fuel for the process. In front of the coke zone is a mobilized, upgraded oil that is then produced by a horizontal well through gravity drainage. The THAI process does not require gas or water [as in steam-assisted gravity drainage (SAGD)] and, in fact, the produced water is appropriate for sale to SAGD producers.

CAPRI is a catalytic process that takes advantage of the in situ conditions created by THAI (i.e., temperature and pressure) to further upgrade the oil as it passes through a catalyst-filled annulus in the production liner (Figure 3). Laboratory results indicate that THAI will provide an upgrade of approximately 10° American Petroleum Institute (API) gravity and CAPRI a further upgrade of approximately 7° API. A more detailed discussion on the THAI and CAPRI processes can be found in Chapter 1 of this book.

Conditioning of Seismic Data for Time-lapse Analysis

Four seismic volumes (Figure 4) will be used in the following discussion:

1) the 2003 coarsely sampled (15 × 20 m bins) P-wave data shown in the western portion of Figure 4

2) the 2005 coarsely sampled (15 × 20 m bins) P-wave data shown in the eastern portion of Figure 4

3) the 2008 true baseline high resolution (15 × 20 m bins) multicomponent data in the northern portion of Figure 4

Chapter 23: *Using Time-lapse Seismic to Monitor the THAI™ Heavy-oil Production Process* 277

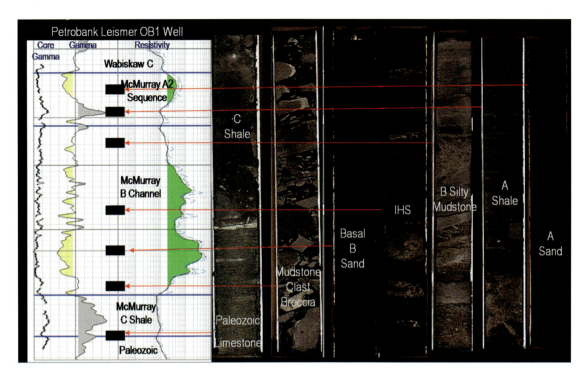

Figure 2. Log-to-core correlation for the Whitesands area. The THAI combustion zone is contained within the Basal B and IHS.

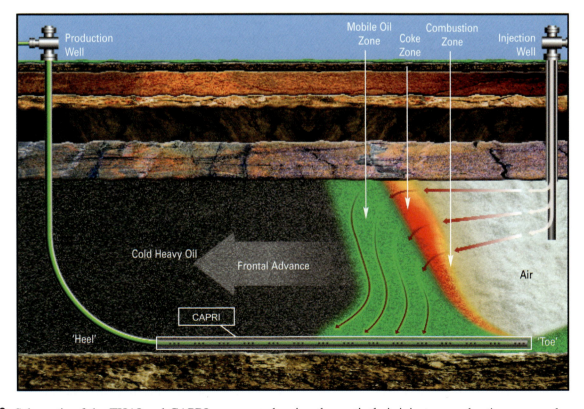

Figure 3. Schematic of the THAI and CAPRI processes showing the vertical air injector, combustion zone, coke zone, mobilized oil zone, and the catalyst-lined horizontal production well.

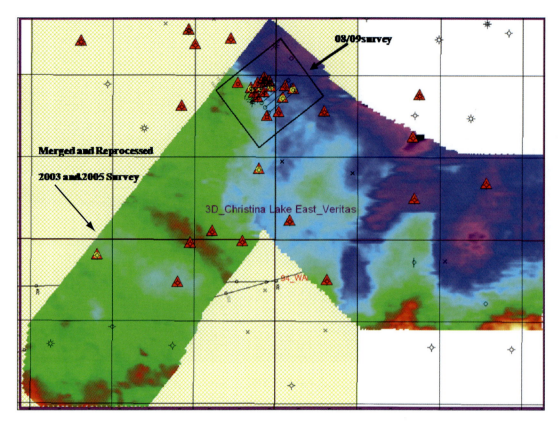

Figure 4. The two coarsely sampled 3Ds (east and west) were merged and then interpolated and regularized to the geometry of the 2008 high-resolution multicomponent 3D.

4) the 2009 monitor survey that has identical acquisition parameters to the 2008 data (Figure 4).

The 2003 and 2005 data were merged, processed (Figure 5), and interpreted for regional mapping purposes. The merged volume was then interpolated and regularized in the area that overlaps with the 2008 data. Following a scheme similar to Trad et al. (2008), new shots and receivers were built using a five-dimensional (offset, azimuth, inline, crossline, and frequency) interpolator based on Fourier reconstruction. By taking advantage of the multidimensional aspect of the process, information from different dimensions can be used simultaneously to infill missing data. Simultaneously interpolating in the four spatial dimensions of offset, azimuth, inline, and crossline instead of just the latter two fully exploits the redundancy of the 3D data, significantly improving the spatial sampling. In this particular case, we created approximately 48 times as much data through interpolation (Figure 6). In turn, these interpolated data, although obviously not ideal, were used as a quasi-baseline survey for time-lapse. The 2008 baseline multicomponent survey was used as a monitor survey in conjunction with the interpolated data. In 2009, the survey was repeated so that we had an interpolated baseline and two high-resolution monitor surveys for the PP mode and a high-resolution and one monitor survey for the PS mode. To test the robustness of the interpolation, we compared the 4D results using the interpolated and noninterpolated flows.

- *Flow 1.* Independent trace equalization, interpolation, and band-limited data.

$$(2009 - 2003) - (2008 - 2003) = 2009 - 2008$$
$$(C - A) - (B - A) = C - B$$

- *Flow 2.* High-resolution noninterpolated data.

$$2009 - 2008 = C - B$$

As it turns out, the interpolated data provided surprisingly good results when used as a baseline. Figure 5 shows the processing flows that were used after interpolation in the case of the 2003 data and directly to the 2008 and 2009 data.

Velocity Anomalies

The P- and S-wave velocities of heavy oil are very dependent on temperature (Han et al., 2006). Figure 7 left shows the decrease in P- and S-wave velocities from 0°C to 200°C. Although the P-wave velocities drop dramatically, the S-wave velocities are even more dramatic and in fact go to zero at approximately 80–90°C. Figure 7

- **2008/2009 Whitesands vertical processing flow**
- **Apply Tilt Corrections**
  - Select Vertical
- Manual Trace Edits
- TOMO refraction statics
- Spreading gain recovery
- Pre-decon NA – shot domain
- Surface consistent scaling
- Surface consistent decon – 60ms op, 0.01% prew.
- Velocity analysis 1 – 250m grid – DSR
- Surface consistent statics 1
- Shot domain NA
- Surface consistent scaling
- Final velocity analysis – 250m grid – DSR
- Spectral whitening – only for non-AVO version
- Surface consistent statics 2
- Shot domain NA
- **TRIM statics – 2008 model used for both 2008 and 2009**
  - widow to derive statics taken above zone of interest.
- **Apply non-common shot/stn/spread edits**
  - to both 2008 and 2009 surveys
- Mute
- **Single function dB scalar**
  - for amplitude decay recovery and matching 2008 and 2009
- 100ms AGC – only for non-AVO version
- Stack 1/n
- SAFA (footprint removal)
- FXY decon
- 3D Kirchhoff migration
- **Bulk match 2009 to 2008**
  - 0.23ms up and -1.0 degrees phase rotation

- **2008/2009 Whitesands radial processing flow**
- **Apply Tilt Corrections**
  - rotate to radial and transverse, select radial
- Use same geometry as vertical with ACP binning
- Manual Trace Edits
- Spreading gain recovery
  - TV function using single function velocity point from vertical
- Pre-decon NA – shot domain
- Surface consistent scaling
- Surface consistent decon – 60ms op, 0.01% prew.
- **Apply SHOT component statics from vertical data**
- **Calculate horizon based STN drift statics**
- Gamma analysis 1 – 250m grid – DSR
- Surface consistent statics 1
- Shot domain noise attenuation
- Surface consistent scaling
- Final gamma analysis – 250m grid – DSR
- Spectral whitening – only for non-AVO version
- Surface consistent statics 2
- **CDP based LW static to match 2009 to 2008**
- **TRIM statics – each survey used individual models**
- **Apply non-common shot/stn/spread edits**
  - to both 2008 and 2009 surveys
- Mute
- **Single function dB scalar**
  - for amplitude decay recovery and matching 2008 and 2009
- 100ms AGC – only for non-AVO version
- CCP binning and Stack 1/n
- SAFA (footprint removal)
- FXY decon
- 3D Kirchhoff migration
- **Bulk match 2009 to 2008 – 43 degrees phase rotation**

Figure 5. Multicomponent seismic processing flow. Bold steps indicate processes that were unique to the time-lapse aspects of the processing.

right shows the phase change of heavy oil with temperature. Because the heavy oil in bitumen sand acts as the matrix, especially at low temperatures, the properties of the heavy oil will dominate the rock properties of the bitumen sand. As such, bitumen sand behaves as a non-Newtonian fluid such that its shear stress and viscosity are not constant at all shear rates. Furthermore, Gassmann-style fluid substitution is not valid below approximately 60°C (Schmitt, 1999; Han et al., 2007).

Time-lapse Results

Time-lapse or 4D seismic is analogous to time-lapse photography (Figure 8). It is accomplished by recording surveys over the same field at different stages of production. The technique assumes that the production process changes the elastic properties of the reservoir and that seismic signals related to the reservoir geology are common to all of the surveys. The crossequalization and normalization flow is shown in Figure 9 and was the same for the PP and PS volumes. However, because the 2003 data were single-component P-wave data, we could not do the PS interpolation or difference the converted-wave data back to 2003.

PP Time-lapse Results

The first 4D map was produced using the 2008 high-resolution 3D and the 2003 interpolated data. The time-delay anomalies are shown in Figure 10 and they clearly indicate that the lower velocity/high-temperature bitumen sand is concentrated toward the toe of the producing wells and moving toward the heel. The thermocouples near the toes of the wells all indicate temperatures in excess of 600°C compared with the native state temperatures before combustion of approximately 12°C.

These results encouraged us to proceed with the 2009 survey. With the acquisition of the 2009 high-resolution survey, we were able to test the robustness of the interpolation and get our first look at the converted-wave (PS) time-lapse results. Figure 11 shows the 2009–2003 time-delay anomalies, and Figure 12 shows the difference of the

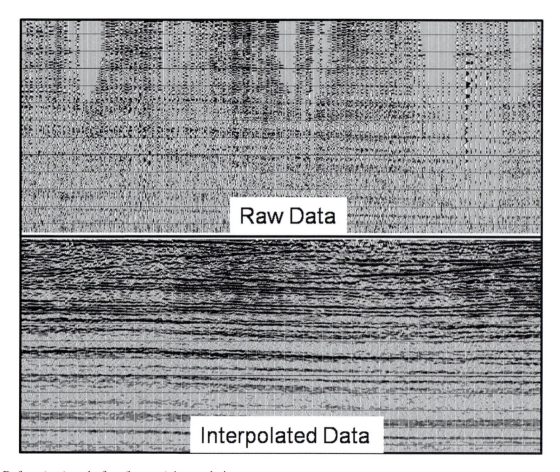

Figure 6. Before (top) and after (bottom) interpolation.

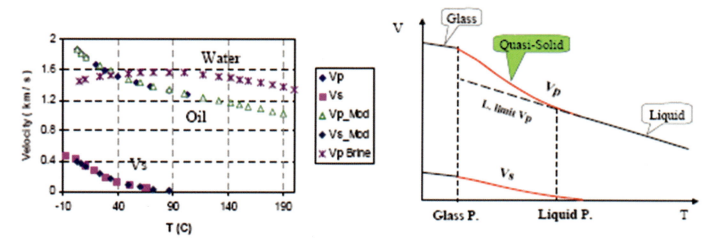

Figure 7. Measured and modeled velocities of heavy oil and water (left) and schematic of phase change of heavy oil with temperature. From Han et al., 2007.

2009–2003 and the 2008–2003, which is effectively the time-delay anomalies for the 2009–2008 period. The differences suggest that some parts of the reservoir were actually cooling during 2009. The areas of blue on Figure 12 indicate that the bitumen got cooler between 2008 and 2009 whereas the areas of white suggest no change in the temperature and the areas of red a further warming. Figure 13 shows the time-delay anomalies for the high-resolution 2009–2008 surveys. Clearly the two flows give similar answers yet the 2009–2008 results are much more detailed. Figure 14 shows the amplitude differences for the PP 2009–2008 data. Although we do not show the full suite of comparable amplitude anomalies, they do agree with the time-delay anomalies but in a less dramatic way.

Chapter 23: *Using Time-lapse Seismic to Monitor the THAI™ Heavy-oil Production Process* 281

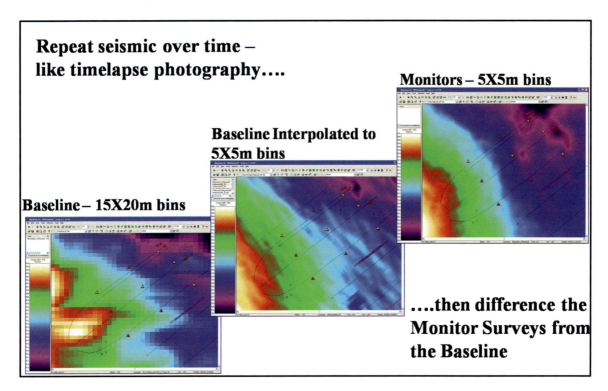

Figure 8. Time-lapse seismic is analogous to time-lapse photography. It can show subtle changes in the reservoir, while everything else remains unchanged.

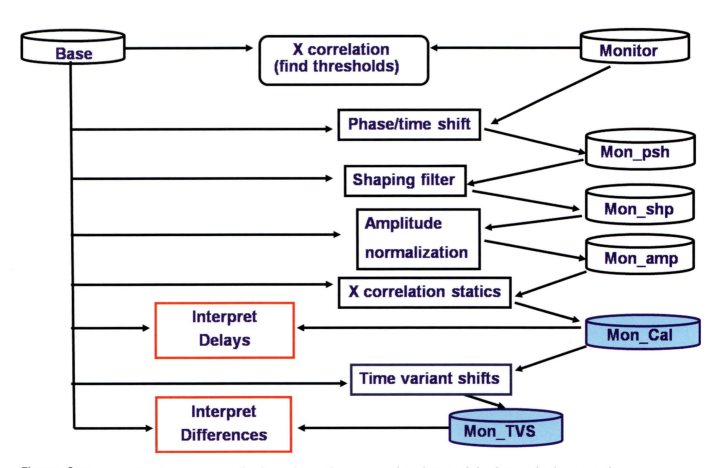

Figure 9. Crossequalization and normalization scheme that was used to detect subtle changes in the reservoir.

282 Heavy Oils: Reservoir Characterization and Production Monitoring

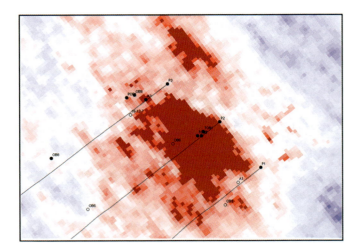

Figure 10. Results from the first time-lapse analysis (2008–2003) using an interpolated (15 × 20 m to 5 × 5 m) and regularized "baseline" survey and a high-resolution multicomponent "repeat" survey.

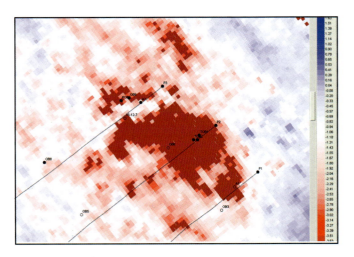

Figure 11. Time-delay anomalies using 2009–2003 data.

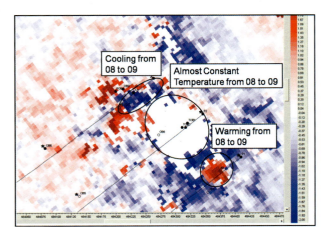

Figure 12. Difference of the 2009–2003 and 2008–2003 time-delay results. Note the areas of cooling in blue.

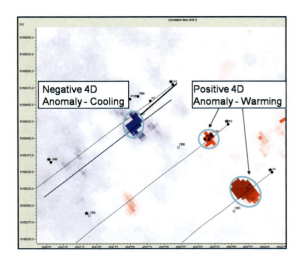

Figure 13. 2009–2008 time delay anomalies. The high resolution seismic data show greater detail but are still in good agreement with the results of Figure 12.

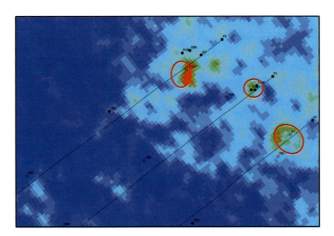

Figure 14. 2009–2008 PP amplitude anomalies with time-delay anomalies outlined in red. There is an excellent agreement between these two results, yet the time delay method appears to be more robust.

We monitor the combustion front using downhole thermocouple strings in adjacent observation wells, as well as in the production liners. Figure 15 and 16 show temperature contours in black superimposed on the seismic time delay anomalies. The downhole thermocouple recording can be used to calibrate the 4D results, while the seismic provides valuable information away from the well bore.

PS Time-lapse Results

Figure 17 shows the time-delay anomalies for the 2009–2008 PS data, and Figure 18 is the amplitude differences for the PS data. It is important to go back to

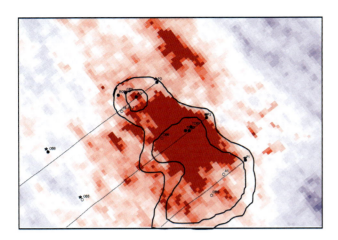

Figure 15. Temperature contours superimposed on the 2008–2003 time-delay anomalies.

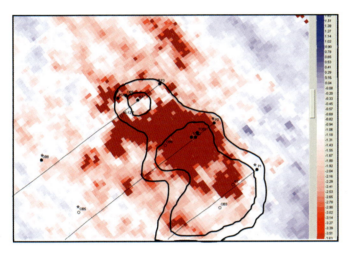

Figure 16. Temperature contours superimposed on the 2009–2003 time-delay anomalies.

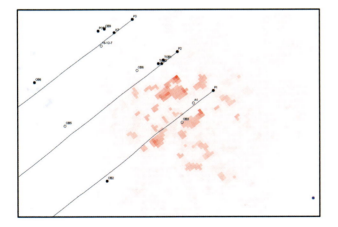

Figure 17. 2009–2008 PS time-delay anomalies.

Figure 7 to understand why we might not be seeing similar time-delay anomalies on the PP and PS. We know that the THAI combustion process can exceed temperatures of 600°C and that the temperature profiles show a

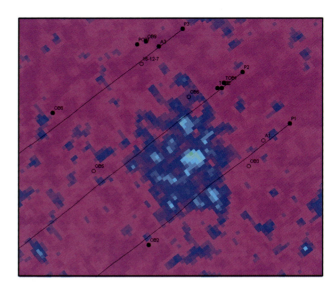

Figure 18. 2009–2008 PS amplitude anomalies. Note the good agreement with the PS time-delay anomalies.

focusing of the effectively warmed area. The pre-ignition-heating-cycle (PIHC) and the initial start-up conditions of the THAI process will increase the temperatures well beyond the point where the shear velocities of bitumen go to zero. If we were able to go back to the native-state temperatures (approximately 12°C) with the baseline PS, then it is conceivable to imagine a very sharp line coinciding with the subtle change in temperature from 12°C to 90°C that the PP data would not be as sensitive to. The critical temperature for THAI combustion to occur is at approximately 90°C.

Conclusions

The compressional and shear velocities of bitumen sand decrease dramatically with increasing temperature. The shear velocities and hence the PS data are most sensitive between 12°C and 90°C. The THAI process can be monitored using time-lapse seismic, and the results indicate that the combustion front is moving from the toe area of the producing wells toward the heel area.

Multidimensional interpolation can be used to dramatically reduce the bin size and hence increase the trace count of coarsely sampled 3Ds. The integrity of the interpolated data is sufficient for monitoring THAI combustion-style production using time-delay 4D methods. The implications for production monitoring of enhanced oil recovery are substantial using interpolation.

Acknowledgments

We thank Petrobank Energy and Resources Ltd., and Whitesands Insitu Partnership for allowing the presentation of these results and CGGVeritas for seismic acquisition and processing.

References

Han, D., J. Liu, and M. Batzle, 2006, Acoustic properties of heavy oil measured data: 76th Annual International Meeting, SEG, Expanded Abstracts, 1903–1907.

Han, D., H. Zhao, and M. Batzle, 2007, Velocity of heavy oil sand: 77th Annual International Meeting, SEG, Expanded Abstracts, 1619–1623.

Schmitt, D., 1999, Seismic attributes for monitoring of a shallow heated heavy oil reservoir: a case study: Geophysics, **64**, 368–377.

Trad, D., M. Hall, and M. Cotra, 2008, Merging surveys with multidimensional interpolation: 2008 CSPG-Canadian Society of Exploration Geophysicists-Canadian Well Logging Society Convention, Expanded Abstracts, 172–176.

Section 5

Geomechanical Aspects

Chapter 24

Geomechanics Effects in Thermal Processes for Heavy-oil Exploitation

Maurice B. Dusseault[1] and Patrick M. Collins[2]

Introduction

There is a great need for high-quality monitoring data from thermal enhanced oil recovery processes (THEOR; Figure 1) applied to heavy-oil reservoirs. We need to understand the physical mechanisms, to improve our attempts at mathematical simulation of these processes, and to provide information for "real-time" project optimization.

THEOR processes (e.g., steam injection) generate large changes in amplitude and traveltime in the complicated signature of a reservoir's seismic response. Conventionally, pressure, temperature, and saturation changes induced by thermal recovery are examined to study their complex and often counteracting effects on the seismic attributes of the reservoir and bounding rocks. The effects of geomechanical changes in stress and stiffness are less often considered, although they may dominate seismic property changes. Thus, seismic data interpretation to deconvolve in situ rock properties is nonunique and it varies with time as these physical parameters change with the maturity of the THEOR process. A fuller understanding of the effects of these changes will lead to better interpretation of changes of seismic attributes; hence, better project management.

Considerable enthusiasm exists for application of seismic monitoring techniques ranging from time-lapse vertical seismic profiles (VSP), X-hole tomography, and 4D seismic surveys to microseismic monitoring. Indeed, these methods are becoming integrated as we learn more and more about how to use spatiotemporally "random" microseismic events as sources and as we learn how to introduce various physical constraints into inversion methods.

Nevertheless, there are interpretation difficulties with using only seismic methods and conventional monitoring well data (e.g., p, T, near-well geophysical response).

This brief chapter is intended to point out some geomechanic issues arising in THEOR monitoring and to advocate incorporation of more monitoring dimensions.

Thermally Induced Stress Changes

In steam THEOR processes (Figure 1), temperatures, typically 200°, but as high as 350°C are feasible, giving a ΔT of as high as 325°C. In combustion processes such as Toe-to-Heel Air Injection (THAI^{TM}), temperatures can reach 500°C to 750°C in the combustion zone, depending on the amount of water vapor present (wet or dry combustion) and the rate and oxygen-richness of the injected oxidant. The question becomes just how large of a stress change could be expected from such a process.

Assuming uniform heating of a flat-lying tabular sandstone reservoir with a coefficient of thermal expansion of 10^{-5} °C^{-1} and a confined Young's modulus of 5 GPa, the increase in horizontal effective stress ($\Delta\sigma'_h$) can be estimated to be approximately 20 MPa for $\Delta T = 300$°C. As an example, consider a steam process at a depth of 500 m; the overburden stress and original hydrostatic pressure are approximately $\sigma_v = 11$ MPa and $p_o = 5$ MPa, respectively, giving an effective overburden stress of $\sigma'_v = 6$ MPa. This may not change much with the heating of a laterally extensive reservoir because the thermal expansion will be accommodated by surface heave. However, the large increase in σ'_h is enough to lead to generalized shearing of the reservoir, particularly because the high injection pressures reduce the confining stress. The resultant drop in formation stiffness may be as large as an order of magnitude.

Of course, the configuration of heated zones is complex, and induced stress distributions around heated zones

[1]University of Waterloo, Department of Earth and Environmental Sciences, Waterloo, Ontario, Canada
[2]Petroleum Geomechanics, Inc., Calgary, Alberta, Canada

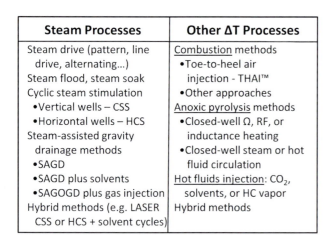

Figure 1. THEOR processess for heavy-oil extraction.

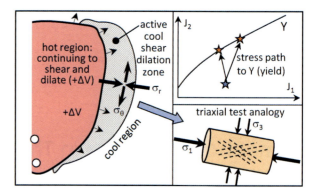

Figure 2. Stress changes in advance of a steam chamber (e.g., SAGD).

show massive increases or decreases in the local σ'_3 and σ'_1, depending on the geometry and the sharpness of the thermal front. If σ'_3 were reduced, as would be expected in any high-pressure injection process such as cyclic steam stimulation (CSS) or high-pressure steam-assisted gravity drainage (SAGD), the tendency to yield in shear would be greatly enhanced.

Mathematical modeling shows that yielding can be expected to be a consequence of any process that involves ΔT values greater than perhaps 80–100°C, even in materials such as oil sands that are less stiff than highly competent rock such as a carbonate. In softer rocks, stress changes are less than in stiff rocks, but the stresses needed for yield are far lower because soft rocks — unconsolidated sandstones — have much lower cohesive strength than stiff rocks such as competent carbonates.

Indeed, there is much independent evidence for shear. Among other phenomena, early attempts at microseismic monitoring of firefloods in 1981–1982 showed clear shear events (Nyland and Dusseault, 1983), casing shear is endemic in CSS projects, laboratory work confirms shearing in oil sands at stress states commensurate with those predicted by simulation, and ground surface uplifts related to shear dilation are regularly measured.

Shearing, Dilation, and Mechanical Damage of the Rock

Shear and dilation

THEOR induces shearing, which is the failure mode that occurs in rock when the anisotropic stresses in the rock exceed its frictional and cohesional strengths. This results in numerous shear zones within the rock along which sliding occurs. Because of the confinement conditions and the shape of the heated reservoir zones, shear within the reservoir does not occur only on isolated local planes. In unconsolidated formations, this sliding results in the translation, rotation, and displacement of individual sand grains that invariably results in a permanent disturbance of the original granular structure and an increase in bulk volume known as dilation. Shearing is prevalent in shallow unconsolidated reservoirs undergoing THEOR because of the high differential thermal stresses induced by the process and by the reduction in effective confining stress with high-pressure injection that reduces the frictional strength of the rock.

What are the consequences for geophysics?

As the thermal front propagates laterally into the reservoir (Figure 2), σ'_h is increased inside and outside of the thermal zone because of laterally constrained thermal stress within the growing thermal zone. At the same time, σ'_v is decreased outside of the shoulders of the thermal zone because of thermal jacking. Vertical extensional strains due to thermal expansion in the heated zone unload some of the stress within the cold reservoir. For these shallow reservoirs, not only does this make shearing inevitable, it means that any remnant stiffer (unyielded) rock will attract higher stresses by virtue of its higher stiffness and eventually be forced into a yielding condition.

Thus, in advance of the front, almost all of the rock shears at low confining stresses. In competent quartzose sands (i.e., most of the world's oil sands), grain crushing is minimal, and the shear dilation potential is great. For a typical Athabasca oil sand at a confining stress of 1–2 MPa, it is likely that the dilation would approach 5–6%; in other words, the porosity would increase from 30% to approximately 35% as shear distortion takes place (Figure 3). Much of this increase in porosity and permeability will occur within a network of induced shear planes within the oil sands. The consequences on transport properties are huge. Not only does the increase in absolute porosity result in an increase of absolute permeability by a factor of 2–10, in typical cases with high oil saturation the

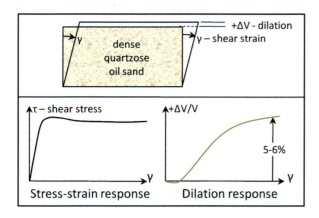

Figure 3. Shear dilation of an unconsolidated sandstone.

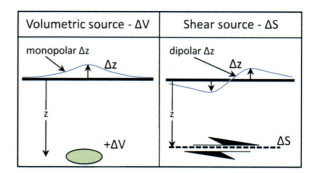

Figure 4. Deformation mode and surface response.

relative permeability to the mobile water phase is increased by orders of magnitude. Furthermore, shear will break up the thin clay dustings found on bedding planes and between bed sets; macroscopic flow impedance is thus greatly reduced. The porosity increase is filled with the most mobile phase, water, which increases the transport rate of pressure and fluids. The success of SAGD and CSS in bituminous sandstone reservoirs can in significant part be attributed to the effects of shear dilation that enhance the transport properties (Collins 2007a, 2007b).

Effective stress

Seismic characteristics will be affected significantly as an indirect result of the increase in relative permeability caused by imbibition due to dilation. The increase in relative permeability ahead of the thermal zone will propagate full injection pressures tens of meters ahead of the thermal zone into the cold reservoir. These elevated pressures will reduce the effective stress on the rock, within the cold reservoir, thus reducing the seismic transit times.

The changes in pressure and stress cause shear yield, as evidenced by the endemic and continual microseisms associated with THEOR processes, and the stiffness of yielded rock is far lower than for intact rock, drastically increasing the transit times and attenuation, resulting in marked differences in seismic characteristics even in areas beyond the thermal zones. Note that increases in gas saturation during injection are unlikely in these cold areas because the increase in pressure tends to drive the pore liquids toward undersaturated conditions. However, solution gas drive is expected to play a dominant role during drawdown.

Shear at the shale caprock

What else happens? The shales overlying a THEOR project are essentially impermeable in comparison to the reservoir. As thermal expansion and dilation take place, a huge shear stress concentration develops between the reservoir and the nonexpanding overlying shales, leading to shearing along the interface between the two, and possibly within very weak planes within the shale caprock. This shearing is not favorable because it is sufficiently large to collapse any penetrating well. Projects such as Imperial Oil Cold Lake CSS project experience dozens of well shear events yearly (Dusseault et al., 2001); these wells must be repaired or replaced at considerable cost. Methods of reducing the incidence and the impact of shearing have been suggested (e.g., horizontal wells or changes in the process sequence).

Reservoir Deformations

Deformations in the reservoir cause deformations at the surface. Horizontal deformations are generally small, but vertical deformations can be of the order of 300–600 mm. The shape of the surface deformation reflects the nature of the reservoir-level deformation. For example, a pure volumetric dilation source gives rise to a monopolar Δz distribution, whereas a shear distortion gives rise to a dipole distribution (Figure 4). For reasonably well-constrained cases (e.g., a single flat-lying reservoir at constant depth), the surface deformation field can be inverted to give a distribution of volumetric and shear distortions at the depth of the reservoir horizon. In practice, this has been used to help understand production mechanisms and even, in one unpublished case, to predict well shearing arising because of steam injection. Hence, another process-monitoring procedure is available that is complementary to seismic monitoring and provides a broad examination of the entire project as opposed to localized measurements at observation wells.

Cyclic injection THEOR projects as well as SAGD projects evidence large vertical surface deformations (Δz). In THEOR processes taking place at depths of 200–500 m, cumulative Δz values as large as 600 mm have been measured after a few years (Figure 5). In fact, for cyclic THEOR processes, Δz is economically important because recompaction is a dominant drive mechanism during late

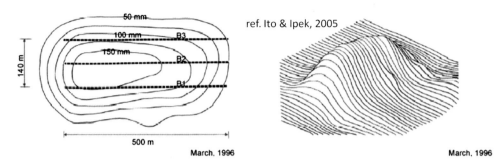

Figure 5. Surface uplift contours above a SAGD well group.

production cycles. This recompaction drive effect increases oil recovery rates and likely the ultimate recovery factor.

To what processes can a total vertical heave of 600 mm be attributed? Sandstones typically have a coefficient of thermal expansion of approximately 10^{-5} °C^{-1}; assuming only thermal expansion, a 200°C average increase in a 30-m reservoir should give Δz of approximately 60 mm, one-tenth of the measured amount. All of the rest of the permanent deformation can only be explained by dilation; there is no other viable explanation. Hence, if we can track volume changes in the reservoir by some indirect means such as deformation inversion coupled with seismic data analysis, we can also track important process parameters such as the homogeneity of the process, changes in permeability, and the distribution of swept zones. The implications for learning and process control are clearly quite large and economically interesting.

Impact of Geomechanics on Seismic Monitoring

What happens when a seismic wave passes through a sandstone and what changes might arise because of changes in the fabric, saturations, or intrinsic conditions? This turns out to be a remarkably complex process in unconsolidated sandstones subjected to THEOR for the following reasons, in increasing order of importance:

- Massive changes in saturation take place, with free gases replacing liquids in the pores.

- Density changes take place because of increasing gas saturations, but also because of dilation.

- Stress changes take place because of the thermally induced volume changes and any changes arising because of injection and production.

- Vital but less appreciated, reduction in rock stiffness occurs because of the reduction in effective confining stress with increased pressure, and the drastic loss of stiffness once the rock experiences yield and dilation.

These changes greatly affect all seismic parameters such as velocity, amplitude, frequency content, and birefringence, thus changing all calculated seismic attributes.

For example, it is common for the shear-wave amplitude to disappear in pressurized regions during the injection phase of CSS above fracture pressure because grain-to-grain contacts and effective stresses disappear. In strongly dilated zones, compressional wave amplitudes are reduced, high frequencies are totally filtered out (e.g., if a microseismic event is used as the source), and any initial horizontal shear-wave anisotropy is likely eliminated. However, during drawdown periods, effective stresses are reestablished, and shear-wave transmission becomes possible again, albeit in a rock material that is much softer than it was originally.

Unfortunately, there is no rigorous model that allows the relationships among seismic parameters and various changes to be quantified to a level that will permit deconvolution of the stress field or the dilation field, as examples. Also, there are severe challenges and limitations that arise in attempting to reduce uncertainty through careful laboratory tests. This is because of core disturbance in unconsolidated sands, scale effects, boundary conditions that are different in the field, difficulties in testing at elevated temperatures, and because of issues such as wavelength scale and dispersion.

Suggestions

We believe that careful reservoir and process monitoring pays for itself many times over in terms of enhanced understanding and process control. Also, if it were possible just to optimize injection and production rates, temperatures, and times based on independent monitoring data, recovery factors in the cyclic steam process would undoubtedly rise. The way forward requires several actions.

- Integrate geomechanics simulation explicitly with thermal-flow simulation, including the effects of dilation and stress changes (Dusseault 2008; Yin et al., 2009).

- Include deformation monitoring (Figure 6) and coupled inversions with seismic monitoring (Du et al., 2005; Ito and Ipek, 2005; Hohl et al., 2009; Maxwell et al., 2009). With time, the data from deformation measurements will help constrain and therefore refine seismic attribute extraction.

- Develop better seismothermomechanical models so that it becomes feasible to do forward modeling of seismic

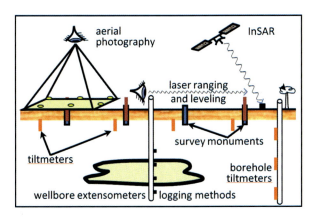

Figure 6. Some methods of measuring deformations.

parameter changes, which can then be verified in the field, calibrated, and used to give better constraints on seismic monitoring data inversion in THEOR cases.

- Consider incorporating other independent measurements that can have a synergetic benefit to data analyses. Some suggestions are:
  - extension of the precision of deformation inversion by including more precise deformation measurements at the surface and with depth in multipurpose observation wells
  - gravity field measurements through use of surface and borehole gravimeters
  - electrical impedance tomography using a three-dimensional array of electrodes
  - better downhole pressure measurements at high temperatures

Different disciplines in the geosciences intersect only irregularly. In geomechanics, we too often ignore the value that geophysics and flow simulation can bring. Conversely, we are strong in measurement methods that are not commonly used in petroleum engineering or geophysics (e.g., deformation measurements), and we understand aspects such as dilation and stress. Disciplinary integration has clear benefits to all parties, and the result will be better interpretation, greater efficiencies, and more value for the investment in monitoring.

References

Collins, P. M., 2007a, The false lucre of low-pressure SAGD: Journal of Canadian Petroleum Technology, **46,** 20–27.

Collins, P. M., 2007b, Geomechanical effects on the SAGD process. SPE Reservoir Evaluation and Engineering, 367–375.

Du, J., S. J. Brissenden, P. McGillivray, S. Bourne, P. Hofstra, E. J. Davis, W. H. Roadarmel, et al., 2005, Mapping reservoir volume changes during cyclic steam stimulation using tiltmeter-based surface deformation measurements: Proceedings of the SPE International Thermal Operations and Heavy Oil Symposium, Calgary, SPE/PS-CIM/CHOA 97848.

Dusseault, M. B., M. S. Bruno, and J. Barrera, 2001, Casing shear: causes, cases, cures: SPE Drilling and Completion Journal, **16,** 98–107.

Dusseault, M. B., 2008, Coupling geomechanics and transport in petroleum engineering: Keynote paper, Proceedings of the SHIRMS International Conference on Geomechanics.

Hohl, D. F., J. Lopez, R. Bos, and K. P. Maron, 2009, Reservoir surveillance technologies for thermal EOR projects: Proceedings of the 15th European Symposium on Improved Oil Recovery.

Ito, Y., and G. Ipek, 2005, Steam-fingering phenomenon during SAGD Process: Proceedings of the SPE International Thermal Operations and Heavy Oil Symposium, Calgary, SPE/PS-CIM/CHOA 97729.

Maxwell, S. C., J. Du, J. Shemeta, U. Zimmer, N. Boroumand, and L. G. Griffin, 2009, Monitoring SAGD steam injection using microseismicity and tiltmeters: SPE Reservoir Evaluation and Engineering Journal, **12,** 311–317.

Nyland, E., and M. B. Dusseault, 1983, Fireflood microseismic monitoring: results and potential for process control, Journal of Canadian Petroleum Technology **22,** 62–68.

Yin, S., B. F. Towler, M. B. Dusseault, and L. Rothenburg, 2009, Numerical experiments on oil sands shear dilation and permeability enhancement in a multiphase thermo-poroelastoplasticity framework: Journal of Petroleum Science and Engineering, **69,** 219–226.

Chapter 25

Passive Seismic and Surface Monitoring of Geomechanical Deformation Associated with Steam Injection

Shawn C. Maxwell,[1] Jing Du,[2] and Julie Shemeta[3]

Introduction

Monitoring injections is important for optimal reservoir management of any enhanced oil recovery. Monitoring is of particular importance in the case of steam injections, which often have small economic margins and where significant material property changes take place. Imaging can be used to optimize the injection, make informed reservoir engineering decisions, and help understand the mechanics of the injection.

Monitoring the steam chamber growth is critical in optimizing heavy-oil recovery; it ensures that the stimulation is confined to the reservoir and helps identify bypassed regions. Steam injection results in geomechanical strains associated with increased pore pressure, thermal stress changes, and dramatic changes in material properties associated with heating the reservoir sufficiently to mobilize the heavy oil/bitumen. This geomechanical deformation may be marked by seismic deformation and the release of seismic energy as fractures adjust to the strain field. Deformation may also be evident in surface expansion or subsidence. Monitoring the microseismic activity with sensitive seismometers and surface deformation with precise tiltmeters could allow the steam injection to be tracked with complementary technologies that respond to different expressions of the geomechanical deformation.

In some fields, this geomechanical deformation also leads to casing deformations and well integrity problems, which may result in operational problems. The combined monitoring of passive seismic and surface deformation provides insight into the mechanisms that lead to casing deformations and will also potentially identify the circumstances that may lead to casing failures. The combined monitoring can also track fluid movements in the reservoir, allowing optimum well and pattern design and subsequent operational improvements, including optimization of steam volumes, rates, and cycle timing. Finally, the passive seismic and surface deformation monitoring can be used to track unwanted steam breakouts. Thus, during steam injection, the combination of passive seismic and surface deformation monitoring offers critical information for several reservoir engineering and management issues.

Many steam injections are at a relatively low injection pressure, which may be below the "frac" pressure required to create tensile hydraulic fractures. Nevertheless, fracture activation may still occur as increased pore pressures and stresses induce shear movement along pre-existing fractures. Previous studies have reported microseismic activity and surface deformations for cyclic steam injections (CSS). Steam-assisted gravity drainage (SAGD) typically uses lower injection pressures and rates compared with CSS and results in less seismic and surface deformation. Surface deformation can be monitored with various techniques, including InSAR and GPS monuments, although tiltmeters offer the highest precision for monitoring small deformation changes. With SAGD applications, the slow injections of relatively small steam volumes point to the use of the most sensitive surface deformation measurements.

Tiltmeters are an interesting geophysical technology. Although the technology arguably sits in the "gray area" between geophysics and engineering, reservoir engineers have utilized the technology for several years to map hydraulic fractures and monitor surface and borehole deformations (e.g., Davis et al., 2000). Tiltmeters are extensively used in geotechnical projects for applications including slope stability. They are also often combined with seismometers to monitor volcanic eruptions. A key feature of tiltmeters is the ability to monitor small

---
[1]Formerly at Pinnacle Technologies, Calgary; now at Schlumberger, Calgary, Canada
[2]Pinnacle Technologies, Houston, Texas, U.S.A.
[3]Formerly at Pinnacle Technologies, Calgary; now at MEQ Geo, Denver, Colorado, U.S.A.
This paper appeared in the September 2008 issue of THE LEADING EDGE and has been edited for inclusion in this volume.

Figure 1. Tiltmeter measurement of the solid earth tide.

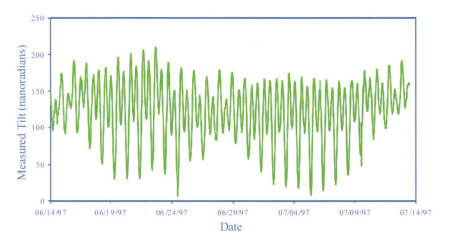

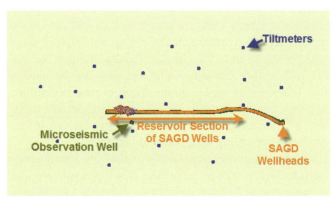

Figure 2. Map view showing the locations of the SAGD wells, microseismic observation well, microseismic events, and surface tiltmeters. The borehole is approximately 1000 m long through the reservoir.

deformations by measuring the gradient or tilt of the deformation, which is significantly more precise than the absolute deformation itself. The heart of the tiltmeter is a glass tube containing an air bubble in a conductive liquid, and when the bubble moves the electrical resistance changes (see SEG's *Encyclopedic Dictionary of Applied Geophysics* for further description). To understand the sensitivity, consider the tiltmeter's measurement of one of the sources of noise associated with the solid earth tide. These tidal strains are very small in amplitude but can be routinely observed in tiltmeter data (Figure 1) and they are also easily filtered out. Another key feature of a tiltmeter is its frequency bandwidth, which responds to deformations with time scales ranging from a few seconds (including passage of seismic waves) down to months and years. This allows measurement of uplift over various time scales.

In this chapter, we present a case study of monitoring a steam injection with passive microseismic and surface tiltmeter deformation. We first describe the site and then the results of monitoring a "warm-up" phase of an SAGD well pair in Western Canada. The microseismic and tiltmeter results are then combined to provide an integrated interpretation of the geomechanical response of the system.

Case Study

Combined microseismic and surface tiltmeter mapping was used to monitor the initial steam injection corresponding to the warm-up phase of an SAGD well pair. The wells were at a nominal depth of approximately 500 m with the horizontal sections extending approximately 1000 m (Figure 2). An existing vertical well, approximately 60 m from the toe of the lateral and 175 m offset, was used for the microseismic monitoring during approximately six weeks of steam injection. An array of 20 surface tiltmeters was also deployed to monitor the surface deformation before and after the microseismic monitoring period.

The microseismic system used a retrievable fiber-optic-wireline-based array, consisting of eight triaxial geophones spaced at 10 m. An earthquake detection algorithm was used to detect or "trigger" seismograms with coherent energy on a preset number of channels. These triggers were transmitted via satellite Internet connection for offsite processing. Noise recording and trigger attributes were also stored and transmitted.

The tiltmeter array consisted of 20 surface tiltmeters in shallow (7 m) boreholes to avoid surface noise and thermal effects. Each tiltmeter was solar powered with radio telemetry, using data transmission integrated with the microseismic communication system. Each tiltmeter consists of a pair of orthogonal bubble levels with a precise curvature capable of resolving tilt as small as 1 billionth of a radian (or 0.00000005°). A solid-state magnetic compass provides the tool orientation so tilt direction can be determined.

Microseismic Results

Over 2000 microseismic events were recorded during the six-week recording period. Many were concentrated in specific regions around the injection wells. The moment magnitudes ranged from −4.2 to −3.6. Extremely low levels of background seismic noise had to be achieved to detect the microseisms. However, only events within a few hundred meters of the geophone array have a sufficient signal-to-noise ratio to be recorded.

Figure 3 shows a vertical section view of the microseismic locations. Note that the events are clustered near the toe of the well but are concentrated toward one side of the observation well. No activity was recorded on the other side of the well; however, if the microseisms had occurred over a distance equivalent to the seismically active zone, the detection sensitivity would be adequate to record them from this region. The lack of seismicity from this region will be discussed later. Because the detection range is limited to a few hundred meters because of the small event magnitude, it is believed that microseisms may be occurring along the borehole but simply are not detected from greater offsets. For example, seismicity is probably occurring at the heel of the well, but it is too far away from the observation well.

Two colors are used in Figure 3 for the microseisms. Each relates to a different type of event with distinct signal characteristics. Figure 4 shows representative seismograms of the two event types. Notice that type I (blue events in Figure 3) is characterized by relatively small amplitude P-wave relative to the S-wave amplitude as compared with the type II (red events in Figure 3). Furthermore, type II events have a relatively larger shear wave on the vertical geophone.

Several high-quality seismograms were selected, and the ratio of the amplitudes of the P-waves to the horizontal shear (p/sh) was measured along with the ratio of horizontal-to-vertical S-wave amplitudes (sh/sv) for the two classes of events (Figure 5). For these plots, events in a small region containing both types of events were selected, and only the arrivals on one sensor at the hypocentral depth of the events are included. Thus, there are no variations in the source-receiver geometry, and the differences in the signal attributes indicate a distinct radiation pattern difference between the two types of events. One possible interpretation is that two different fracture surfaces are being activated by the steam injection. However, the first cluster (blue symbols in Figure 3) are believed to be related to fracture movement in the reservoir from the steam injection, whereas the second cluster (red events in Figure 3) may be associated with a different source type.

The radiation pattern characteristics in Figure 5 are also consistent with the theoretical radiation pattern of a casing failure, which is distinct from the radiation pattern of shear failure in the reservoir. Although the energy release of these events is much smaller than the signals expected for a complete parting of the casing, the events could be a result of deformation of the borehole completion associated with the thermal expansion of the metal liner. It is interesting to note that Imperial Oil uses these same signal characteristics to detect casing failure events

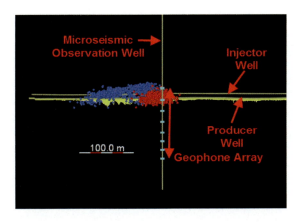

Figure 3. Cross-sectional view of the toe of the wells. The lower vertical well displays a gamma-ray log. The vertical observation well displays the geophone array. The microseismic events are represented by the blue and red symbols.

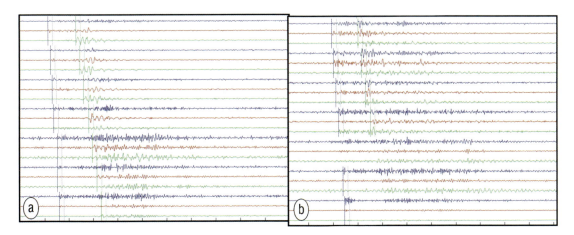

Figure 4. (a) Typical seismogram for the type 1 events. Shallowest geophone on top and blue, red, and green are the vertical, and two horizontal geophones. (b) Typical seismogram for the type II events.

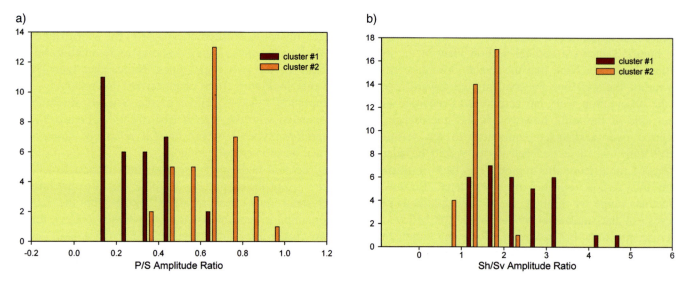

Figure 5. (a) Histogram of the amplitude ratio of P- to S-waves for the two classifications of events. (b) Histogram of the amplitude ratio of the horizontal-to-vertical shear waves.

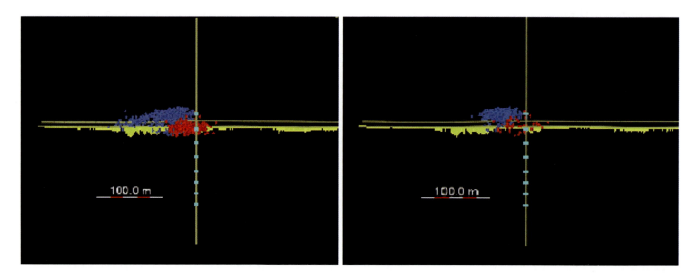

Figure 6. Locations of microseisms recorded in the first 40 h of recording (left) and another 40-h period after 10 days of steam injection (right). See Figure 2 for further details.

at Cold Lake (Smith et al., 2006). However, at Cold Lake, a complete casing failure, which was a consequence of the tensile parting of a casing joint, causes a signal of substantially greater energy. Otherwise these small-magnitude events are consistent with the deformation mechanism of the casing failure and are believed to represent small deformations resulting from thermal expansion as the uncemented slotted liner in the well expands. Through the rest of the chapter, these will be referred to as "completion deformation events" to distinguish from those believed related to fracture movement in the reservoir, which will be referred to as "reservoir deformation events."

In Figure 3, these completion deformation events are highly localized in a region where both types of events initiate, although the reservoir deformation activity temporally migrates along the borehole. Figure 6 illustrates the temporal variation in location, showing the events recorded during the first 40-h period, and another 40-h period starting 10 days into the monitoring period. Between these particular time periods, the reservoir deformation events migrate between the regions defined by these clusters. In contrast, the completion deformation events remain essentially in the same region.

Steam enters the toe region of the well via injection through tubing that ends near the toe. A deformation nucleation point occurs at some distance from the actual steam entry point, which suggests a zone of weakness that is exploited by the steam. The common nucleation may suggest that the mechanism of one of the deformations may be providing a nucleation point for the other. To speculate, the completion deformation may indicate a liner flaw that allows easier steam penetration into the reservoir and initiates the reservoir deformation in the same region. However, over time, the temporal migration of the reservoir deformation events moves toward the toe of the wells, which suggests that the prolonged steam injection causes the steam to continue to enter the reservoir from this nucleation point back toward the steam entry point (at the end of tubing near the toe of the well). Further details of the microseismicity will be discussed in the context of an integrated interpretation with the tiltmeter data.

Tiltmeter Results

Figure 7 shows the surface deformation integrated from the tilt data measured during the six weeks of microseismic monitoring. There are two discrete regions of significant surface uplift: one near the toe of the well and another with larger uplift (approximately 15 mm) near the heel. In addition to the steam injection, through tubing that enters the toe portion of the well (approximately 10 mm of uplift), steam is also injected into the annulus outside of the tubing. The region of uplift near the heel roughly corresponds to the location of the casing shoe where steam could enter the reservoir. There is also a region of subsidence in the upper portion of the tilt image, presumably related to a net fluid withdrawal in the region from other neighboring well pairs (these additional well pairs are not shown). However, over the SAGD well pairs examined in this study, essentially no relative elevation change was observed over the central section of the well.

The observed tilt data were inverted to deduce the volumetric reservoir strain associated with the uplift. In the inversion process, the reservoir region along the borehole was chosen as the inversion domain for two reasons. First, this monitoring period is the warm-up phase of the SAGD project, and the stream is not expected to migrate laterally away from the well to any significant extent. Second, the specific monitoring objective of this project is to focus on this specific well pair. Limiting the inversion domain helps us focus the objective by eliminating any noises in the data (deformation data which are not associated with the process monitored) coming from other wells. Figure 8 shows the resulting strain down the length of the boreholes. Notice that strain is confined to the extremities of the boreholes, as expected from the distribution of the surface uplift. Also note the distribution of the microseismic events near the strained region at the toe of the well (shown in more detail in Figure 9). Microseismicity also may occur at the heel of the well, but the observation well is too far away to detect it because of the small event size and seismic attenuation. However, both data sets suggest that the warm-up phase has not provided uniform steam coverage along the length of the well pairs. Although this limited conformance may not hold true during the actual production SAGD phase, this case study highlights the potential to track conformance, which enables the injection design to be modified for complete conformance.

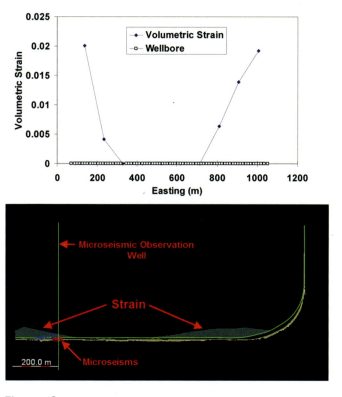

Figure 8. Inverted stain along the wellbore (top). Cross-sectional view of the entire SAGD borehole (bottom), showing strain anomalies (dark green) at toe and heel of the well.

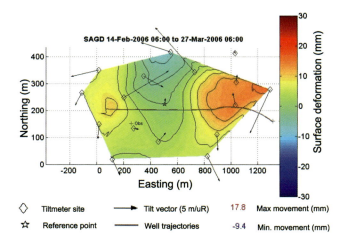

Figure 7. Map view of tiltmeters and observed tilt over the six-week period, contours of a best-fit deformation surface, and the SAGD well pairs.

Discussion and Conclusions

To compare the seismic and surface deformations, the total cumulative seismic moment (logarithm of seismic moment is proportional to moment magnitude) was computed. Figure 10 shows the time line of the injection details (rate, pressure, and cumulative volume) for the steam injected through the tubing into the toe of the well, along with cumulative seismic moment for each of the two event types and the surface uplift over the toe region of the well (corresponding to the red arrow in Figure 11). The cumulative moment for the completion deformation (cluster 2) shows an approximate constant increased rate up to about 27 February 2006, after which the rate decreases. This effect could be related to the thermal expansion of the liner slowing as the temperature equilibrates with the steam.

For cluster 1, the moment increases rapidly for the first few days and then slows until about 25 February when it again accelerates, until after approximately 27 February when the rate of increase of moment slows significantly. Notice that a few days after both time periods of rapid moment release, there is a period of rapid surface uplift. One possible interpretation is that the period of rapid reservoir seismic deformation corresponds to the development of a fracture network, which is then permeated with steam, resulting in a sudden advance of the steam into the previously unheated reservoir. This steam could then heat a relatively large region of the reservoir over the next few days, resulting in a surface uplift response that corresponds to the delay associated with the steam entering the fractures, consequently heating the surrounding reservoir.

This hypothesis points to an important aspect of the induced seismicity. We presume the seismic deformation in the reservoir is associated with stress- or pressure-induced shear failure on preexisting fractures because the injection pressures are believed below frac pressure. The induced fracturing, whether associated with activation of preexisting fractures or creation of new fractures, could then provide a stimulated region of enhanced permeability. If the seismic deformation corresponds to new fractures, this would enhance the permeability over the unfractured state. Alternatively, if the seismicity is associated with the shear deformation of preexisting fractures, the shearing would likely result in fracture dilation associated with shear offset of mismatched opposite sides of the fracture surfaces. Regardless of the mechanism, the occurrence of the seismic deformation suggests enhanced permeability and that the steam chamber will advance more rapidly. This appears to be supported by the subsequent rapid uplift observed in this region with the tilt array.

To further examine the stress changes associated with the steam injection, elastic stress changes were computed from the volumetric strain. Total effective stresses were computed based on a tensor summation of uniform in situ stresses and incremental stress changes from the volumetric strain. The following properties were used (Du et al.,

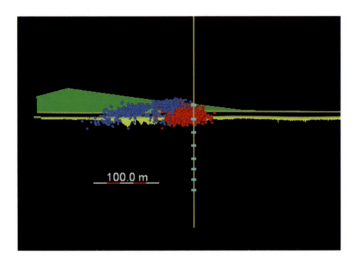

Figure 9. Zoomed in cross-sectional view of the strain and microseismic events at the toe of the well pair.

Figure 10. Time line of the injection data, total recorded seismic moment/deformation, and surface uplift over the toe of the well.

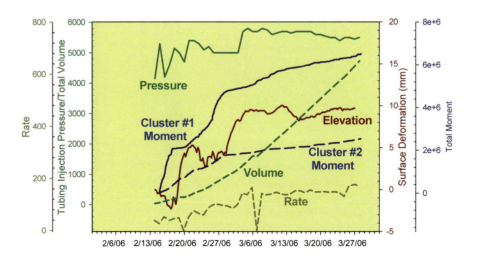

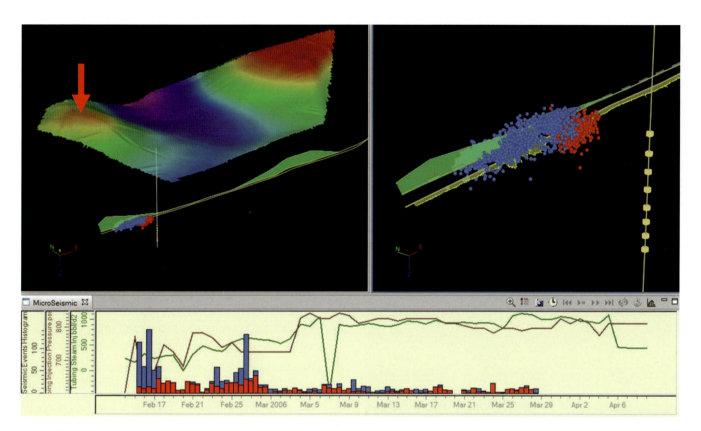

Figure 11. Illustration of exaggerated surface elevation change, volumetric strain, and mircoseismic data. Event histogram and injection data are displayed at the bottom.

2007). The minimum horizontal in situ stress gradient of 20.4 kPa/m and its orientation of S49E are assumed to be the average values from the published data at a similar location and depth. The vertical in situ stress gradient is estimated to be 22.54 kPa/m, using mean density of 2300 kg/m^3. For the depth of interest, the vertical stress is likely to be the intermediate principal stress and close to the maximum horizontal stress. Because of the lack of maximum horizontal stress data, it is taken to be 50 psi (0.344 kPa) larger than the vertical stress. Pore pressure is assumed to be hydrostatic, and a Young's modulus of 800 MPa was used.

Figure 12 shows contours of the calculated dimensionless normal stress, and Figure 13 shows the dimensionless shear stress. Note that two discrete stress anomalies are associated with the individual strain regions near the toe and heel of the well. Also note that the shear-stress anomaly extends further above and below the borehole, compared with the more contained normal stress. To examine the effect of this on potential shear failures, a Mohr–Coulomb stability calculation was performed. A random grid of points was generated, and at each the Mohr–Coulomb failure criteria were examined, assuming no cohesion and a friction coefficient of 0.5. Figure 14 shows the resulting points, which were calculated to experience shear failure. Note that the objective

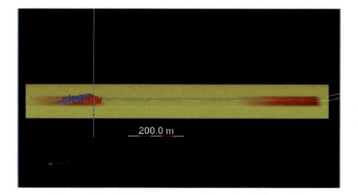

Figure 12. Dimensionless normal stress on a scale from yellow (0.0075) to red (0.025).

here is to examine the location of the microseisms and not the number, and the locations predicted to be experiencing shear failure are in good general agreement with the spatial location of the observed microseisms. Similar agreement was found with a sensitivity test varying the maximum horizontal stress. Therefore the microseisms are consistent with induced shear deformations associated with stress changes associated with the steam injection. However, pore-pressure changes are also likely a factor in the microseisms, and further effort is focused on examining the detailed relationship between the steam chamber

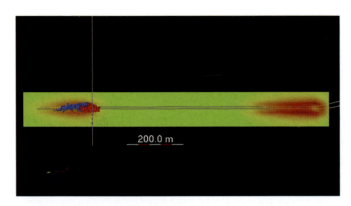

Figure 13. Dimensionless shear stress on a scale from yellow (0.0015) to red (0.007).

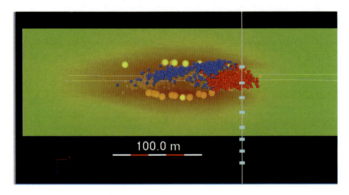

Figure 14. Zoom-in view of the toe of the well with shear stress, observed microseisms, and the location of estimated shear failure (large orange/yellow symbols).

and associated thermal and pore-pressure changes and the microseismicity and shear stability.

The integration of microseismic events and strain can also potentially be used for the calibration of a geomechanically linked reservoir simulator. The deformation associated with the steam heating can result in alterations to the permeability and porosity. Therefore, incorporating deformation analysis into a flow simulator is critical for interpreting the reservoir performance. The volumetric strain could also be modified by including the microseismic locations to constrain the strain by matching the observed points of shear failure with the passive seismic deformation. The inverted strain and observed failure regions can be used to constrain the deformation model results and, along with conventional temperature and pressure measurements, to validate the simulator.

In summary, the case study in this paper demonstrates:

- Significant seismic deformation can be recorded during an SAGD steam injection; however, a highly sensitive passive seismic array is required to record the small magnitude activity.
- Two distinct types of microseismic events were observed and are believed to correspond with seismic-deformation-associated thermal expansion of the uncemented liner and induced fracturing around the steam chamber.
- Surface tiltmeters recorded substantial uplift at the toe and heel of the SAGD well.
- The surface deformation was inverted for the variation in volumetric strain, which was constrained to the region around the well and showed strain increased at the toe and heel of the well.
- The microseismicity and volumetric strain indicated that steam was not uniformly distributed through the well.
- Periods of increased seismic deformation were found to precede periods of rapid surface uplift by a few days.
- Geomechanical analysis indicated stressed areas near the strained regions at the toe and heel of the well.
- Shear stability based on Mohr–Coulomb failure indicates increased shear stress shear instabilities in the region of the microseismicity.
- Integration of volumetric strain inverted from tiltmeter data and microseismic fracturing regions can potentially be used to calibrate a geomechanically linked reservoir simulator.

References

Davis, E., C. Wright, S. Demetrius, J. Choi, G. Craley, 2000, Precise tiltmeter subsidence monitoring enhances reservoir management: SPE Paper 62577.

Du, J., S. C. Maxwell, and N. R. Warpinski, 2007, Fluid production and injection-induced stress changes using reservoir volume changes inverted from tiltmeter-based surface deformation measurements: SPE Paper 110832.

Smith, R. J., C. M. Keith, and J. R. Bailey, 2006, Passive seismic monitoring for casing integrity at Cold Lake, Alberta: EAGE Workshop on Passive Seismic.

Chapter 26

Using Multitransient Electromagnetic Surveys to Characterize Oil Sands and Monitor Steam-assisted Gravity Drainage

Folke Engelmark[1]

Introduction

Heavy oil and bitumen constitute the largest easily accessible remaining hydrocarbon deposits in the world, with the greatest potential resources found in Canada, Venezuela, and the former Soviet Union. There are many ways to produce these assets, starting with surface mining of the shallow oil sands to various in situ recovery methods.

Heavy oil can sometimes be produced cold, whereas bitumen requires heating or injection of solvents to be mobilized. Cold production of heavy oil may or may not be successfully monitored by electromagnetic surveys (EM), depending on the recovery technique. Slow drainage by pumping the oil is no different from other oil production in terms of EM and should hence be successful, but for the process known as cold heavy-oil production with sand (CHOPS), it is unclear whether this can be successfully monitored by EM because the recovery rate is only 10% and most of the oil and sand is produced from so-called "wormholes," which affect only restricted parts of the reservoir.

The preferred methods of thermal in situ recovery are cyclic steam stimulation (CSS), also known as "huff and puff," and steam-assisted gravity drainage (SAGD). CSS is better suited to reservoirs that are fairly thin and have good horizontal permeability in which the steam spreads primarily laterally within the reservoir. The reservoir is first injected with steam for a period of time, followed by soaking, followed by a production phase in which the mobilized oil is produced by the same wells in which the steam was injected. When production declines, the cycle starts all over again. The recovery rate for CSS is approximately 25%.

SAGD is better suited to thicker reservoirs with good vertical permeability in which a balloon-shaped steam-filled chamber is created by injecting steam along a horizontal well bore placed low in the oil-charged reservoir sand, as shown in Figure 1. The mobilized oil and water are then recovered by gravity drainage in a production well bore running parallel with and directly below the injection bore 5 m deeper down. The two parallel well bores typically run horizontally for 600–1000 m. The steam rises through buoyancy and the dynamically changing shape of the steam chamber is governed by the local geology. The horizontal permeability in the homogeneous rock is likely to be a factor five larger than the vertical permeability for the homogeneous reservoir sand. However, thin tight streaks that are more or less horizontal can aid in developing the width of the chamber by temporarily inhibiting the vertical migration of steam, whereas dipping tight streaks can severely deform the symmetry of the steam chamber and leave large volumes of the oil-charged sands untouched. It is one of the primary tasks of a monitoring technology to outline the shape of the steam chamber. The recovery rate for SAGD can exceed 60%.

It is important to characterize the spatial distribution of oil-charged sands in place to find the optimal locations for the well tracks, but ideally we would like to also know the reservoir quality in terms of storage capacity (porosity), connectivity (permeability), and hydrocarbon saturation S_{hc}. Armed with this knowledge, we can estimate the hydrocarbon pore volume (HCPV; defined as the product of effective porosity Φ_{eff} and hydrocarbon saturation S_{hc}), recoverable reserves, and grade high the areas subjected to initial development.

Direct hydrocarbon detection of heavy oil by means of seismic is not possible because of the small contrast in the bulk modulus of water and oils at similar density, corresponding to an American Petroleum Institute (API) gravity of approximately 8–10°. The distribution of sand and shale may be possible to map on the basis of consistent differences in acoustic impedance, and this is of considerable value especially when combined with other information.

The optimum technology for mapping subsurface resistors by means of electromagnetic data requires a

[1]Petroleum Geo-Services, Houston, Texas, U.S.A.

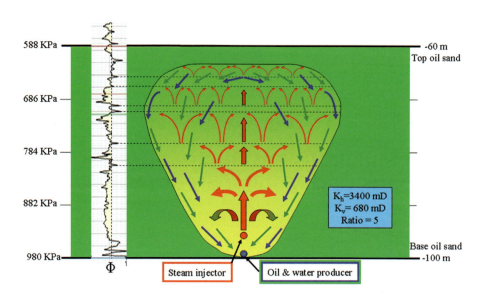

Figure 1. Vertical cross section through an SAGD chamber perpendicular to the horizontal well tracks. The steam injector and producer wells are shown at the bottom with the producer approximately 5 m below the injector. Rising steam is shown as red arrows with gravity draining of mobilized oil as green arrows and water condensation shown in blue arrows. The porosity column to the left shows there are a series of tight streaks (dashed lines) that temporarily inhibit the rise of steam and broaden the steam chamber, allowing a larger subsurface volume to be drained.

galvanic dipole source in which current is injected directly into the ground and an array of receivers are measuring the voltage inline with the dipole. Multitransient EM (MTEM) is such a technology that implements a pseudorandom binary sequence (PRBS) as source signal. The received voltage is deconvolved for the measured input current to recover the impulse response of the subsurface with the highest signal-to-noise ratio (S/N) possible given a particular source strength and acquisition time. Further benefits are gained in field efficiency by emulating 2D seismic acquisition with a line of 40 receivers and the required S/N at target depth can be achieved quickly with patented removal of cultural noise (50/60 Hz) from the data in the field. The resistivity-depth profile is then recovered by inversion in which the impulse responses from forward-modeled resistivity profiles are matched to the recorded impulse responses.

Electromagnetic data can provide the complementary information by mapping the transverse resistance (product of resistivity and thickness) of the oil-charged units. On the basis of a reasonable thickness estimate of the charged zone, the spatial distribution of HCPV can be estimated from EM data.

There is a need to monitor cold and thermal production of heavy oil. Cold production will not render any detectable difference on seismic, whereas the resistivity will change according to changes in water saturation (S_w).

In thermal production by means of steam injection, there are primarily two things we would like to know: Where in the subsurface is the steam migrating and from where is the produced oil mobilized? The live steam and evolved hydrocarbon gas lowers the bulk modulus of the rock, making the affected volumes stand out in a seismic monitoring survey as areas where the acoustic impedance has decreased. On the other hand, electromagnetic methods are sensitive to changes in temperature, S_w, and the salinity of the pore water.

Steam Injection from the Pore Perspective

If we could view a single pore space at the time of steam entry, we would observe the following, assuming some nominal numbers for in situ saturations, as shown in Figure 2:

A nominal pore space has an in situ irreducible water saturation $S_{w\text{-irr}} = 0.15$ and an oil saturation $S_{oil} = 0.85$.

Early in the SAGD cycle when the steam chamber is small, the rate of expansion of the chamber is faster than the heat flow through the rock. The required entry pressure for steam will then be very high because the relative permeabilities for mixed fluids are proportional to the abundance of each particular phase as seen in Figure 3, and with no gas phase present in the pore, the required entry pressure for steam will be very high. This also means that once the pore throat is breached, the pore fills up almost instantaneously by displacing oil until the residual oil level is reached at 0.15 to render a steam saturation of 0.7. As the steam chamber expands over time, the heat flow will exceed the rate of steam expansion. The fact that gas is evolving from the heated oil makes the steam entry easier in the peripheral areas because there is already a gas phase present in the pore space by the time the steam front is close enough to enter. The relative permeability for gas/steam is now much higher easing the steam expansion. This also means that steam will preferentially enter gassy oil-charged pores rather than 100% water-charged pores in which the steam is lost without any production benefit. After steam entry, and because gases

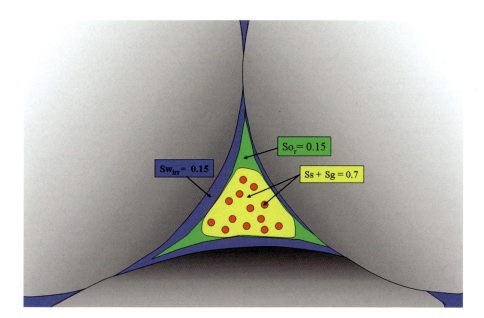

Figure 2. The fluid-filled pore space shown between three clastic grains. The irreducible water saturation is 0.15, and the in situ oil saturation is 0.85. After steam invasion and evolution of hydrocarbon gas, oil is displaced until residual oil remains at a saturation of 0.15. The combined saturation of steam and gas is 0.7. The steam may condense to water, but the gas will remain stable.

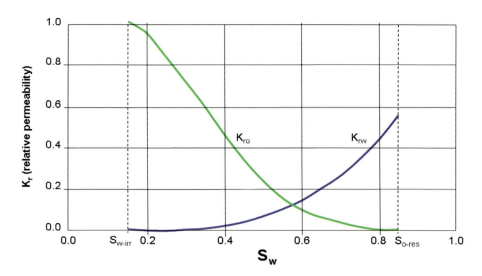

Figure 3. Relative permeability trends for water and oil. A fundamental rule is that once a fluid has been introduced into the pore space, it cannot be completely drained either. Any fluid that has a low saturation will also have a very low relative permeability and hence eventually stop flowing altogether. Irreducible water $S_{w\text{-irr}}$ and residual oil saturation $S_{o\text{-res}}$ are the two end points of the permeability trends.

mix perfectly well, there will actually be a gas phase in the pore space consisting of a mixture of steam and hydrocarbon gas adding up to a saturation of 0.7. Also important to remember is that steam is as nonconductive as oil and hydrocarbon gas. The conductivity of the rock originates exclusively from the more or less saline water in the pore space.

The steam condensates after the initial entry into the cooler pore space, adding a tiny amount of distilled water and also allowing oil to reflux into the pore space, where the amount of reflux depends on the remaining hydrocarbon gas saturation. A particular pore space is likely to go through several cycles of steam charge→condensation→steam charge before the rock is heated sufficiently to support a permanent steam charge. The distilled water added by the steam condensation increases water saturation but lowers the salinity of the pore water. The evolved hydrocarbon gas remains stable but will become mobilized and migrate upward when the gas saturation locally exceeds approximately 0.4, and a gas cap may build up at the top of the chamber.

Recovery rates of 70% have been reached in long-running SAGD processes. However, because the steam is generated with natural gas as fuel, the optimal recovery rate depends on the value of the extracted bitumen and the cost of natural gas. Hence, the optimum economy is currently reached at approximately 55% recovery rate. Evolved gas assists in reaching target recovery because it helps displace the mobilized oil out of the pore space and inhibits reflux into that pore space, allowing the oil to gravity drain down to the production well at a faster rate than would be the case if only steam and condensation water displaced the oil.

Resistivity Monitoring of Thermal Recovery

Cold production of heavy oil is no different to monitor by means of EM than production of light oil, gas, water injection into hydrocarbon, or carbon dioxide (CO_2) sequestration. The only thing that affects resistivity in a given reservoir is S_w as seen in Figure 4, where the resistivity index, defined as R_t/R_0, is shown as a function of saturation for a clay-free clastic rock, where R_t is the true bulk resistivity at any saturation and R_0 is the resistivity for the 100% water-saturated rock. The presence of clay in the rock will push the trend lower down, and if the rock matrix is partially oil wet, as is often the case in carbonates, the trend can become much steeper.

However, thermal recovery such as SAGD is very different to monitor, and by rearranging the Archie equation to express the bulk resistivity R_t in the other parameters, we obtain

$$R_t = \frac{a \cdot R_w}{\Phi^m \cdot S_w^n}, \qquad (1)$$

where a = tortuosity, R_w = the resistivity of the formation water, Φ = porosity, m = the cementation exponent, S_w = water saturation, and n = the saturation exponent. The parameters a, Φ, and m are properties of the rock matrix, whereas the R_w and S_w are fluid properties and n is related to the wettability, which is a mineral, pore geometry, and grain-fluid interface phenomenon.

Steam recovery affects the resistivity in four ways:

1) increased temperature
2) increased S_w
3) decreased salinity
4) change of rock model from weakly cemented to unconsolidated sand with a concomitant increase in porosity

The temperature effect alone is quite dramatic, as shown in Figure 5. The trend is based on the equation shown, which is available in all of the wireline logging company's chart books.

$$R_2 = R_1 \cdot [(T_1 + 21.5)/(T_2 + 21.5)], \qquad (2)$$

where R_1 and T_1 are the resistivity and temperature at in situ conditions, respectively, and R_2 is the resistivity estimated at temperature T_2. Already at a moderate temperature increase from 10°C to 42°C, the resistivity has dropped to one-half of the in situ value. Further heating shows that at 104°C, the resistivity is down to one-quarter and at 230°C it has reduced to one-eighth of the in situ resistivity. Thus temperature increase alone is a major source of resistivity decrease when monitoring steam injection.

When the injected steam condensates, distilled water is added to the pore space, increasing the water saturation but also diluting the salinity of the brine in the process. The increase in the S_w alone is seen in Figure 4. The net effect, at least in the short term, is likely to be a lowering of the resistivity, although repeated condensation and drainage cycles will deplete the original salts in place and increase the resistivity. This effect is expected to be most dramatic close to the injection well, where small amounts of water will precipitate throughout the production cycle as the pressure drops when steam enters the reservoir. This precipitated, distilled water will mix with the irreducible water and continuously be transported away by the continuous flow of steam, leaving salt-depleted pore water behind. The dramatic increase in resistivity at very low salinities is shown in Figure 6.

Finally there will be changes in the rock matrix because of the dissolution effect on grain cements by the hot circulating water, and this will to some extent affect the grains themselves depending on mineralogy. The exact

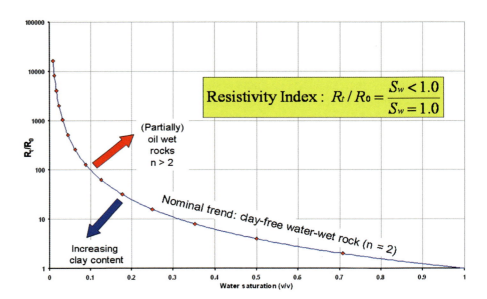

Figure 4. The resistivity index as a function of water saturation for a clay-free water-wet rock. The red diamonds on the trend indicate successive doubling of the resistivity index. The presence of clay will subdue the trend and partially oil-wet rocks will show a more aggressive trend.

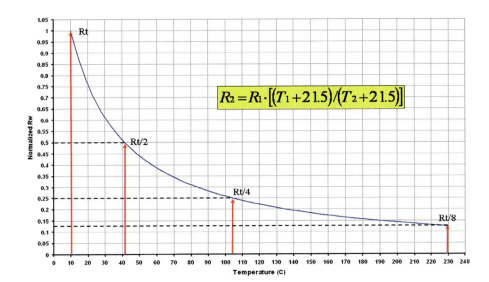

Figure 5. The sensitivity of the pore water resistivity R_w and bulk rock resistivity R_t to increasing temperature from an in situ value of 10°C. The resistivity has been normalized at in situ temperature.

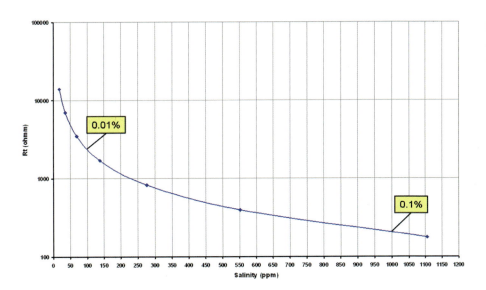

Figure 6. The resistivity of pore water as a function of salinity for very low salinity pore waters.

changes in rock matrix should be evaluated in monitoring wells, but we can illustrate the effect by modeling the in situ sand reservoirs as weakly cemented sands in situ with the nominal values for $a = 0.845$ and $m = 2.02$ as seen in Figure 7. Then, after a long period of exposure to steam and hot water, the steam chamber may be best represented by an unconsolidated sand model in which $a = 0.62$ and $m = 2.15$. This change of model will by itself lower the resistivity 15%. But in addition to mineral loss, the steam chamber also experiences an expansion of the rock volume measurable at the surface with tilt meters as part of the monitoring effort to provide early warning of a potential surface blowout. This expansion can be illustrated by increased porosity in the Archie equation. The change in rock model and the increased porosity can substantially lower the resistivity further by at least 30%, in addition to the temperature, S_w, and salinity effects.

When all of these effects are combined, they result in a nominal resistivity profile as seen in Figure 8. The thermal halo creates a substantial drop in resistivity, and within the steam chamber the resistivity drops further because of increased S_w and increased mineral dissolution. Close to the injection site, the resistivity will then rise sharply because of salt depletion.

The ideal feasibility study to answer the question how well the SAGD process can be monitored by EM would be to run one of the commercial reservoir simulators specifically designed for simulation of the SAGD process. The simulator can track all of the relevant parameters: temperature, $S_w = 1 - (S_s + S_g + S_{oil})$, and salinity. From this information, the resistivity changes can be estimated for every grid cell at different time steps for the entire volume by entering the appropriate dynamic changes in the Archie equation. A model for additional resistivity reduction because of mineral dissolution and volume expansion can be developed based on time-lapse wire-line log data from monitoring wells combined with repeated core sampling.

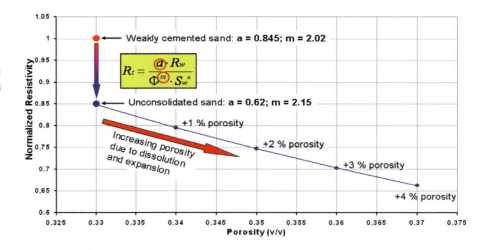

Figure 7. The effect of mineral dissolution modeled as the change from a weakly cemented sand to an unconsolidated sand. The dissolution and volume expansion caused by the growing steam chamber is illustrated by the increasing porosity values.

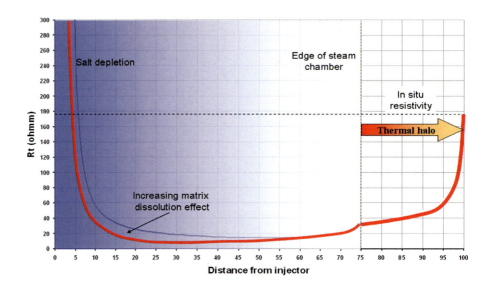

Figure 8. A cross section through a nominal steam chamber with the injection site to the left and the temperature "halo" to the right. The temperature increase will by itself lower the resistivity outside of the steam chamber, and within the chamber there will be a further reduction in resistivity because of the increased water saturation from steam cooling to water. Close to the injection site there will be a sharp rise in resistivity because of the depletion of salts in the pore water.

The resulting resistivity changes over time can then be forward-modeled in three dimensions and inverted to evaluate the resulting spatial resolution that MTEM offers as a SAGD monitoring tool.

Conclusions

Reservoir characterization using EM can determine the approximate recoverable reserves in place before onset of production. Production methods that involve mobilizing the oil or bitumen by means of steam injection are well understood in terms of how the recovery process affects the resistivity. There will be a resistivity decrease within the "thermal halo" radiating beyond the steam chamber. The bulk of the steam chamber will experience a further drop in resistivity because of the combined effects of heating and increased S_w and to some extent mineral dissolution and increased porosity also, whereas the volume closest to the injection well bore will experience a dramatic rise in resistivity because of salt depletion.

Gas evolving from the heated oil aids recovery in three ways:

1) Evolved gas increases the relative permeability for gas/steam, allowing steam entry at a lower pressure.
2) The evolved gas displaces oil faster, adding gas volume to the injected steam volume.
3) The gas is not reassimilated once evolved but stays stable and prevents reflux of oil into pores.

If seismic is used to monitor the SAGD production, the time-lapse changes that will be observed are the steam and/or hydrocarbon gas-charged volumes at the time of data collection. If EM is used to monitor the SAGD production, it is the temperature, S_w, and salinity changes that determine the resistivity changes. In neither case is the parameter of main interest directly evaluated, namely the oil saturation S_o. It has to be inferred from other changes.

The ideal feasibility study to evaluate EM as a SAGD monitoring tool should be based on reservoir simulation of the recovery process where temperature, water saturation, and salinity are tracked through time and the resistivity changes are then estimated for each grid cell. Forward modeling and inversion of the modeled data at different time steps will then show the sensitivity and spatial resolution of the method.

The spatial resolution of the resulting resistivity pattern offered by EM depends on depth of burial of the target.

References

Wright, D., A. Ziolkowski, and B. Hobbs, 2002, Hydrocarbon detection and monitoring with a multicomponent transient electromagnetic (MTEM) survey: The Leading Edge, **21**, 852–864.

Wright, D. A., A. Ziolkowski, and B. A. Hobbs, 2005. Detection of subsurface resistivity contrasts with application to location of fluids: U. S. Patent 6914433.

Ziolkowski, A., B. A. Hobbs, and D. Wright, 2007, Multitransient electromagnetic demonstration survey in France: Geophysics, **72**, no. 4, F197–F209.

Section 6

Environmental Aspects

Chapter 27

Tar Sands: Key Geologic Risks and Opportunities

Jack R. Century[1]

As the price of oil rises and as conventional hydrocarbon resources become scarcer, increased exploration and production activity is occurring in heavy-oil, tar-sands, and bitumen deposits. Although these contribute significantly to the global energy budget, they also contribute a greater share to the global carbon budget and to the detriment of the global environment. The balancing act between economics and environmental concerns is demonstrated on a grand scale in the evaluation of these geologic deposits. This chapter is intended to present the concerns relating to the "carbon footprint" of the development of these deposits in northern Alberta, Canada (referred to hereafter as "tar sands" for brevity) and to outline opportunities for more balanced tar-sands development by improved integration of geoscience and engineering disciplines.

Although the emphasis in this chapter is the Athabasca, Peace River, and Cold Lake districts, the concerns about these deposits are global in nature. Tar sands occur in as many as 70 countries, including Canada, Venezuela, Colombia, Trinidad, the United States, Romania, Albania, Madagascar, the former Soviet Union, Saudi Arabia, and China. Because of declining production of conventional, light, and medium oil, the world is depending on oil supplied from bitumen and heavy oil much faster than had been previously thought.

As an example, an ancient oil seep (Figure 1), with the help of Russian geologists, was the reason the Karamai heavy-oil field was discovered in 1955 and is now the third largest field in China. Karamai Field is in the Junghar Basin in the extreme northwest corner of China. The discovery well had initial production of 75 barrels per day (b/d) from fractured Triassic conglomerates. This had reduced to 25 barrels at the time of the picture (during a visit in 1983). The field is 85 mi in length and has been under water flood over its life.

Currently, heavy oil accounts for 13% of China's total daily production of 3.6 million barrels. Some Chinese estimates predict that heavy oil will be 60% of total Chinese production in a decade or two. Other countries, including Canada, join China in this headlong quest for more production of heavy oil.

Although tar sands (and other heavy oils) represent a small fraction of global oil production, they represent a far greater fraction of the greenhouse gas contribution from liquid hydrocarbons. Overall, this extraction and upgrading generates up to three times the carbon dioxide (CO_2) emissions as conventional oil, mostly through burning natural gas to heat the oil in situ to stimulate flow through the formation. If we replace the current world's supply of 85 million b/d of conventional oil with more costly, heavier oil, we would produce the CO_2 equivalent of nearly one-quarter of a billion barrels of oil per day. Is this the direction our society (and profession) should continue to follow?

The current exploitation of the tar sands is environmentally damaging enough by burning (and depleting resources of) natural gas as the main energy source to extract oil from bitumen. Consider the consequences of adding multibillion-dollar nuclear plants to generate steam or electricity. Some estimates reason these would provide little, if any, improvement in the carbon budget when evaluating the full CO_2 cycle. The cycle starts with uranium mining and milling, power-plant construction and transportation, and proceeds all the way to permanent disposal of radioactive waste and ultimate decommissioning of the nuclear power plant. We may be no better off with nuclear energy just on the issue of full-cycle greenhouse gas emissions.

At a meeting with the Canadian Association of Petroleum Producers (CAPP) in 2007, it was revealed there was no current interest by the oil-sands industry in developing nuclear power, but at the same time CAPP maintained nuclear is not off the table. This position prompted promoters of CANDU reactor sales to change plans from

---
[1] J. R. Century Petroleum Consultants, Calgary, Alberta, Canada
This paper appeared in the September 2008 issue of THE LEADING EDGE and has been edited for inclusion in this volume.

Figure 1. The oil seep associated with discovery of the Karamai heavy-oil field in Junghar Basin, Xinjiang Province, China (photographed in 1983).

building nuclear reactors in the Fort McMurray area to two other Alberta communities, Peace River and Whitecourt. The original location, Lac Cardinal, a shallow lake west of Peace River, was chosen by Energy Alberta Corporation (EAC) and Atomic Energy of Canada (AECL) with some dubious community support. Ontario-based Bruce Power Corporation later bought out EAC. Lac Cardinal lies on top of the Peace River Arch (PRA), potentially the most tectonically active region in Alberta. As a result of Alberta municipal elections later in 2007, voters in Peace River threw out of office their incumbent mayor and other elected officials who supported bringing in the first nuclear reactor to Alberta. Although the current Alberta government supports nuclear power for electric power generation, it is not clear whether nuclear power for production of tar sands can ever gain political traction or solve the carbon problem associated with that production.

The short-term projected growth of Alberta tar-sands production will result in the generation of more greenhouse gas emissions than in any other oil-producing region in the world, including Saudi Arabia. There are two basic reasons for this. First, producing oil from the McMurray area generates up to three times the CO_2 per barrel compared with conventional oil. Second, conventional world oil production appears to have already peaked. When Alberta tar-sands production increases from more than 1 million to 3 million b/d, it will generate the equivalent CO_2 of up to a 9 million b/d conventional oil field. At that point, continued global decline of light- and medium-gravity oil would result in Canadian tar sands becoming the world champion in greenhouse gas emissions generated by oil fields. Does Canada or the United States really want to assume this responsibility and liability by paying the price of long-term pain for short-term gain now readily accepted by our industry and profession? Should we not ask this, as geologists, geophysicists, and engineers? Have we really considered these consequences?

There are other major environmental tar-sands issues such as water supply limitations from the Athabasca River (used for steam generation which, in turn, is injected into the ground to stimulate production). In addition, there is serious contamination of land and freshwater, loss of boreal forests, loss of flora and faunal species, important health issues arising from pollution, tremendous drains on human and civic infrastructure resources, and the disruption of aboriginal rights because of the vast areas impacted on the surface by production. It would make much more sense to slow down extreme development and adverse social consequences already happening. It would be prudent for Alberta to reduce absolute, rather than just localized "intensity," CO_2 emissions by controlling future tar-sands expansion ... or pay a bigger price later.

To minimize our production-related carbon footprint, we do not need to abandon all tar-sands production as some propose. There are positive alternatives we can do as earth scientists and engineers.

The Alberta government has studied the effect of production of oil sands, and a short brochure outlines their findings (see http://environment.gov.ab.ca/info/library/7925.pdf). The Alberta Oil Sands Consultation (see http://www.oilsandsconsultations.gov.ab.ca) communicates progress in research and public policy concerning oil-sands production in Alberta and has published results and maps online (including Figure 2, without the added faults).

Figure 2 is a map of total bitumen distribution with faults determined from aeromagnetic surveying. This map indicates that no bitumen occurs east of the Clearwater fault between the intersection with the McMurray fault and the Virgin River shear zone (VRSZ), whereas the

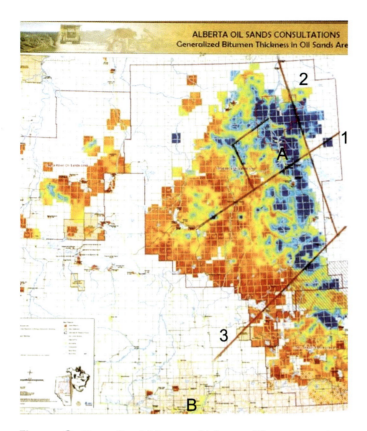

Figure 2. Generalized bitumen thickness. Blue represents greater bitumen thickness and red indicates areas of less thickness within Alberta. Lines are faults added by the author and inferred from published aeromagnetic surveys. (1) Fort McMurray fault. (2) Clearwater fault. (3) Virgin River shear zone. A = Fort McMurray, B = Edmonton.

Mannville sandstone group is up to 100-m thick in this area; it is void of any tar sands. Bitumen apparently has been biodegraded or flushed in the area east of this segment of the Clearwater fault. This fault appears to correlate with salt solution-collapse features as well as with biogenic and geochemical activity affecting bitumen quality distribution.

Most of the cumulative Athabasca tar-sands production is inside of the McMurray faulted structure. This is not just a coincidence. A topographic high inside of this basement "box structure" would account for more recent erosion and deeper tar-sands outcrop exposure. This is where the bitumen areas most accessible for open pit mining were first produced. South of the VRSZ, heavy-oil production in Cold Lake is in the 11–14 American Petroleum Institute (API) gravity range compared with the heavier (6–10 API) in the Athabasca area. The greatest range of 8–23 API gravity occurs in the Peace River oil-sands district. The VRSZ extends for approximately 1200 mi starting in the Precambrian Shield and trends southwest, terminating under the Laramide orogeny of the Canadian Rocky Mountains. The VRSZ also crosses under the major Redwater reef oil field of Upper Devonian age and the nearby city of Edmonton. The Alberta Research Council is planning a huge carbon-capture and storage project in the now depleted Redwater reef with CO_2 injections to begin in 2011. Investigations to monitor microseismicity over the Redwater field for pre-CO_2 injection baseline information have already begun.

Finally, geothermal variability may account for the range of bitumen characteristics. Basement faults and Precambrian geothermal activity in the Athabasca and Cold Lake Oil Sands areas are likely contributing factors in controlling oil gravity distribution and production. The same probably applies to the Peace River District tar/oil sands with additional Mississippian and more recent faulting on the PRA. The PRA is a classic Paleozoic crustal-collapse feature and has been closely compared with the seismically active New Madrid Tectonic Zone.

This analysis should help the exploration and development in areas of better quality oil-sands deposits, with the lowest possible CO_2 footprints.

In conclusion, it is my opinion that production of the Alberta tar-sands resources is environmentally unsound, largely because of the CO_2 released during production. Instead of producing tar sands of low quality, we should concentrate on exploration for oil/tar sands that are higher quality, thereby reducing our impact on global climate due to CO_2 and other greenhouse emissions into the atmosphere.

Petroleum geophysicists, geologists, and engineers often consider they are only doing their professional jobs, whereas the public, commercial, industrial, and government consumers can choose which kind, how much, and in what manner energy is consumed. However, in the case of oil-sands production, we geoscientists and engineers are making that choice ourselves and releasing unacceptable amounts of carbon into the atmosphere as a result. We must improve our professional practices in the oil patch to become more responsible citizens of the world.

Acknowledgments

I thank Wayne Pennington of Michigan Technological University for his assistance in editing and revising this manuscript.

Index

A

Alberta Oil Sands Technology Research Authority (AOSTRA), 47
American Petroleum Institute (API)
    definition of heavy oils, 73
    gravity oil deposits, 89
anisotropic imaging, 176
aromatics, 7, 73
asphaltenes, 7, 73
Athabasca, Alberta, Canada, 8
    deterministic mapping, 165–171
    environment of deposition, 203–204
    mechanisms for formation of deposit, 10
    mudstone clasts and SAGD process, 203–214
AVO analysis
    elastic parameters from, 33
    three-term, 166–167

B

benzene, 7
Berea sandstone, 116
biodegradation, 267–268
Biot-Squirt model. *see* BISQ model
BISQ model, 113–116
bitumen
    compressional velocities, 16
    definition, 8, 73
    density as function of temperature, 17
    exploitation, 161–163
    temperature dependence, 16
    viscosity as function of temperature, 16
    viscosity effects on production, 265–273
bituminous-oil reservoirs, 107–112
    elastic property changes, 121–127
bulk modulus, 14

C

Canada
    heavy oil deposits in, 22–25
Canadian tar sands
    HS modeling, 86–87
    modeling studies, 82–84
CAPRI process, 30–31, 276–277
carbonate triangle
    Alberta, 10
cold heavy-oil production with sand (CHOPS), 31
Cold Lake, Alberta, Canada, 9
cold production, 237–241, 243–249
    collaborative methods, 251–257
    model and methodology, 237–238
    reservoir simulation of, 255–256
    rock physics of, 253–255
    seismic modeling and imaging, 238–240
    seismic resolution of zones, 251–253
common conversion point (CCP), 191
common depth point (CDP), 191
compressional velocity
    oil sand, 18–19
constrained prestack linear inversion, 178–179
conventional seismic profile, 36
core velocity data
    rock physics model and, 215
crosswell imaging, 42–60, 259–263
    background, 259–260
    reservoir features, 260–263
crude oil
    chemical properties of, 6–8
    classification, 8
crystalline solid, 74
cyclic steam injection (stimulation) (CSS), 28, 251
cyclohexane, 7

D

deasphalting migration, 10
density after inversion of data, 35
density reflectivity
    from AVO analysis, 35
depth-variant stack
    radial component data processing, 195–196
derived facies profile, 36
deterministic mapping
    Athabasca, Alberta, Canada, 165–171
diagenetic evolution
    carbonates in Western Canada, 137–138
dilation, 288–289
dry gas, 13–14
dry gas reservoirs
    Western Canada, 144

E

effective stress, 287
elastic moduli *versus* temperature, 18
elastic property changes
    bitumen reservoir, 121–127
    sequential, 126–127
Elgin sandstone, 114, 116
enhanced oil recovery (EOR), 42–60

F

facies modeling, 209
facies prediction, 230–231
fluid heterogeneity
    caused by biodegradation, 267–268
fluid property variations
    impact on recovery, 268–269
fluid substitution
    Gassmann's equation, 243–245
foamy oil
    cold production and, 251
fracturing
    Grosmont Formation, 157–159
frequency attenuation, 52–55
frequency dependence
    oil sand, 20
frequency effect, 79
fuel consumption, 62

G

Gabor deconvolution
    radial component data processing, 196–197
    vertical component data processing, 192
gas effect, 76–77
gas-oil ratio (GOR), 13, 74
Gassmann equation
    fluid substitution, 243–245
    steam injection, 123–124
    Uvalde heavy-oil rock, 84–86
geomechanical deformation
    steam injection, 293–300
geomechanics
    impact on seismic monitoring, 290
    in thermal processes, 287–291
geometry
    radial component, 195–196
    vertical component data processing, 192–193
geotailoring
    reservoirs with viscosity variations, 269–270
glass point, 74
glassy solid, 74
greenhouse gas emissions, 64–65
Gregoire Lake In Situ Steam Pilot (GLISP), 46, 47, 49
Grosmont Formation, 139, 155–163

H

Hangingstone steam-assisted gravity drainage (SAGD), 121–127, 215–226
heavy oils
    Alberta, 10
    challenges for production, 60–66
    cold production, 243–249
    correlating chemical and physical properties, 89–97
    definition, 73
    geology of two major areas, 21–25
    geomechanics in, 287–291
    geophysical characterization of formations, 32–41
    laboratory measurements, 102–105
    modeling studies, 81–87
    multicomponent characterization for formations, 37–42
    physical properties pre- and postproduction, 245
    properties of, 14–18
    recovery, 25–31
    reservoir properties, 99–105
    reservoirs, 1–17
    resource for the future, 1–2
    rocks and, 5–6
    rocks saturated with, 18–20
    samples, 90
    upgrading of, 63
    viscosity effects on production, 265–273
    Western Canada, 143
    worldwide production, 9
horizon slice, 33
horizontal well pair planning, 231–233

horizontal wells, 32, 34
hot production, 243
Hashin-Shtrikman (HS) bounds
    elastic property estimation, 82–87
hydrocarbons, 6
    general phase behavior, 11–12
    low-shrinkage, 13
    mixtures, 13–14
    multicomponent system, 12
    single-component system, 12, 13
hydrology
    Devonian petroleum system, 139–143
hyperbolic normal moveout, 191

I

imaging
    oil-sand reservoirs, 173–181
inclined plate setters (IPS), 27
inelastic losses, 177
in situ recovery techniques, 27–28
in situ recovery zones (ISC), 27
    monitoring fire front, 43–44
inverted velocity profiles, 47–48
iso-butane, 7

J

Jackfish Heavy Oil Project, 191–200
J-well and gravity-assisted drainage steam stimulation (JAGD), 270–273

K

Keg River, 139–141
Kirchhoff migration
    vertical component data processing, 194–195

L

laboratory measurements
    heavy-oil reservoirs, 102–105
    sandstone reservoirs, 114
land surface disturbance and reclamation, 62
large tailings pond, 27
layered mining, 27
light crude reservoirs
    Western Canada, 148–151
liquid-assisted steam-enhanced recovery (LASER), 32
liquid point, 78–79
lithology distribution, 41
log curves, 41
Long Lake South project, 183–190

M

McMurray Valley system, 183
measured P-wave velocity
    oil sand, 20
mechanical damage to rock, 288–290
mega-breccia zones
    Grosmont, 159–161
megaporosity zones
    Grosmont, 158
microseismic results, 294–297
mine tailings disposal, 62
mudstone clasts facies
    effect on SAGD process, 203–214
    permeability of sand, 204–206

mudstone volume modeling, 209–211
multiattribute analysis, 34
multicomponent data, 37–42
multilayer feedforward neural network (MLFN), 187–189
multitransient electromagnetic surveys, 301–307

N

n-butane, 7
near-surface static solution
    radial component data processing, 195–199
    vertical component data processing, 192–194
neural networks analysis
    characterization of heavy oil reservoir, 183–190
Nisku Pools, 144
North Sea sandstone, 113

O

oil sands, 8
oil-sands reservoir
    Alberta, 10
    facies prediction, 230–231
    geologic background, 227–228
    geophysical background, 228–229
    monitoring, 215–226
    optimization of fields development, 227–234
    viscosity effects on production, 265–273
    workflow, 229–230
oil-to-steam ratio (OSR), 28
oil viscosity. see viscosity
oil-water contact (OWC), 267
oxidation, 10

P

passive seismic and surface monitoring, 293–300
Peace River, Alberta, Canada, 9
Peace River Arch, 134
petroleum reserves, 6
phase behavior
    hydrocarbons, 11–12
pore fluid viscosity
    effects on P-wave attenuation, 113–119
porosity and permeability estimations, 211–212
postproduction
    physical properties of heavy oils, 245
PP data, 38–42
    time-lapse results, 279–280
preproduction
    physical properties of heavy oils, 245
pressure dependence
    during steam injection, 121–122
pressure effect, 76–77
pressure pulse technology (PPT), 32
probabilistic neural network (PNN), 187–189
PS data, 37–41
    time-lapse results, 279–280
PSEI
    heavy oil reservoirs, 99–105
P-SV converted-wave 3D monitoring, 215–226
    analysis and interpretation, 222–225
P-to-S converted-wave elastic impedance. see PSEI
P-wave attenuation
    pore fluid viscosity effects, 113–119

pyrolysis-MBMS
    heavy oil properties, 91–92, 95–96

Q

quasi-solid, 74

R

radial filter
    radial component data processing, 195
    vertical component data processing, 192
radial component data processing, 195–199
reflected light view, 19
relative density, 35
reservoir deformations, 289–290
reservoir description
    Whitesands, 275–276
reservoir geology
    Grosmont, 155–157
reservoir heterogeneity
    Athabasca oil sands, 165–171
reservoir modeling
    geostatistical, 203–214
    workflow, 208–212
reservoir properties, 99–105
    Western Canada, 143–151
reservoir simulation
    cold production, 255–256
resins, 7, 73
resistivity monitoring
    thermal recovery, 304–306
retrograde gas condensate, 13
rheology
    heavy oil properties, 90–93
rock physics
    cold production of heavy oils, 253–255
rock physics analysis
    Athabasca oil sands, 165–166
rock physics model
    core velocity data and, 215
rocks
    saturated with heavy oil, 18–20
Russia
    natural bitumen, 4–5

S

sandstone
    pore fluid viscosity in, 113–118
SARA fractionation, 7
saturates, 7, 73
scanning electronic microscope
    carbonate saturated with heavy oil, 19
seismic calibration and interpretation, 215–222
seismic data
    acquisition, 191–192
    attribute selection for facies discrimination, 206–208
    comparison with 2D time-lapse imaging, 111–112
    conditioning of, 276–278
    impact of geomechanics, 290
    implications of monitoring, 116–118
    multicomponent processing, 191–200
seismic inversion
    oil-sand reservoirs, 173–181
seismic resolution
    cold production zones, 251–253
seismic response
    effects of cold production, 237–241

seismic rock physics
  steam injection, 107–112
seismic tomography methods, 56–60
seismic transformation and classification (STAC), 35
seismic wave attenuation, 113
sequential rock physics model, 124–125
shale caprock
  shear at, 289
shearing, 288–289
shear properties, 17–18
shear storage modulus, 18
  heavy oil rock, 21
shear velocity *versus* temperature, 76
shear-wave measurements, 75
slurry fracture injection (SFI), 31
solubility contrast, 159
sonic log, 21
source rocks
  Western Canada, 139
sour gas reservoirs
  Western Canada, 143
spectral decomposition, 222
steam and gas push (SAGP), 30
steam-assisted gravity drainage (SAGD), 4, 28–29, 73, 251
  in bituminous-oil reservoirs, 107–112
  Grosmont, 161–162
  Hangingstone, 121–127, 215–226
  monitoring flooding, 50–52
  mudstone clasts effects, 203–214
  multitransient electromagnetic surveys, 301–307
steam flood monitoring, 45–46
steam injection
  in bituminous-oil reservoirs, 107–112
  elastic property changes in bitumen, 121–127
  geomechanical deformation with, 293–300
  from pore perspective, 302–303
steam-to-oil ratios (SORs), 32
sulfur, 7
surface mining, 25–26
Surmont bitumen reservoir, 173–181
sweet gas reservoirs
  Western Canada, 144
synthetic modeling
  rock physics and reservoir parameters, 109–111

T

tailings oil recovery (TOR), 27
tar mats, 73
tar sands, 8, 311–313
temperature dependence
  during steam injection, 121–123
temperature effect, 77–78
thermal enhanced oil recovery processes (THEOR), 287–291
thermally induced stress changes, 287–288
thermal recovery
  resistivity monitoring, 304–306
tiltmeter results, 297
time difference maps, 54
time-lapse 3D seismic monitoring, 215–226
  monitoring of THAI, 275–282
time-lapse imaging, 2D
  comparison with seismic data, 111–112
time-lapse monitoring, 42–60
  conditioning of seismic data for, 276–278
  results, 279
Toe-to-Heel-Air-Injection (THAI^{TM}), 30–31
  time-lapse seismic monitoring of, 275–283
transducer design, 103
true amplitude recovery (TAR), 193

U

U. S. Geological Survey
  definition of heavy oils, 73
ultrasonic measurements
  heavy oil properties, 91, 93
Uvalde heavy-oil rock
  generalized Gassmann's equation, 84–85
  HS modeling, 83–84
  modeling studies, 81–82

V

vapor extraction (VAPEX) method, 31
velocity, 16
  anomalies, 278
  dispersion, 125–126
  factors influencing, 76–79
velocity models
  light and heavy oils, 74–76
velocity-temperature measurements, 74
Venezuelan asphaltene, 7
vertical component data processing, 192–195
viscosity
  impact on production, 265–273
  pore fluid, 113–119
  versus temperature, 16
  temperature trends, 74
  variations, 269–270
V_P/V_S ratio
  characterization of heavy-oil reservoir, 183–190
  heavy-oil cold production effects, 243–249

W

water consumption, 62
water washing, 10
wavelet stretch correction, 176
Western Canadian Sedimentary Basin, 1, 131–151
  basin evolution and structure, 133–136
  crude oil resources, 132
  diagenetic evolution of carbonates, 137–139
  hydrology and migration, 139–143
  reservoirs, 143–151
  sedimentation and facies, 136–137
  source rocks, 139
wet gas, 13
Whitesands project, 275–282
wide-angle prestack seismic inversion, 173–181
  angle requirements and difficulties, 174–179
  anisotropic imaging, 176
  constrained prestack linear, 178–179
  for density, 175–176
  inelastic losses, 177
  wavelet stretch correction, 176
Winterburn Group, 135–137
Woodbend Group, 135–137
wormholes, 31, 239, 243, 253
  cold production and, 251